# Advanced Microsystems for Automotive Applications 2010

Gereon Meyer · Jürgen Valldorf

# Advanced Microsystems for Automotive Applications 2010

## Smart Systems for Green Cars and Safe Mobility

Springer

Dr. Gereon Meyer
VDI/VDE Innovation + Technik GmbH
Steinplatz 1
10623 Berlin
Germany
gmeyer@vdivde-it.de

Dr. Jürgen Valldorf
VDI/VDE Innovation + Technik GmbH
Steinplatz 1
10623 Berlin
Germany
valldorf@vdivde-it.de

ISBN 978-3-662-51982-0
DOI 10.1007/978-3-642-16362-3
Springer Heidelberg Dordrecht London New York



*Coverdesign:* deblik, Berlin

Printed on acid-free paper

Springer is part of Springer Science+Business Media (www.springer.com)

# Preface

The automobile of the future has to meet two primary requirements: the super-efficient use of energy and power and the ultra-safe transportation of people and goods. Both features are increasingly enabled by smart, adaptive and context aware information and communication technologies (ICT), electrical or electronic components and systems rather than solely by the mechanical means of classic automotive engineering. The most advanced example of this trend is the electrified vehicle combining a full electric powertrain with completely electronic controls like smart power and energy managers, steer-by-wire technologies and intelligent networking capabilities allowing all providers and consumers of energy to work in efficient synergy.

In the course of this year the first series production electric vehicles will finally come into the market. Automakers – unsure if electric vehicles would really sell – have long time been hesitant to make the necessary changes of their product portfolios. In the coincidence of economic crisis and growing concerns about global warming and energy security companies and public authorities jointly succeeded to overcome many obstacles on the path towards electrification.

It has been the mission of the International Forum on Advanced Microsystems for Automotive Applications (AMAA) for more than twelve years now to detect paradigm shifts and to discuss their technological implications at an early stage. Previous examples from the fields of active safety, driver assistance, and power train control can be found in many passenger cars today. The enabling technologies of the electrified vehicle thus nicely fit into the scope of the conference. The heading of the 14th AMAA held in Berlin on 10-11 May 2010 is "Smart Systems for Green Cars and Safe Mobility". Among the co-organizers is the European Technology Platform of Smart Systems Integration (EPoSS), which in cooperation with the European Road Transport Research Advisory Council (ERTRAC) and the SmartGrids platform is now playing a major role in the Public Private Partnership European Green Cars Initiative.

The book at hand is a collection of papers presented by engineers from leading companies and world-class academic institutions at the AMAA 2010 conference. They address ICT, components and systems for electrified vehicles, power train efficiency, road and passenger safety, driver assistance and traffic management. Highlights of the contributions include energy management and power train architectures of electrified and optimized conventional vehicles, autonomous collision avoidance, safety at intersections, as well as a number of features of car-to-car and car-to-infrastructure communication applications including the use of Galileo navigation data for traffic management. Being typical for the AMAA the presented applications are complimented by recent sensor and actuator developments, e.g. active engine management sensors, advanced camera systems, and active vehicle suspensions.

We would like to thank all authors for making this book an outstanding source of reference for contemporary research and development in the field of ICT, components, and systems for the automobile of the future. The time and effort that the members of the AMAA Steering Committee spent on making their assessments are particularly acknowledged. We like to thank also the European Commission, EPoSS, and all industrial sponsors for their continuous support of the AMAA.

In our role as the editors and conference chairs we would like to point out that preparing a book like this is a serious piece of hard work built upon the endurance and enthusiasm of a multitude of great people. Our particular thanks goes to Laure Quintin for running the AMAA office as well as to Anita Theel, Michael Strietzel, and David Müssig for technical preparations of the book, and to all other involved colleagues at VDI|VDE-IT. We also want to express our deep gratitude to Wolfgang Gessner for his leadership and for continuous support of the AMAA project.

Berlin, May 2010

Dr. Gereon Meyer
Dr. Jürgen Valldorf

## Funding Authority

European Commission

## Supporting Organisations

European Council for Automotive R&D (EUCAR)

European Association of Automotive Suppliers (CLEPA)

Advanced Driver Assistance Systems in Europe (ADASE)

Zentralverband Elektrotechnik- und Elektronikindustrie e.V. (ZVEI)

Mikrosystemtechnik Baden-Württemberg e.V.

Hanser Automotive

mst/news

enabling MNT

## Organisers

European Technology Platform on Smart Systems Integration (EPoSS)

VDI|VDE Innovation + Technik GmbH

## Honorary Committee

| | |
|---|---|
| Eugenio Razelli | President and CEO,<br>Magneti Marelli S.P.A., Italy |
| Rémi Kaiser | Director Technology and Quality<br>Delphi Automotive Systems Europe, France |
| Nevio di Giusto | President and CEO<br>Fiat Research Center, Italy |
| Karl-Thomas Neumann | Executive Vice President E-Traction<br>Volkswagen Group, Germany |

## Steering Committee

| | |
|---|---|
| Mike Babala | TRW Automotive, Livonia MI, USA |
| Serge Boverie | Continental AG, Toulouse, France |
| Geoff Callow | Technical & Engineering Consulting, London, UK |
| Bernhard Fuchsbauer | Audi AG, Ingolstadt, Germany |
| Kay Fürstenberg | Sick AG, Hamburg, Germany |
| Wolfgang Gessner | VDI\|VDE-IT, Berlin, Germany |
| Roger Grace | Roger Grace Associates, Naples FL, USA |
| Klaus Gresser | BMW Forschung und Technik GmbH, Munich, Germany |
| Horst Kornemann | Continental AG, Frankfurt am Main, Germany |
| Hannu Laatikainen | VTI Technologies Oy, Vantaa, Finland |
| Günter Lugert | Siemens AG, Munich, Germany |
| Roland Müller-Fiedler | Robert Bosch GmbH, Stuttgart, Germany |
| Paul Mulvanny | QinetiQ Ltd., Farnborough, UK |
| Andy Noble | Ricardo Consulting Engineers Ltd., Shoreham-by-Sea, UK |
| Pietro Perlo | Fiat Research Center, Orbassano, Italy |
| Detlef E. Ricken | Delphi Delco Electronics Europe GmbH, Rüsselsheim, Germany |
| Christian Rousseau | Renault SA, Guyancourt, France |
| Patric Salomon | 4M2C, Berlin, Germany |
| Florian Solzbacher | University of Utah, Salt Lake City UT, USA |
| Egon Vetter | Ceramet Technologies Ltd., Melbourne, Australia |
| David Ward | MIRA Ltd., Nuneaton, UK |
| Hans-Christian von der Wense | Freescale GmbH, Munich, Germany |

**Conference Chairs:**

| | |
|---|---|
| Gereon Meyer | VDI\|VDE-IT, Berlin, Germany |
| Jürgen Valldorf | VDI\|VDE-IT, Berlin, Germany |

# Table of Contents

## Electrified Vehicles

## Power Train Efficiency



## Safety & Driver Assistance

## Components & Systems

## Traffic Management

## Appendices

# Electrified Vehicles

# Embedded Systems: The Migration from ICE Vehicles to Electric Vehicles

M. Ottella, P. Perlo, Centro Ricerce Fiat
O. Vermesan, SINTEF
R. John, Infineon Technologies
K. Gehrels, NXP Semiconductors
H. Gall, austriamicrosystems
J. Aubert, FICOSA International

## Abstract

Future generations of electric vehicles (EVs) will require a new level of convergence between computer and automotive architectures, with the electric power train being a mechatronic system that includes a multitude of plug and play devices, embedded power and signal processing hardware, software and high level algorithms. This paper discusses the current advances in the computing devices, communication systems and management algorithms embedded in the EV building blocks used for implementing the distributed energy and propulsion architectures required for high efficiency, reduced complexity and safe redundancy.

## 1 Vehicle Architecture and Systems

The aim of the new electric vehicle architectures which are based on distributed embedded computing and power electronic systems, is to achieve significant energy saving capabilities, an enhanced fun-to-drive experience, increasing safety and comfort while decreasing the overall complexity of the vehicle. Further objective includes reducing the overall system design cost.

At present the electronic devices in conventional ICE-propelled vehicles [1] can be divided into four domains: Power train, Chassis, Body and Infotainment, each offering different functional and computational characteristics:

- **Power train controllers** require high computational power and stringent time constraints to run complex control algorithms in short sampling periods (ranging from 100 µs to 1 ms).
- **Chassis controllers** still require high computational power, but also multi-task operation capabilities within strict trigger timing constraints,

i.e. running safety critical tasks (steering, braking). Sampling periods are usually in the order of some milliseconds.

- **Body controllers** run fewer critical tasks with less stringent time constraints and sampling periods (in the order of some seconds). They also run some safety-critical applications, e.g. airbag ignition.
- **Infotainment controllers** perform no critical functions but require operational features such as upgradeability, plug-and-play capabilities, high and fluid bandwidth and security.

a)

b)

Fig. 1. Complexity in the vehicle - (a) the vehicle today (b) the vehicle tomorrow.

In addition, the overall **signal distribution system** serves each domain by performing the necessary communication services, inside and outside the vehicle (Vehicle-to-Vehicle – "V2V" and Vehicle to Infrastructure – "V2I").

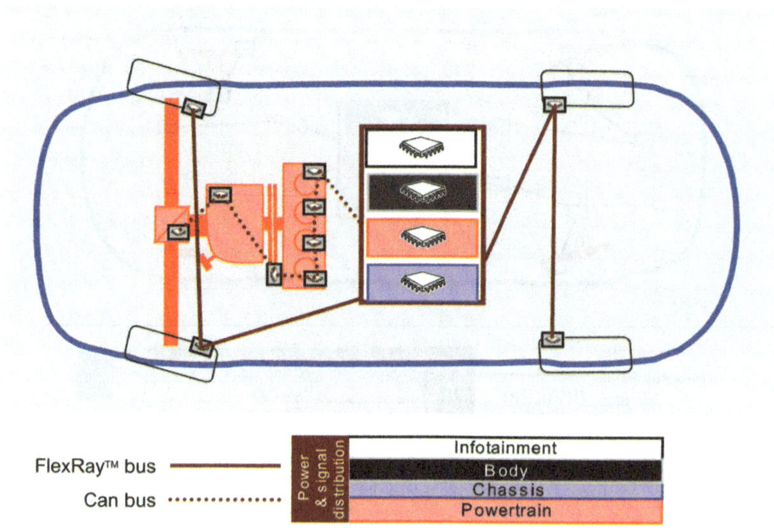


Fig. 2. ICE vehicle architecture, building blocks and domain partition.

In EVs the overall architecture can be thought of as a multi-layered hierarchical structure with similar domains having different boundaries (Fig. 2):

- **Energy sources:** Power electronic modules, control hardware and firmware, and high level algorithms performing the functions of energy/power storage (battery/super capacitors) overall management (cell charge/discharge/equalisation functions, failure management functions, etc) and recharging (grid connection, range-extension, photovoltaic)
- **Propulsion:** Control modules and control algorithms, power modules and electrical motors performing core power functions (traction torque distribution, energy recuperation and conversion).
- **Chassis:** Power modules and electric actuators performing steering, braking and active suspension. Control algorithms performing stability control, ABS, ESP, etc.
- **Vehicle body and on-board control:** Performing master supervision, human machine interface (HMI), comfort, infotainment and driver assistance functions.

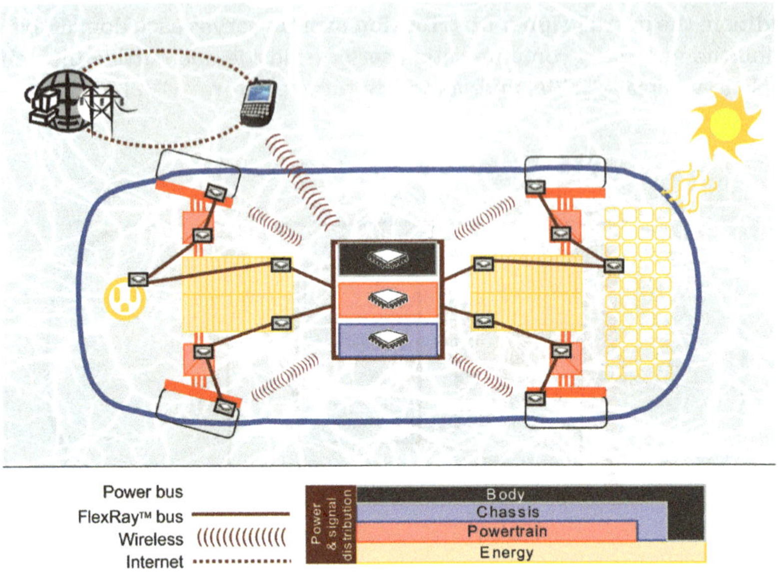


Fig. 3. EV architecture, building blocks and domain partition hierarchy.

The interconnection among the various domains of the vehicle is demanded to the **Power and Signal Distribution system** also featuring V2V, V2I, Vehicle to Grid and Internet Connection (V2G+I).

## 2 Embedded Systems Computing Platforms

**Embedded systems** are penetrating the automotive industry very rapidly with software and electronics currently accounting for some 35% of the cost of premium vehicles [2]. A significant growth is expected for full EVs because of distributed energy sources, storage units, power converters, motors, electro-mechanical chassis devices and novel high-speed power and signal distribution systems.

ICE vehicles are characterised by a low number of concentrated high-power computing units, whereas new generations of EVs are characterised by a high number of low computational power units.

For instance, modern Engine Control Units (ECUs) require up to 500 Mega instructions per second necessary to handle complex critical fail-safe tasks. Incremental evolutions of ICEs, including combustion chamber pressure control,

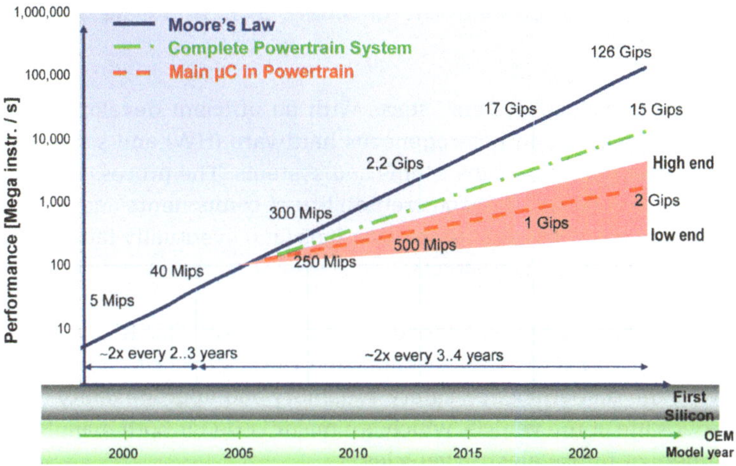


Fig. 4. Trends in computational power in conventional (ICE) powertrains.

valve position control, oil condition control and exhaust gas composition sensors and control, are expected to require even greater computational power. It is expected that the EV architecture will be more computational intensive than the ICE one, in that most functions will be performed through electrical and electronic devices. However, in the newer generations of EVs, in analogy with biological systems, the complexity will be minimised by evolving into an optimal trade-off between local and central computing/sensing while preserving safety and reliability. This will require significant improvements in the standardisation and implementation of fail safe computing platforms demanding for high speed time triggered communication and computation tasks.

Fig. 5. The four-leg principle for sprinter and runner: distributed propulsion, distributed power.

## 3 AUTOSAR Reference Platform for Smart, Swift and Safe System Integration

The design of embedded systems starts with an efficient development process in which dedicated and heterogeneous hardware (HW) and software (SW) modules are integrated into subsystems and systems. The process must follow guidelines that enable simultaneous reusability of components and algorithms by standardisation and traceability of functions (and eventually faults) through consistent multi-layered abstractions.

Specifically for automotive embedded systems the AUTOSAR[1] concepts and methodologies [3] paves the way for managing the electronic system complexity, improving the cost-efficiency, safety, reliability and upgradeability throughout the service life of the vehicle, which is a major issue for both manufacturers and consumers of future electric vehicles.

The implementation of the layered architecture presented in figure 6 ensures that the applications and the high level functions are decoupled from the underlying hardware and software services. This enables the design of highly distributed systems: the runtime environment (RTE) ensures communication between hardware and software components regardless of the topology.

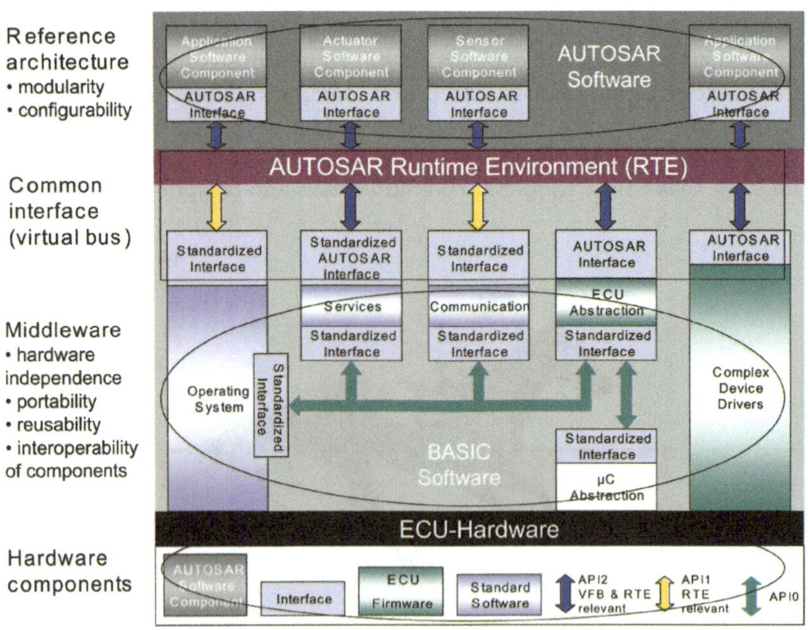


Fig. 6. AUTOSAR layered reference platform.

In addition the future evolution of AUTOSAR is the key for high level of safety-integrity levels (ASIL[2] D: Failure rate < $10^{-8}$ / h) [4], as it enables the development of a methodology that can easily be mapped on those defined by the ISO 26262[3] standard.

## 4 Power Distribution and Communication Systems: Smart Power Cables, Wireless.

EVs architectures are demanding a new level of safety and reliability to be deployed for in- and off-vehicle communication where mechanical systems will be substituted 100% by electrical systems. As a consequence the in-vehicle networks are becoming the backbone of EVs. In addition the evolution towards a distributed embedded system platform is closely related with the development of high speed automotive communication systems: the characteristics of high bandwidth (>10 Mbps) time triggering and fault-tolerance make FlexRay the preferred candidate protocol for the distributed electric propulsion. Moreover the interaction of the vehicle with the forthcoming infrastructures will also require new fast and secure wireless implementations (WAVE, IEEE 802.11p, LTE, etc.).

The presence of high electrical and magnetic fields, with the related electromagnetic compatibility and interference (EMC-EMI) issues, pose a new level of design complexity because of the tight interaction between high power distribution lines and high speed wired and wireless communications media. The major issue is then to address networking architectures and dedicated components to achieve high radiation immunity.

Moreover to perform in-vehicle networking immune from interferences, a viable option is foreseen in the implementation of smart shielded power cabling which will also reduce the potentially harmful magnetic field emissions.

## 5 Conclusions

In summary the migration from ICE based vehicles to the full electric ones, will change the overall domain partitioning and their boundary conditions together with the distribution of the core system functions. With respect to the conventional vehicle, more functions will be performed by distributed embedded systems. Communication protocols will play a crucial role in that high speed protocols (i.e. FlexRay) will be required for a safe and reliable management of the architecture.

[1] AUTOSAR (Automotive Open System Architecture) is an open and standardized automotive software architecture, jointly developed by automobile manufacturers, suppliers and tool developers. It is a partnership of automotive OEMs, suppliers and tool vendors whose objective is to create and establish open standards for automotive E/E (Electrics/Electronics) architectures that will provide a basic infrastructure to assist with developing vehicular software, user interfaces and management for all application domains. This includes the standardization of basic systems functions, scalability to different vehicle and platform variants, transferability throughout the network, integration from multiple suppliers, maintainability throughout the entire product lifecycle and software updates and upgrades over the vehicle's lifetime as some of the key goals.

[2] Automotive Safety Integrity Level (ASIL) is defined in ISO 26262 as a relative level of risk-reduction provided by a safety function, or to specify a target level of risk reduction. In simple terms, ASIL is a measurement of performance required for a Safety Instrumented Function (SIF).

[3] International standard IEC 61511 was published in 2003 to provide guidance to end-users on the application of Safety Instrumented Systems in the process industries. This standard is based on IEC 61508, a generic standard for design, construction, and operation of electrical/electronic/programmable electronic systems. Automotive sector has its own standard based on IEC 61508: ISO 26262 (for road vehicles, currently a draft international standard).

## References

[1] Simonot-Lion, F., The design of safe automotive electronic systems: Some problems, solution and open issues, Keynote speech, IES - IEEE Symposium on Industrial Embedded Systems, Antibes, France, Oct 18-20, 2006.

[2] Charlette, R. N., This car runs on code, IEEE Spectrum online edition http://spectrum.ieee.org/green-tech/advanced-cars/this-car-runs-on-code/1, Feb 2009.

[3] Furst, S., AUTOSAR – An Open Standardized software architecture for the automotive industry, 1st AUTOSAR Open Conference & 8th AUTOSAR Premium Member Conference, October 23rd, Detroit, MI, USA, 2008.

[4] Lafouasse, B., AUTOSAR for Electric Vehicles at SVE – Dassault Group, Vector Congress, Oct 2008.

**Marco Ottella, Pietro Perlo**
Centro Ricerche Fiat
Strada Torino, 50
10043 Orbassano
Italy
marco.ottella@crf.it
pietro.perlo@crf.it

**Ovidiu Vermesan**
SINTEF
Forskningsvn. 1
0314 Oslo
Norway
ovidiu.vermesan@sintef.no

**Reiner John**
Infineon AG
81726 Munich
Germany
reiner.john@infineon.com

**Kees Gehrels**
NXP Semiconductors
Building FT3.135
Jonkerbosplein 52
6534AB Nijmegen
The Netherlands
kees.gehrels@nxp.com

**Harald Gall**
austriamicrosystems
Tobelbaderstr. 30
8141 Unterpremstaetten
Austria
harald.Gall@austriamicrosystems.com

**Jordi Aubert**
FICOSA International S.A.
Pol Ind Can Magarola, Ctra C-17. Km 13
Mollet Del Vallès – Barcelona
Spain
ftr.jaubert@ficosa.com

**Keywords:** embedded systems, E/E architectures, electric vehicles

# Green Combustion Cars Drive on Electric (BLDC) Motors

D. Leman, Melexis


## Abstract

Vehicle manufacturers recently have demonstrated impressive fuel economy improvements in combustion engines thanks to the adoption of a range of electrical motors. When the benefits of electrical motors are fully leveraged throughout the engine development cycle, they allow a significant gain in fuel economy without compromising performance and comfort. Through engine down sizing, weight reduction, improvement in aerodynamics and on-demand operation (including starter-alternators for stop-start technology) electrical motors have applied a silent (r)evolution on the combustion engine. The electrified combustion engine is fully equipped to hold back the break through of hybrid and plug-in electric vehicles for a significant number of years to come. This paper explains why brushless DC (BLDC) motors are both technically as well as cost wise an ideal solution to implement any type of fan or pump. First the back ground of the fuel savings is analyzed to better understand where the different operating conditions originate from. Furthermore the challenges that these operating conditions impose on sensorless BLDC motor control are being elaborated.


## 1 Fuel Savings and Their Origins

The interest in green cars is strongly driven by legislation. Legislation imposes specific driving cycles when measuring emission rates. City cycle emissions are largely influenced by idle mode, deceleration and acceleration, and low speed operation, while the highway cycle is essentially defined by high speed operation and aerodynamics. We will now further break down these modes exposing the influence of electric motors, and the challenges they impose on sensorless BLDC motor control to ensure there is no loss in performance or comfort.

## 1.1 Legislation

Since the 1960's and the introduction of rudimentary emissions controls there has been a continuing trend in driving down the level of pollutants emitted from passenger cars and light trucks. This evolution has progressed differently in the different world markets. In Europe for example Euro 5 standards are effective since September 2009 and become 100% mandatory for all vehicles by 2011. The Euro 6 standard represents the design target for automotive manufacturers and demands a 50% reduction over Euro 5 limits. In the US there have been ongoing State and Federally defined standards. One of the first announcements of President Obama was the shortened introduction time of the CAFE standards which will be effective in 2016.

| | Particulates | NOx | Combined |
|---|---|---|---|
| Euro 4 | 6,25 | 250 | 300 |
| Euro 5 | 5 | 180 | 230 |
| Euro 6 | 5 | 80 | 170 |

-80% -57% -68% -57%

Fig. 1. Overview of EURO norms for Diesel engines

## 1.2 Idle Mode and Low Speed Operation.

The most fuel efficient idle mode is "no operation". Stop start technology imposes a number of indirect consequences to the operating modes of the pumps and fans:

- The need for cooling of the engine and the cabin during stop function imposes that water pumps and cooling fans, previously operated from the belt, have to be operated by electric motors.
- Additionally the slightest noise, created by the pumps and in particular the fans in the cabin area, becomes audible as the back ground noise of the combustion engine is gone. Lowest noise levels can only be achieved by sinusoidal driven BLDC motors, so called Permanent Magnet Sinusoidal Motors (PMSM).

- In order to avoid draining the battery all pumps and motors have to be operated at minimum speed, or are halted completely to limit their current consumption. In case the electric motors are stopped, for instance pumps that generate a hydraulic pressure, they have to start up as fast as possible when the engine is started again.

Pumps and fans are dimensioned to operate at maximum performance, but these performances are hardly ever reached during the life of a vehicle. Therefore even without the implementation of a stop start, on demand operation of pumps and fans has a significant impact on the fuel efficiency in idle and low speed mode. On demand operation of fuel pumps and HVAC fans allow to save 1.9g $CO_2$/100km, Electric Power Steering (EPS) saves 5.9g $CO_2$/100km, while an electric water pump even saves as much as 7.1g $CO_2$/100km [1].

## 1.3 Acceleration: Weight Savings

Weight reduction need to result in the same sensations using a smaller engine, or at least with burning less fuel.

Belt driven water pumps have to ensure sufficient cooling at low engine speed and maximum towing torque. To avoid excessive pressures at high engine revolutions, the cooling ducts have to be over dimensioned, leading to larger and heavier engines. Applying on demand water pumps allows optimizing the cooling ducts leading to smaller and lighter engines and thus reducing weight.

Brushless motors are more efficient than brushed DC motors and space as well as weight savings run up to typically 30%. In high temperature environments like the power train this delta increases to 50% because removing the heat from the rotor requires oversizing of the commutation brushes.

HVAC blowers were previously driven in a linear way burning all excessive energy as heat. This large heat dissipation implied the need for heavy heat sinks. Driving the pumps with a more efficient PWM (pulse width modulation) scheme is more complex, especially towards achieving EMC.

During acceleration, on demand pumps allow the engine ECU to temporally reduce the power drawn by the auxiliaries, reducing the load on the alternator and therefore boost the engine performance.

### 1.4 Aerodynamics

In low end cars, a simple method to increase aerodynamics is by closing the grill shutter at higher speeds when no cooling is required. Such a grill shutter can be implemented with a 150mA brushless motor, similar to an HVAC flap.

Higher end cars can afford electro-hydraulic pumps to lower the chassis by a few cm when they reach a certain minimum speed.

Fig. 2. HVAC flap and grill shutter module

## 2 Automotive Challenges for Sensorless BLDC Motor Control

As previously mentioned BLDC motors offer higher efficiency than there brushed counterparts. Additionally they offer lower noise operation and longer longevity. The lower raw material content however does not imply a lower cost yet as they are less wide spread and therefore less subject to competition. Recently several vehicle manufacturers have put large volume RFQs on the market. This was the trigger needed for the supplier market to start investigating this technology, trying to resolve the challenges they bring along.

Based on its ASIC experience in automotive waterpumps since 2005, Melexis brought its TruSense technology to the market in 2009. TruSense ensures robust and reliable sensorless operation for any type of motor under any kind of load condition. This technology combines reluctance sensing at lower speeds with Back EMF (BEMF) sensing at higher speeds. This technology is applicable independent from the motor construction, and allows any kind of motor current shaping, whether a simple block commutation is applied, or overlapping motorstates with extensive lead angle adjustment or even when a full sine wave motor current is applied. Additionally TruSense only requires a 16 bit microcontroller in combination with some dedicated hardware modules to achieve performances previously only possible with Field Orient Control (FOC) using expensive DSPs.

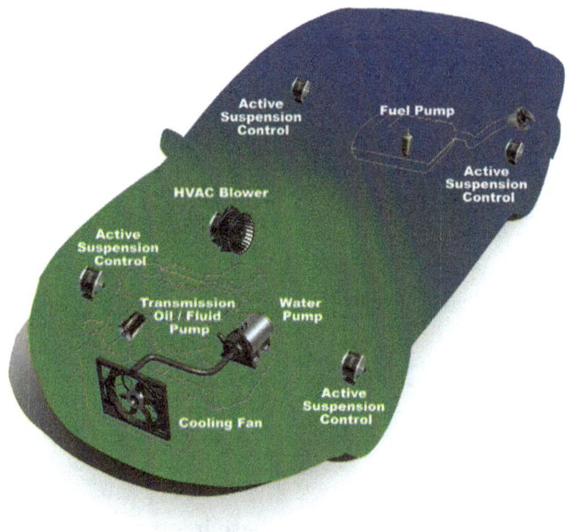


Fig. 3. Overview pump and fan applications

## 2.1 Closed Loop Start Up

As the driver presses the gas pedal a stop-start system has to shift into gear and start driving. Under these conditions every ms counts. Applying the pressure in for example a hydraulic transmission is therefore crucial. Hydraulic systems leverage the control capabilities of Brushless DC motors one step further by rendering an expensive pressure sensor obsolete. From motor speed and torque output information corrected for temperature effects and combined with specific pump information it is possible to predict and control the hydraulic pressure.

A key challenge to realize hydraulic pumps in a sensorless way is to ensure reliable and fast start up under a wide range of loads. For instance a 500W transmission pump should start up to 12 bar in less than 50ms, and this at -40$^{o}$C with highly viscous fat as load, as well as at maximum engine temperature with liquid oil. TruSense technology has demonstrated on multiple types of motors the possibility to reach 12 bar pressure in the system within as low as 50ms, independent of the applied load condition.

TruSense is able to sense the rotor state at stand still through reluctance sensing, avoiding rotor pre-alignment. Thanks to the closed loop start up the motor starts up regardless of the load condition. As long as no BEMF is detected reluctance sensing is applied to track the rotor position. And as soon as a Back EMF signal is detected the robust BEMF tracking is applied to ensure the motor is operated at its most efficient point regardless of sudden load changes.

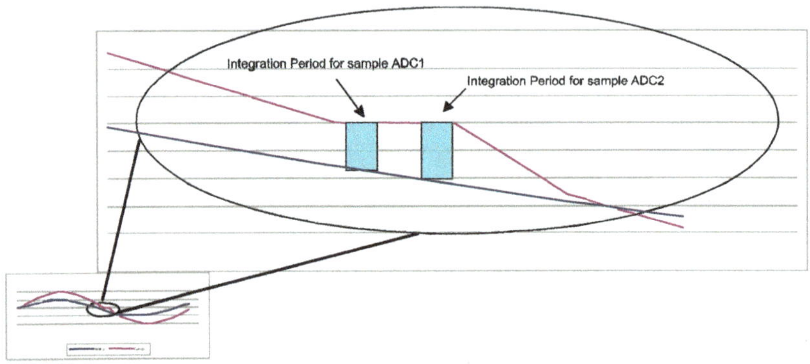


Fig. 4. BEMF sensing scheme using phase integrators

## 2.2 Robust On Demand Operation

On demand operation implies the need for reliable operation over a wide dynamic range, regardless of sudden load variations. A water pump is a typical application that requires a wide range of operating speeds. At very low speed it allows spreading the limited engine heat at start up evenly over the engine to ensure minimum friction as soon as possible. If the water flow is too high it would remove the heat instead. A water pump has to run as efficient as possible, but exceptionally at maximum towing torque it may temporally be boosted to maximum speed by applying a large lead angle at cost of some efficiency. A standard parameter defining the On Demand system is the dynamic range = [Minimum speed / Maximum speed]. Typically TruSense technology achieves a dynamic range of less than 5%. For instance a fuel pump with a maximum speed of 8000 rpm can be operated reliably down to 400 rpm.

### 2.3 Low Noise Operation

Blowers inside the cabin or water pumps in electric vehicles require low noise operation. Sinusoidal operation of so called PMSM motors is a challenge for BEMF operated controllers since there is no free running coil, and thus there is zero-crossing detection window. Additionally to achieve maximum performance the actual motor current is out of phase with the applied PWM. This phase shift influences the resulting voltage on the coils. The Melexis TruSense technology provides similar performance as FOC implementations, without the requirement of position resolvers for closed loop start up.

Fig. 5. BLDC Water pump with integrated electronics

## 3 Summary

The need for improved fuel economy is generating a large need for cost effect and reliable solutions to drive electric motors under very challenging conditions. With its smaller raw material cost and its inherent higher reliability brushless motors are the obvious way of the future. The today's capabilities of sensorless control bring the future already a significant step closer. New developments in hard- and software will continue to increase performance and reduce cost.

## References

[1] VDI Berichte 2007, Nr. 2000.

**Dirk Leman**
Melexis Belgium
Transportstraat 1
3980 Tessenderlo
Belgium
dlm@melexis.com

**Keywords:** power train efficiency, driving cycle, sensorless BLDC motor control, PMSM sine wave motor control, water pump, fuel pump, oil pump, cooling fan, HVAC blower, power steering

# Achieving Efficient Designs for Energy and Power Systems of Electric Vehicles

E. Larrodé, J. Gallego, S. Sánchez, Universidad of Zaragoza

## Abstract

With the advent of the electric car as a new transport model for the twenty-first century, this paper attempts to give an analysis of this technology and justify the need for efficient energetic designs for this kind of vehicles. It is necessary to reach the best compromise of performance with the least waste energy possible. This is the reason why private demand studies based on the type of vehicle and where the vehicle is going to be used are indespensable. In order to be aware of the magnitude of the obtained results, also a comparison of benefits over a conventional diesel drive vehicle, sized under the same constraints, is made. The analysis has been done through ADVISOR (Advanced Vehicle Simulator), a software which vehicles are modelled conceptually in the environment of Matlab / Simulink. This study is a step towards the implementation of a smart management energy system in electric logistic vehicles.

## 1 Introduction

The society, on the one hand due to past excesses and lack of foresight, and on the another hand due to the increased demand caused by economic growth of new countries like China, India and Brazil [1], must face the end of cheap fossil energy resources, whose maximum production, according to some experts, has already been exceeded [2]. Furthermore, the Kyoto Protocol signed by the industrialized countries in 1997, has meant a turn in global energy policies, which have led to commitments with the aim of reducing greenhouse emissions and preventing climate change. These measures greatly affect the transport sector, because in Europe, the transport sector is responsible for 28% of total $CO_2$ emissions. The road transport contributes with 84% of this amount of $CO_2$ [3]. Moreover, private transportation is 95% dependent on fossil fuels. In 2005 this private transportation reach 47% of world oil consumption, which is predicted to increases to 52% in 2030 [4], and therefore this dependence, as the total amount of oil consumed maintain an increasing trend. For all this, electricity must play a significant role both in the transportation and in the exploitation and processing of renewable energies.

## 2 Powertrain Technologies

The powertrain system or a vehicle is made up by the set of elements that provide the mechanical power to the axle. Depending on the nature of the energy sources used to provide the power, traction technologies can be divided into three main groups:

- Internal combustion powertrain: vehicle motion is produced by an engine that converts thermal energy of fuel into mechanical energy that is transmitted to the axle. As fuel can be used a liquid or gas fossil fuel: gasoline, diesel, natural gas or LPG, a biofuel: biodiesel, biogas or ethanol, or other fuels such as hydrogen.
- Electric powertrain: vehicle motion is produced by an electric motor that converts electrical energy into mechanical energy by electromagnetic interactions. The electrical energy has its origin from batteries, fuel cells or solar panels that are mounted on the vehicle.
- Hybrid powertrain: vehicle motion is achieved through a combination of thermal and electrical systems of those seen in previous cases. Depending on who provides the power to the axis, we have two different types of hybridization, traction hybridization and power supply system hybridization. The complexity of the term 'hybridization' when applied to a vehicle will be discussed in the next paragraph.

### 2.1 Hybrid Vehicles

Vehicle hybridization can be viewed from two perspectives; both share as a basic principle the existence of electric and thermal systems that will be supplemented to obtain the movement of the vehicle. These perspectives will differ too, because in the first one is only the electric motor responsible for providing traction to the vehicle, but in the second one both, the electric motor and the heat engine can together propel the vehicle. Thus we have:

- Traction system hybridization: vehicles that have both, an electric drive system and one based on a heat engine, and both have the capacity of propelling the vehicle, either independently or in combination.
- Power supply system hybridization: vehicles that have more than one type of production or storage energy system; at least one of them must be electric. To simplify the case, traction will be provided in any case by an electric motor.

On vehicles that meet the first condition, hybridization is used to make a better use of fuel (increase performances, reduce consumption, or both). An example of this kind of vehicles are those hybrid cars that have engine and electric motor, but only use the electric one for starting and maintaining the vehicle at very low speed for short distances. In the second type, the combination of an electrical system and a fuel one serves to increase the autonomy, and is the case, for example, of one hybrid car that has engine and electric motor, but uses as a system only the electric motor. The function of the thermal traction is to recharge the batteries when they are running low.

## 3 Energy Efficiency in Transportation

The incorporation of new electric-drive systems is linked to efficient and flexible system designs. In these new system designs it is fundamental to know the quantity of energy that the vehicle requires during its operation. In these cases it is necessary to obtain the best compromise of performance with the least possible waste energy. In order to do this it is indispensable to make private demand studies based on the type and the use of the vehicle. Incorporating the concept of energy efficiency also involves the incorporation of a new variable to the transport equation: the operating cycles. An operating cycle characterizes the route of a vehicle; a vehicle must cover a certain distance maintaining at all times previously set speeds.

To pass each of the foreseen cycles, the vehicle needs a certain amount of energy. This energy will vary depending on factors such as vehicle weight or efficiency. In order to achieve the efficient sizing of the powertrain and energy system of an electric vehicle, first of all it is necessary to know the use of the vehicle, the second step is to determine the operating cycle that characterizes it, next step is to determine the amount of energy that the vehicle is going to demand and final step is to select the most appropriate energy supply system.

## 4 Research: Electric Vehicles Architecture

With the objective of performing a comparative study between conventional and electric vehicles, with the help of software ADVISOR (Advanced Vehicle Simulator) the performance and emissions of three prototypes designed for use in three different cases have been analyzed: tourism of private use, tourism of public use (taxi) and delivery commercial vehicle. All of them have a similar outward appearance and have been simulated both with a conventional diesel powertrain system, and with an electric powertrain system:

### 4.1 Private Use

The tested vehicle's powertrain system consists of a 83 kW electric motor, with 18 modules of batteries with a capacity of 91 Ah, providing a nominal voltage of 222 V. The mass of this vehicle is 1,400 kg. The operating cycle in which the vehicle has been tested is characterized for 36,000 seconds of duration by continuous stops, and starts, and parking operations. The range of the vehicle is 10 hours and the maximum speed reached in the cycle is 90 km/h. The cycle is an urban nature driving cycle. In figure 1 are shown: full cycle on the left and a detail of it on the right.

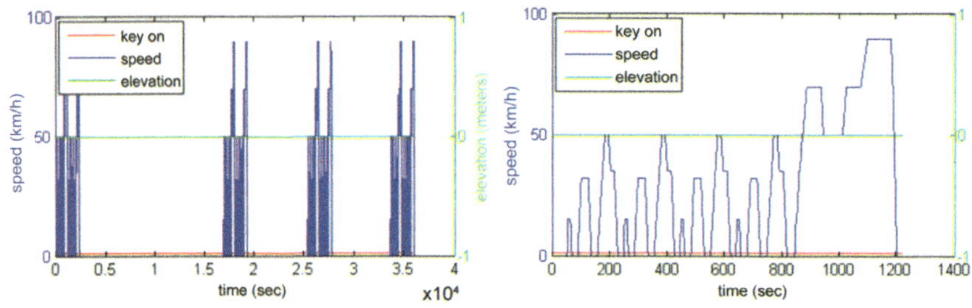


Fig. 1. Urban operating cycle for private use

In order to do a comparison, also has been tested in the same cycle of operating a vehicle propelled by a conventional diesel engine. The vehicle architecture is the same as in the electric vehicle with the exception of the powertrain system. Is this case, the powertrain system consists of a diesel internal combustion engine of 57 kW (75 HP approximately).

### 4.2 Public Use

These cycles are performed by taxis, vehicles for street cleaning or theme park vehicles. They are characterized by continuous stops and starts, short parking operations and a top speed of 50 km/h, since they are limited to an urban environment. The selected cycle has duration of 10 hours and has also been tried without a refuelling stop simulation. In figure 2 are shown: full cycle on the left and a detail of it on the right. Although the vehicle architecture is the same as in the previous case of vehicle for private use, the powertrain system is not the same. In this case the tourism requires a 75 kW electric motor and 25 modules of batteries with a capacity of 91 Ah, providing a voltage of 308 V.

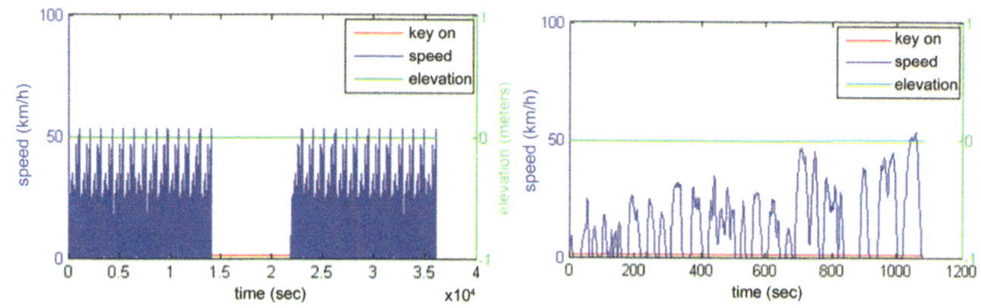


Fig. 2. Urban operating cycle for public use

This new configuration creates a total vehicle weight of 1,556 kg. Performances of this electric vehicle have been compared with a diesel vehicle of 57 kW (75 HP approximately) performances, under the same operating cycle.

### 4.3 Vehicle 3: Delivery Vehicle

In this third case, the vehicle is bigger and heavier and has been tested to its maximum permissible weight: 3,500 kg. The operating cycle is based on numerous stops and starts without parking operations, with a top speed of 45 km/h and autonomy of 10 hours. In figure 3 are shown: full cycle on the left and a detail of it on the right.

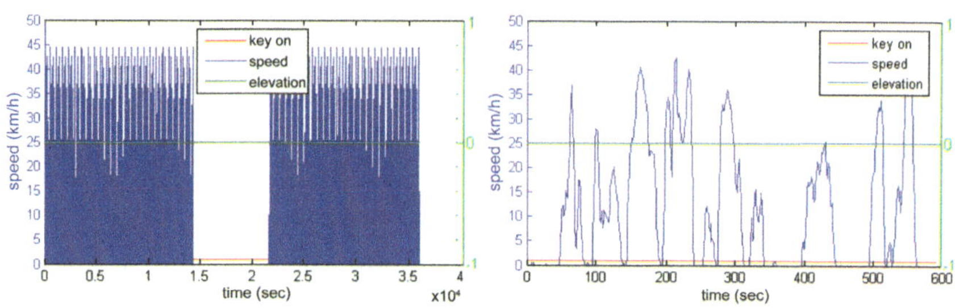


Fig. 3. Urban operating cycle for delivery use

The powertrain system is composed of a 75 kW electric motor, 40 modules of batteries with a capacity of 91 Ah, providing a nominal voltage of 493 V. With this configuration, the car weighs a total of 2,613 kg. Performance of this vehicle has been compared with that of a vehicle powered by a diesel internal combustion engine of 73 kW (98 HP approximately), which has been subject to the same cycle of operation.

# 5 Results

## 5.1 Weight of Powertrain System

In conventional vehicles, the weight of the powertrain system is essentially due to the engine. In electric vehicles, in addition to the electric motor and other components, the propulsion system has also multiple battery modules, so that the total weight of the electric car is higher. For this reason, electric vehicles have a higher energy demand than a conventional vehicle for the same operating cycle. Figure 4 shows in blue the weight of these three electric vehicles, depending on application, while in red is represented the weight of those respective vehicles if they had a diesel engine instead of the electric powertrain system.

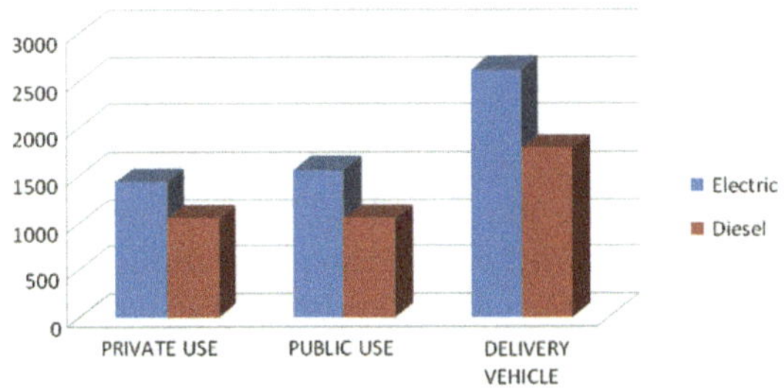


Fig. 4. Total vehicle weight (kg) without load

## 5.2 Fuel Consumption

Electric vehicles have the support of the batteries in times of peak energy requirement; it makes these vehicles have a lower consumption in liters of gasoline equivalent than a conventional vehicle. In figure 5 are represented the consumption of conventional and electric vehicles in blue and red respectively.

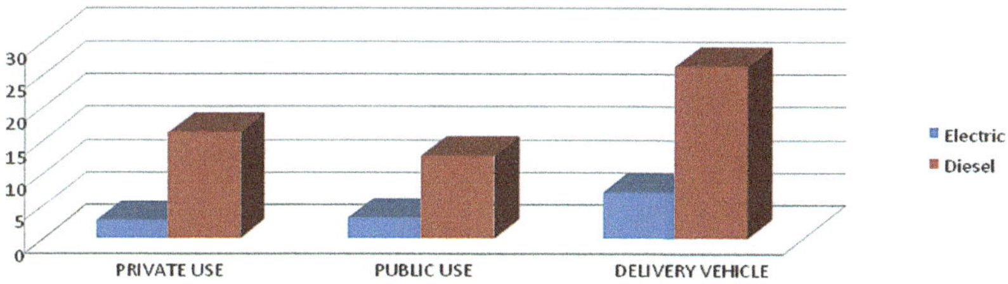


Fig. 5. Consumption of each vehicle in equivalent liters of gasoline per 100 kilometers

## 6 Conclusions

The electric motors are more energy-efficient that conventional engines, so that these vehicles have an energy saving which is determined by fuel consumption measured in liters of equivalent gasoline consumed. Figure 6 shows the percentage of fuel saved when an electric vehicle is used instead of a conventional one for complete the same work.

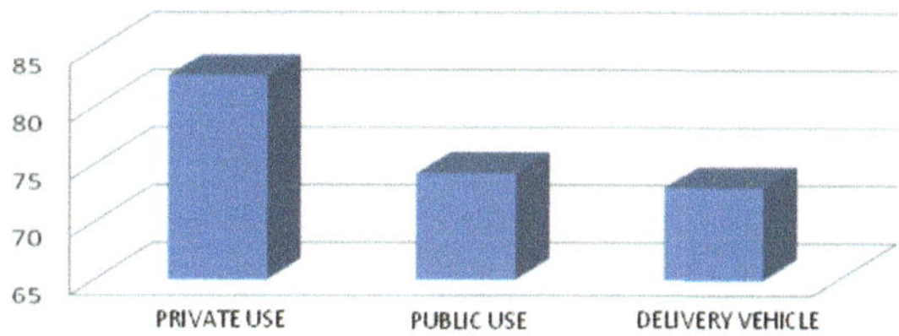


Fig. 6. Decrease in percentage of fuel consumption of electric vehicles in comparison with conventional ones

Electric vehicles have additional weight because they add in their powertrain system an auxiliary power system, such as batteries. Figure 7 represents the increased weight in percentage of electric vehicles vs. conventional ones.

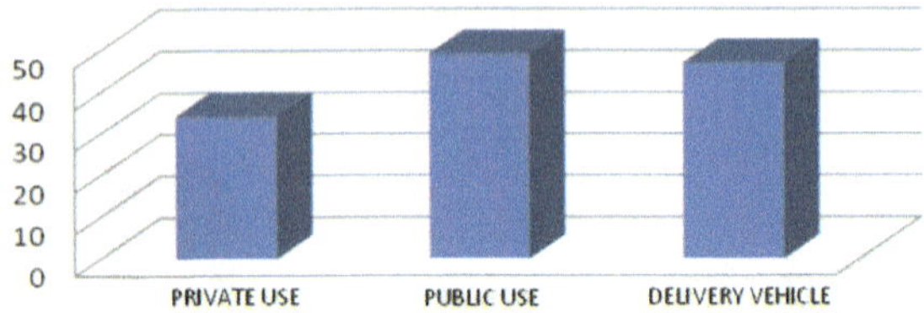


Fig. 7. Weight percentage increased of the electric cars without load with respect to conventional vehicles

## References

[1] Krishna, V., Rapid economic growth and industrialization in India, China & Brazil: At what cost?, William Davidson Institute Working Paper No. 897, University of Michigan, 2007.

[2] Tsoskounoglou, M., Ayerides, G., Tritopoulou, E., The end of cheap oil: Current status and prospects, Icaria Editorial, 2008.

[3] European Commission, 2003.

[4] IEA, World Energy Outlook 2007.

**Emilio Larrodé, Jesús Gallego, Sara Sánchez**
Universidad de Zaragoza
C/María de Luna s/n, 50018
Zaragoza
Spain
jgallego@unizar.es
elarrode@unizar.es
ssanchez@unizar.es

**Keywords:** electric vehicle, hybrid vehicle, energy efficiency, vehicle architecture, operating cycle

# Inverter Losses Reduction Control Techniques for Plug-In HEV and FEV Traction Drive

F. Cheli, D. Tarsitano, F. Mapelli, Politecnico di Milano

## Abstract

Plug-in Hybrid Electric Vehicle (PHEV) or Full Electric Vehicle (FEV) can be very useful in order to meet fuel economy and to reduce pollutant emissions in transport in particular for use in the urban area. This kind of vehicles needs a medium to large battery stack in order to assure a good range in pure electric with zero emissions (30-40 km for PHEV and 120-150 km for FEV). The global power train efficiency is an important factor that has to be improved for a better utilization of the energy stored into the batteries. The electronic power converter (usually an inverter) in the electrical traction drive is one of the key components in the power train. In this paper a control method for reducing the inverter power losses and, consequently, for improving the efficiency and the motor exploiting has been studied. The proposed control method is a Modified Space Vector Field Oriented Control for an induction motor. Furthermore, a better inverter efficiency allows a size reduction of the electrical drive cooling system.

## 1 Introduction

In the PHEV the power train is designed to have an All Electric Range (AER) of 30-50 km. This range is operated at zero emission. In this case batteries that are oversized compared to the conventional HEV are needed and the space available for the whole power train is reduced. The efficiency of all drive components is an important issue for increasing the PHEV AER. In FEV the expected range has to be more than 120 km and consequently, an efficient power train can help to achieve this goal. For both kind of vehicles, PHEV and FEV, the inverter efficiency and dimensions are a key point. This paper shows how, applying a modified Field Oriented Control to the induction motor traction drive, it is possible to decrease the inverter power losses and, as a consequence, to reduce its size especially referring to the cooling system. With the proposed method, the total electrical drive efficiency is increased, especially in the field weakening region. The proposed method adopts a reduced losses space vector

modulator for the constant torque operating region and a six-step modulation for the field weakening region [1, 2, 3]. The two different control methods are linked by means of a bump-less transition algorithm. Finally a simulation analysis referring to a case of a experimental Plug-In-HEV vehicle that was realized at Politecnico di Milano in a project co-sponsored by Regione Lombardia [1, 9] has been performed to test the control system.

## 2 The Reduced Losses Field Oriented Control (RLFOC)

In PHEV and FEV ac drives are used as an electrical traction system adopting induction or permanent magnet synchronous motors [4]. The inverter is usually IGBT-based and controlled by means of a space vector modulator (SVM) or a pulse width modulator (PWM) [4, 5]. Considering an induction motor the typical field oriented control scheme is reported in Fig. 1 [5]. The induction motor model is reported in eq. 1 [5, 6] where the rotor flux $\overline{\Psi}_r$ space vector and stator current $\bar{i}_s$ has been assumed as state variables and a rotating reference frame $d$, $q$ synchronous with the rotor flux has been adopted. The Field Oriented Control (FOC) method is based on a indirect torque and flux regulation operated by means of the $d$, $q$ stator current components control. In fact, in the control scheme of Fig. 1, two current regulators are acting for controlling the flux ($i_d$ regulator) and the torque ($i_q$ regulator). An additional decoupling term and feed-forward terms are added to the output of $i_{d,q}$ regulators in order to make the two closed loop independent [5]. This kind of control is very common for ac electrical drive applied to road vehicles.

The control in field weakening region becomes quite complicated because the flux has to be reduced in order to maintain the motor voltage magnitude compatible with the maximum voltage deliverable by the inverter in relation to the battery dc voltage value. Since the motor voltage depends on stator current, torque, rotor flux and speed the control becomes more complicated and, in literature, a lot of control scheme has been presented [6]. A very simple and effective control method especially in field weakening region (where six step modulator is adopted) is the Direct Self Control (DSC) [1, 7]. DSC has previously been studied [1], demonstrating a good inverter efficiency.

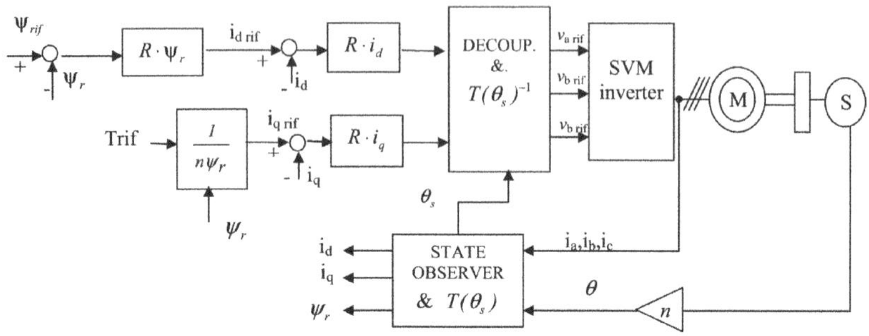


$$
\begin{aligned}
v_{sd} &= (R_s + R_r)i_{sd} + L_{ks}pi_{sd} - \dot{\theta}_s L_{ks} i_{sq} - \frac{R_r}{M}\psi_r \\
v_{sq} &= (R_s + R_r)i_{sq} + L_{ks}pi_{sq} + \dot{\theta}_s L_{ks} i_{sd} + \dot{\theta}\psi_r \\
0 &= R_r\left(\frac{\psi_r}{M} - i_{sd}\right) + p\psi_r \\
0 &= -R_r i_{sq} + \dot{\theta}_r \psi_r \\
T - Tc &= Jp\dot{\theta}/n; \qquad T = n\psi_r \cdot i_{sq}
\end{aligned}
\tag{1}
$$

Fig. 1. Field Oriented Control Scheme.

In the proposed method, the six step inverter modulation is adopted and only the torque is controlled acting on stator voltage space vector position, since the amplitude is fixed and equal to the maximum deliverable.
In this configuration, some drawbacks are unfortunately present [5]:

- Managing a current limit protection is quite complicated.
- The inverter switching frequency is related to the regulator torque hysteresis bandwidth and is not constant in the operating range.
- Is not possible to control easily the flux at zero speed with zero torque reference.

These conditions can be tolerated in railway applications, but not for PHEV or FEV application where, during a city drive cycle, the vehicle starts and stops continuously (zero torque and zero speed condition) and it is important to maintain the induction motor fluxed in order to perform a quick start. On the other hand, FOC has good performance, but it is quite complicated in field weakening and adopts high switching frequency in all operative range, that means inverter losses. In order to improve inverter efficiency maintaining a good control at low speed, FOC has to be modified introducing some DSC feature. For example, different methods based on SVM has been presented [2, 3] in order to reduce switching frequency. The two-phase modulation space vector technique [3] allows to reduce losses and can be adopted together with FOC. For the field wakening region, the minimum losses method for controlling the inverter is to pass to a six step operation adopting a sort of DSC control method [1]. A secondary benefit, adopting the six step modulation, is a power and torque maximum value increase in the field weakening region since the maximum value of the inverter output voltage is greater than the SVM obtainable one. As recall in SVM, if $V_{dc}$ is the inverter input dc voltage, the maximum phase to phase r.m.s. voltage is $V_{dc} / \sqrt{2}$ while in six step operation the maximum is $\sqrt{6} * V_{dc}/\pi$ the difference is about 15 % [5]. An important issue to investigate is the transition from SVM modulation to six step operation since in the first case the voltage vector $\bar{v}_s$ can be controlled in magnitude and position while in the second case $\bar{v}_s$ can be controlled only in position by means of six discrete values [5, 7]. The connection between SVM and six-step is operated by a control in over-modulation mode [5] with a smoothed transition. For field weakening region, the control scheme adopted is reported in Fig. 2. The control architecture is derived from the DSC. In fact considering the induction motor model referred to a fixed reference frame (2) – (4) in steady state condition, and neglecting the stator resistance voltage drop (hypothesis valid under high speed condition such as field weakening region), the stator equation becomes (5) and the torque expression (6). The steady state vector diagram is reported in Fig. 3. The motor torque can be regulated controlling the angle $\gamma$ between stator voltage vector and rotor flux as shown in Fig. 2. The proposed control method is a combination of the architecture of Fig. 1 for constant field region and the scheme of Fig. 2 for field weakening region. A smoothed transition algorithm is introduced for managing the passage between the two control schemes.

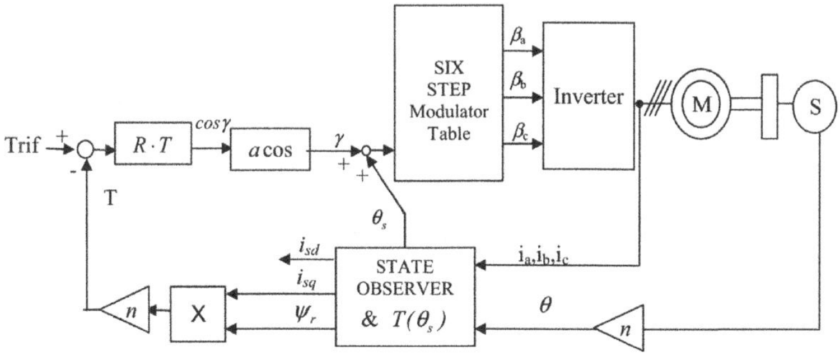


Fig. 2. Modified field oriented control scheme for field weakening.

$$\bar{v}_s^s = R_s \bar{i}_s^s + p\bar{\psi}_s^s \approx p\bar{\psi}_s^s \tag{2}$$

$$0 = R_r \bar{i}_r^s + p\bar{\psi}_r^s - j\dot{\theta}\bar{\psi}_r^s \tag{3}$$

$$p_{fwD} = A_{fwD} \cdot i_f + B_{fwD} \cdot i_f^2 \tag{8}$$

$$p_{swT}(i_c) = (B_{onT} \cdot i_c + C_{onT} \cdot i^2{}_c) \cdot f_s + (B_{offT} \cdot i_c + C_{offT} \cdot i^2{}_c) \cdot f_s \tag{9}$$

$$p_{recD}(i_f) = (B_{recD} \cdot i_f + C_{recD} \cdot i^2{}_f) \cdot f_s \tag{10}$$

$$p_{inv} = 6(p_{swT} + p_{recD} + p_{fwT} + p_{fwD}) \tag{11}$$

## 4 Simulation Analysis

This section presents a simulation analysis performed on full torque-speed operating range for both a standard FOC- control scheme and the purposed reduced losses one RLFOC. The control scheme, the motor and inverter have been implemented in a *Matlab / Simulink* based simulation model. The electrical drive system data referred to the PHEV prototype [9] are reported in Tab. 1.

| Motor Data | | Inverter Data | |
|---|---|---|---|
| **Rated power** | 10 kW | **Input Voltage Range** | 150 – 400 V |
| **Rated torque** | 33 Nm | **Rated current** | 300 A |
| **Overload** | 2.4 p.u. | **Max. current** | 400 A |
| **Rated Voltage** | 105 V | **Max. frequency** | 400 Hz |
| **Rated Current** | 70 A | **Overload** | 2 p.u. |
| **Rated efficiency** | 0.9 p.u. | **Cooling** | Water |
| **Rated Speed** | 2980 r.p.m. | **IGBT** | 2MBI600NT – 060 |
| **pole number** | 4 | - | - |

Tab. 1. Induction motor and inverter data.

| 0.5 p.u. speed | FOC | RLFOC | |
|---|---|---|---|
| **inverter + motor efficiency** | 0.688 | 0.693 | [p.u.] |
| **inverter efficiency** | 0.94 | 0.954 | [p.u.] |
| **inverter losses** | 887 | 749 | [W] |
| **motor power** | 12 | 12 | [kW] |
| **motor torque** | 76 | 76 | [Nm] |
| **1 p.u. speed** | **FOC** | **RLFOC** | |
| **inverter + motor efficiency** | 0.803 | 0.822 | [p.u.] |
| **inverter efficiency** | 0.968 | 0.978 | [p.u.] |
| **inverter losses** | 879 | 650 | [W] |
| **motor power** | 22,4 | 24,1 | [kW] |
| **motor torque** | 71 | 76 | [Nm] |
| **1.5 p.u. speed** | **FOC** | **RLFOC** | |
| **inverter + motor efficiency** | 0.79 | 0.83 | [p.u.] |
| **inverter efficiency** | 0.967 | 0.983 | [p.u.] |
| **inverter losses** | 862 | 517 | [W] |
| **motor power** | 20,89 | 25,2 | [kW] |
| **motor torque** | 44 | 53 | [Nm] |

Tab. 2. Simulation results summary.

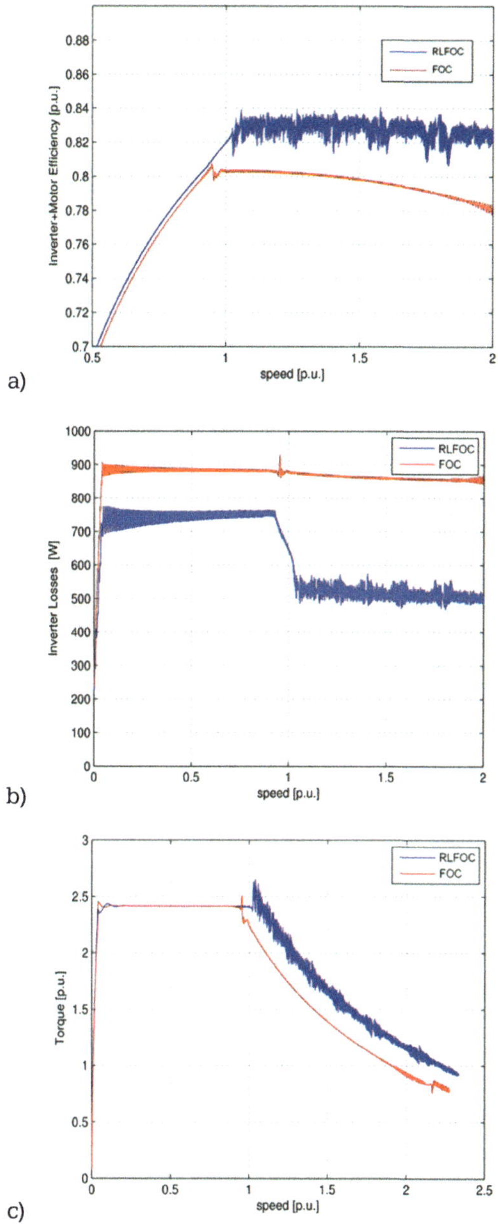


Fig. 4. a) Global electric drive efficiency, b) inverter losses, c) torque capability.

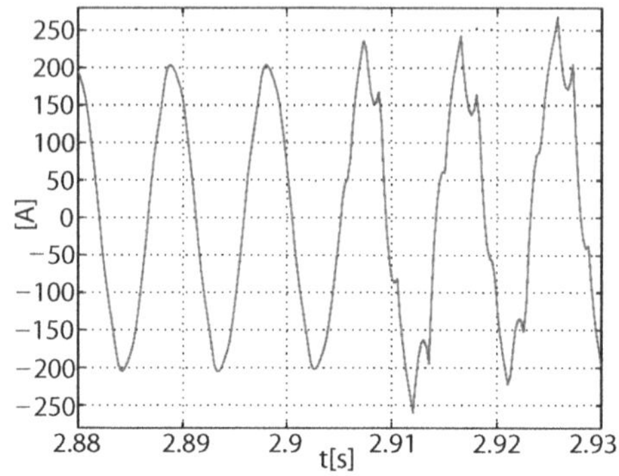


Fig. 5. Stator phase current during transition from constant to field weakening region.

A reference simulation has been chosen for comparison where torque reference $T_{ref}$ has been fixed to the motor maximum overload value. In this condition the motor speed starts from zero and reaches a value of 200% of the rated speed (2 p.u.) and the torque in constant flux region is equal to 240% of rated torque (2.4 p.u.). In this way it is possible to evaluate the efficiency, the inverter losses and torque capability on the overall speed range. In Fig. 4c is shown the maximum torque capability comparison between the standard and reduced losses control. In the second case an extension of the constant torque speed range of about 15% is visible. Fig. 4b shows the instant inverter losses comparison. A losses reduction, especially in field weakening region, can be noticed. In the RLFOC, the field weakening region is operated with six step voltages and the motor current harmonic content is increased respect to the FOC control. As a consequence, the copper motor losses are increased. In order to analyze this drawback of the proposed method, the overall electrical drive efficiency (motor + inverter) has been evaluated and reported in Fig. 4a where an improvement of efficiency is still noticeable. In this way it is demonstrated than the increase of motor copper losses is less than inverter losses reduction so the overall efficiency in RLFOC is better than conventional FOC. Finally, for the RLFOC the transition between the two control modalities has been studied in detail in order to individuate possible discontinuities. Fig. 3 shows the stator phase current during transition reporting a smoothed passage that is also visible in the torque vs. speed curve represented in Fig. 4c. Table 2 summarizes the comparison results referred to the motor maximum torque capability in the whole operating speed range. The same table reports the comparison performed at speed of 0.5, 1 and 1.5 p.u. As is possible to notice, the RLFOC allows to obtain a better inverter efficiency, an inverter losses reduction, a whole electrical drive efficiency improvement and an increase of torque and power capability especially for speed above of 1 p.u. (the field weakening region).

## 5 Conclusions

A control method for reducing inverter losses (RLFOC) in PHEV and FEV and for extending the constant torque region has been analyzed and validated by means of numerical simulations. The evaluation of performances has been performed in comparison with the conventional FOC-SVM control. The RLFOC has been developed joining the classical FOC (for constant torque region) with the DSC control for the field weakening region by means of a "bump-less" transition algorithm that has been validated too. The inverter losses has been modeled by an instant "circuit like" model. The reduction of inverter losses obtained adopting the RLFOC control is very interesting in the whole speed operating range. This could allow to reduce the physical dimension of the cooling system of this component, saving space that can be used for install bigger batteries and extending the AER that is one of the most important feature of the PHEV. The further developing of work will be the implementation on a laboratory test bench for electrical drive for a experimental validation. A subsequence simulation analysis, using a full PHEV vehicle model previously developed [9], will be performed for evaluating the benefit of inverter efficiency improvement on the total energy consumption in an urban drive cycle.

## List of Symbols

| Symbol | Description | Symbol | Description |
|---|---|---|---|
| $\dot{\theta}$ | rotor angular speed | $d, q$ | rotating reference frame |
| $\dot{\theta}_r$ | angular slip speed | $n$ | pole pairs number |
| $\overline{\Psi}_r$ | rotor flux linkage space vectors | $T$ | motor torque |
| $i_{sd}, i_{sq}$ | stator current space vector component | $T_{rif}$ | motor torque reference |
| $v_{sd}, v_{sq}$ | stator voltage space vector component | $T_c$ | load torque |
| $\theta_s$ | rotating reference frame position | $i_{a,b,c}$ | phase currents |
| $L_{ks}, M$ | leakage and magnetizing inductance | $v_{a,b,c}$ | phase voltages |
| $p$ | derivative operator (d /dt) | $v_{a,b,c\ ref}$ | phase voltages reference |
| $T(\theta), T(\theta)^{-1}$ | direct and inverse park transform operator | $R_s, R_r$ | stator and rotor resistance |
| $\overline{v}_s, \overline{\Psi}_s$ | stator voltage and flux linkage space vectors | $s$ | apex for fixed frame referred space vectors |
| $\alpha, \beta$ | fixed reference frames | $p.\ u.$ | per unit values |

Tab. 3. List of symbols.

## References

[1] Manigrasso, R., Mapelli, F., Mauri, M., Tarsitano, D., Inverter loss minimization for a plug-in hybrid vehicle traction drive using dsc control, in SPEEDAM 2008 - International Symposium on Power Electronics, Electrical Drives, Automation and Motion, Ischia, pp. 889–894, 2008.

[2] Trentin, A., Zanchetta, P., Wheeler, P., A new investigation on space vector modulation technique for voltage source inverter in ac drive, in IECON Proceedings (Industrial Electronics Conference), pp. 1627–1632, 2006.

[3] Trzynadlowski, A. M., Legowski, M., Minimum-Loss Vector PWM Strategy for Three-phase Inverters, IEEE Transactions On Power Electronics, Vol. 9, No. 1, 1994.

[4] Zhu, Z. Q., Howe, D., Electrical machines and drives for electric, hybrid, and fuel cell vehicles, Proceedings Of The IEEE, vol. 95, no. 4, pp. 746–765, 2007.

[5] Manigrasso, R., Mapelli, F. L., Mauri, M., Azionamenti elettrici, Voll I-II, P. Editrice, Bologna, 2008.

[6] Casadei, D., Mengoni, M., Serra, G., Tani, A., Zarri, L., Field-weakening control schemes for high-speed drives based on induction motors: a comparison, Power Electronics Specialists Conference, PESC 2008 IEEE, pp. 2159 – 2166, 2008.

[7] Buja, G., Kazmierkowski, M., Direct torque control of pwm inverter fed ac motors - a survey, IEEE Transactions on Industrial Electronics, vol. 51, no. 4, pp. 744–757, 2004.

[8] Manigrasso, R., Mapelli, F., Design and modelling of asynchronous traction drives fed by limited power source, in 2005 IEEE Vehicle Power and Propulsion Conference, VPPC, vol. 2005, Chicago, IL, pp. 522–529, 2005.

[9] Cheli, F., Mapelli, F., Manigrasso, R., Tarsitano, D., Full energetic model of a plug-in hybrid electrical vehicle, in SPEEDAM 2008 - International Symposium on Power Electronics, Electrical Drives, Automation and Motion, Ischia, pp. 733–738, 2008.

**Federico Cheli, Davide Tarsitano, Ferdinando Mapelli**
Politecnico di Milano
Via La Masa 1
20156 Milan
Italy
federico.cheli@polimi.it
davide.tarsitano@mecc.polimi.it
ferdinando.mapelli@polimi.it

**Keywords:** inverter efficiency, field oriented control, space vector modulation, inverter losses, plug-in hybrid electric vehicle, full electric vehicle

## References

[1] Manigrasso, R., Mapelli, F., Mauri, M., Tarsitano, D., Inverter loss minimization for a plug-in hybrid vehicle traction drive using dsc control, in SPEEDAM 2008 - International Symposium on Power Electronics, Electrical Drives, Automation and Motion, Ischia, pp. 889–894, 2008.

[2] Trentin, A., Zanchetta, P., Wheeler, P., A new investigation on space vector modulation technique for voltage source inverter in ac drive, in IECON Proceedings (Industrial Electronics Conference), pp. 1627–1632, 2006.

[3] Trzynadlowski, A. M., Legowski, M., Minimum-Loss Vector PWM Strategy for Three-phase Inverters, IEEE Transactions On Power Electronics, Vol. 9, No. 1, 1994.

[4] Zhu, Z. Q., Howe, D., Electrical machines and drives for electric, hybrid, and fuel cell vehicles, Proceedings Of The IEEE, vol. 95, no. 4, pp. 746–765, 2007.

[5] Manigrasso, R., Mapelli, F. L., Mauri, M., Azionamenti elettrici, Voll I-II, P. Editrice, Bologna, 2008.

[6] Casadei, D., Mengoni, M., Serra, G., Tani, A., Zarri, L., Field-weakening control schemes for high-speed drives based on induction motors: a comparison, Power Electronics Specialists Conference, PESC 2008 IEEE, pp. 2159 – 2166, 2008.

[7] Buja, G., Kazmierkowski, M., Direct torque control of pwm inverter fed ac motors - a survey, IEEE Transactions on Industrial Electronics, vol. 51, no. 4, pp. 744–757, 2004.

[8] Manigrasso, R., Mapelli, F., Design and modelling of asynchronous traction drives fed by limited power source, in 2005 IEEE Vehicle Power and Propulsion Conference, VPPC, vol. 2005, Chicago, IL, pp. 522–529, 2005.

[9] Cheli, F., Mapelli, F., Manigrasso, R., Tarsitano, D., Full energetic model of a plug-in hybrid electrical vehicle, in SPEEDAM 2008 - International Symposium on Power Electronics, Electrical Drives, Automation and Motion, Ischia, pp. 733–738, 2008.

**Federico Cheli, Davide Tarsitano, Ferdinando Mapelli**
Politecnico di Milano
Via La Masa 1
20156 Milan
Italy
federico.cheli@polimi.it
davide.tarsitano@mecc.polimi.it
ferdinando.mapelli@polimi.it

**Keywords:** inverter efficiency, field oriented control, space vector modulation, inverter losses, plug-in hybrid electric vehicle, full electric vehicle zero emission vehicle (ZEV). In this case, the performances have a maximum speed of about 85 km/h with a range of about 40 km, evaluated with an ECE (Economic Commission for Europe Test Cycle) drive cycle and with experimental tests;

- **Start&Stop mode:** this is a hybrid use of the two previously listed modes. In facts, it has been implemented into the vehicle a control strategy by which under a certain speed threshold the vehicle is propelled by the Electrical Motor (EM) and above this threshold it is propelled by the ICE.

In Fig. 1.1 the proposed configuration is schematically represented, pointing out the elements that have necessarily been added for the transformation from a conventional vehicle to a plug-in hybrid one, while in Fig. 1.2 the prototypal vehicle is shown during a competition of the FIA Alternative Energies Cup [5].

## 2 The Numerical Simulation Model

In order to correctly design the components of the electrical drive train that has been added to the standard vehicle, a numerical simulation model has been developed with an object oriented approach. This approach allows a quick model reconfiguration for analyzing different drive train architectures, for testing the sizing of the components and different control algorithms. *Matlab Simulink* simulation tool has been chosen for implementation and in Fig. 2 the main parts of this model are shown.

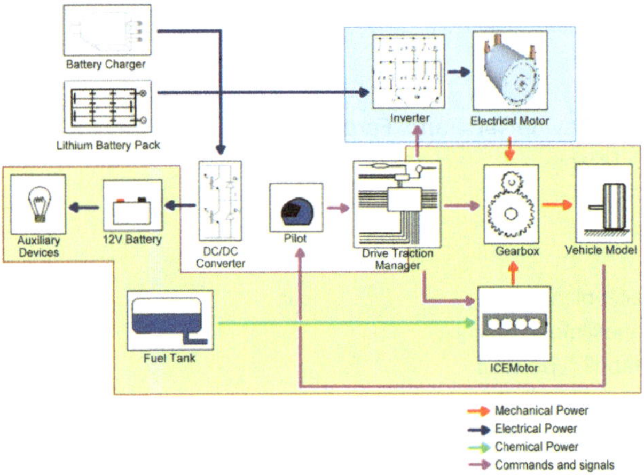


Fig. 1.1 Block diagram of the proposed plug-in HEV

Fig. 1.2 The prototypal plug-in HEV

The objects set, schematically depicted in Fig. 2, represents the whole vehicle model. All the single devices represented have been considered: driver, vehicle control system, Li-ion battery, inverter and electric motor, gear box, clutch, ICE motor, fuel tank, auxiliary on board electrical loads and the longitudinal dynamic of the vehicle. The numerical simulation model has so been used in order to correctly size the performances of the electrical drive train.

The numerical simulation model considers as input the drive mode and the reference speed which is given to the pilot model. This one gives an equivalent signal of accelerator and footbrake to the drive traction manager which splits these signals to the two power drives (EM and ICE) and to the mechanical brake, taking into account the energy control management strategy implemented. The model of both the two motors considers as input the torque reference from the traction manager, the shaft speed and gives as output the effective torque supplied to the input shaft of the gearbox and the motor consumption (gasoline or current from battery). The torque supplied by each motor is summed into the gearbox model which reduces this value to the wheel for the vehicle's longitudinal dynamic calculation (vehicle speed and acceleration). At last the model calculates the gasoline level in the tank and the State Of Charge (SOC) of the battery, taking into account the temperature of the battery pack and the variation of the internal resistance with the SOC [6].

The prototypal vehicle shown in Fig. 1.2 has been instrumented with a large number of transducers in order to have a wide analysis of vehicle's performances under different conditions (for example, urban or suburban driving, enabling or disabling regenerative braking) and to validate the numerical simulation model. For this reason, different comparisons between experimental data and numerical simulation results have been performed; in particular paragraph 2.1

refers to the vehicle propelled by the ICE motor, while paragraph 2.2 refers to the vehicle propelled by the electrical motor. The simulation results have been compared with the experimental ones obtained on the prototype where the same drive cycle has been adopted.

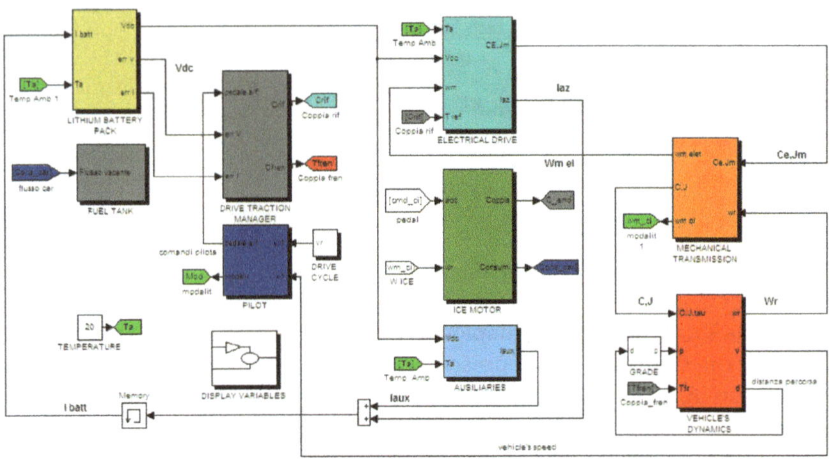


Fig. 2. Vehicle's object oriented numerical model of the Plug-In HEV.

## 2.1 Experimental Validation of the Model Propelled by the ICE Motor

In order to validate the ICE motor model, it is necessary to make a comparison between experimental data and the data obtained from the global model: the same experimental cycle driven in ICE mode was reproduced with the model. Furthermore, provided that our prototypal vehicle has a manual transmission, the moments when the driver has changed the gear ratio and when he has pushed the clutch have been imposed to the model.

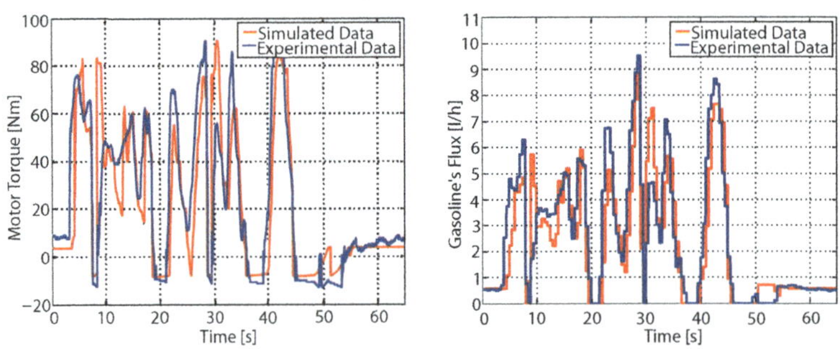


Fig. 3. Comparison between experimental and simulated data for ICE torque (left) and gasoline flux (right).

It has to be noted that there are some differences between the simulation and the experimental results. In particular, the pilot model has to be improved for what concerns the starting from zero speed, in which the clutch effects are not satisfactory simulated. As consequence of these phenomena, a little difference of gasoline flux at the start of the vehicle from zero speed is observed. It is also possible to calculate the cumulative gasoline consumption from the data represented in Fig. 3. It has been obtained a consumption of 45ml from the experimental data and a consumption of 40ml from the simulation. The main cause of the different values obtained in Fig. 3 is due to the pilot's clutch pedal action: in fact in experimental data every change of gear ratio is different from the other one and it is very hard to find a general strategy for modeling this phenomenon.

## 2.2 Experimental Validation of the Model Propelled by the EM Motor

In order to validate the electrical drive train system it has been requested to the model to follow the same drive cycle executed using prototypal vehicle during experimental tests in pure electric mode. In this way it is possible to validate the battery performance in terms of voltage available $V_{\mathrm{batt}}$ and in terms of current supplied $I_{\mathrm{batt.}}$ The comparison between the model performances and the experimental ones is shown in Fig. 4. In the figure it is also reported the energy consumption $Q$ evaluated through the acquired data and through the output of the vehicle's model. The comparison shows a good correspondence between the simulation and experimental data; as consequence the kilometric energy consumption is also well estimated by the model.

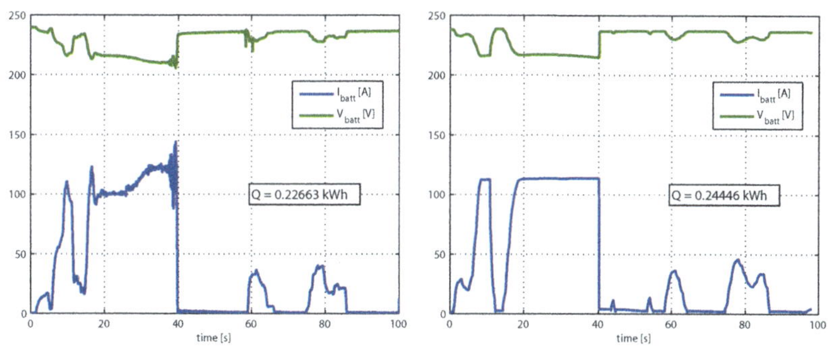


Fig. 4. Comparison between battery experimental data (left) and numerical results (right).

## 3 The Start&Stop Strategy

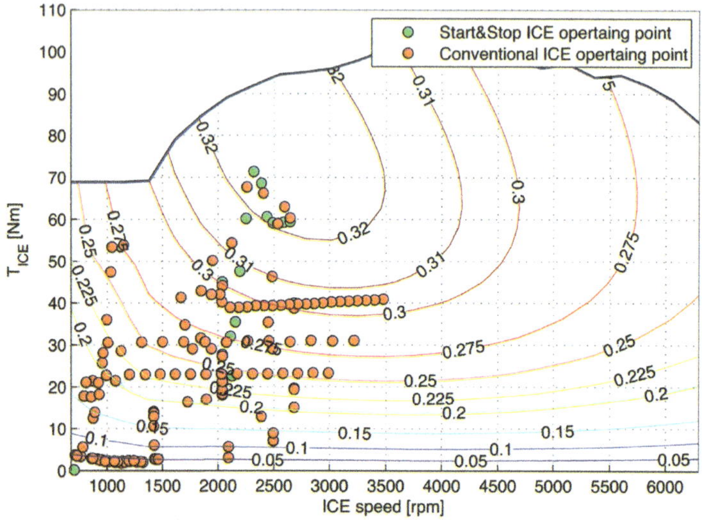


Fig. 5. Operating points of ICE with Start&Stop on ECE drive cycle.

At last the Start&Stop strategy has been studied, by which the vehicle is propelled by the electrical motor under a certain speed threshold. Above this threshold it is propelled by the Internal Combustion Engine. To identify the speed threshold which minimizes the fuel consumption, an optimization procedure has been executed [7]. This approach consist in an a priori optimization procedure since the drive cycle used for simulation is the ECE drive cycle. Using the numerical simulation model it is also possible to analyze the operation point of ICE on its efficiency map considering the original vehicle (equipped only with ICE) and the prototypal one (equipped also with an electrical power train). The results, reported in Fig. 5, highlight that the strategy adopted remove the biggest part of low efficiency operating point, substituting them with the EM operation. This operation leads to an overall improvement of vehicle efficiency and consequently a fuel consumption reduction.

To establish the threshold value which minimizes the overall fuel consumption and that allows to cover a range of about 80 km before the battery has completely discharged, a global analysis on vehicle performances has been executed. In particular in the following table the performance indexes obtained from numerical analysis on ECE drive cycle and from experimental data on a real urban drive cycle are reported. The first index reported is the fuel use drive the ECE or real urban drive cycle considering that the battery is fully charged using renewable sources; the value has been normalized on 100 km as it is usual for automakers. The second performance index, defined as

$$C_{eq-ICE} = c_{fuel} + \Delta SOC \cdot c_{rech} \tag{1}$$

is the equivalent fuel consumption $C_{eq\text{-}ICE}$: this value considers the amount of fuel necessary to feed the ICE and to produce the energy absorbed by the battery using ICE as prime motor for 100 km travel. The third index considered is the equivalent fossil energy consumption at the power station $C_{eq\text{-}PS}$.

$$C_{eq-PS} = c_{fuel} H_{LHV} + \frac{\Delta SOC \cdot E_{batt} 3{,}6}{\eta_{PS} 1000} f \tag{2}$$

In eq. (2) the first part $C_{fuel}H_{LHV}$ represents the chemical energy used to propel the vehicle with the ICE, while the second part refers to the amount of energy necessary to produce the energy absorbed by the battery, considering the European average percentage $f$ of electrical energy produced by fossil sources and the average power station efficiency $\eta_{PS}$. The fourth index

$$CO_2 = c_{fuel} \rho_C \frac{Mm_{CO_2}}{Mm_C} \phi + \Delta SOC \cdot E_{batt} Q_{CO_2-spec} \tag{3}$$

considers the $CO_2$ emission produced during fuel combustion through the molar mass of carbon dioxide and carbon taking into the incomplete combustion of fuel through $\phi$. Also the calculus of the $CO_2$ emissions produced into the power station to replace the energy used from the battery is included. At last, the driving cost €$_{km}$ has been considered as sum of gasoline and electrical energy fee.

$$€_{km} = c_{fuel} €_{fuel} + \frac{\Delta SOC \cdot E_{batt}}{\eta_{ch}} €_{ElEnergy} \tag{4}$$

| Performance indexes | | | | | |
|---|---|---|---|---|---|
| | *ECE Cycle Conventional vehicle* | *ECE Cycle Prototypal vehicle* | *Real Cycle Conventional vehicle* | *Real Cycle Prototypal vehicle* | *Unit* |
| *Fuel consumtpion $c_{fuel}$* | *7,07* | *2,32* | *8,56* | *2,67* | *l/100km* |
| *Equivalent Fuel consumption $C_{eq\text{-}ICE}$* | *7,07* | *7,33* | *8,56* | *8,43* | *l/100km* |
| *Equivalent Fuel consumption $C_{eq\text{-}PS}$* | *2,08* | *1,18* | *2,52* | *1,57* | *MJ/100km* |
| *$CO_2$ emission* | *164* | *95* | *199* | *128* | *g/km* |
| *Driving costs* | *8,5* | *4,9* | *10,27* | *5,66* | *€/100km* |

Tab. 1 Performance indexes.

Finally, in Fig. 6 the experimental data recorded during the test performed on an urban drive cycle and the relative status of ICE motor are reported.

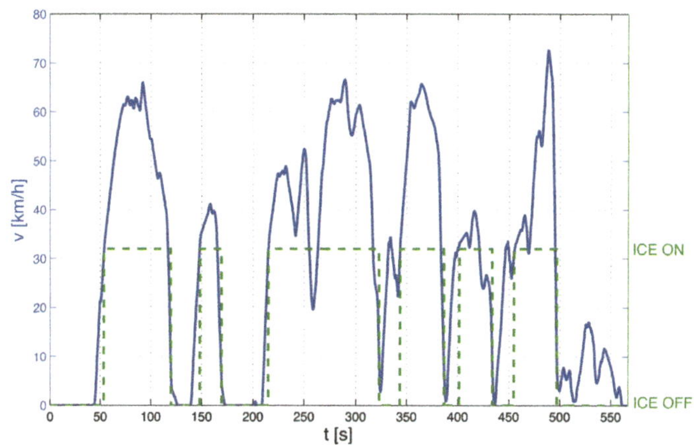


Fig. 6. Operating points of ICE with Start&Stop in real drive cycle.

| *List of Symbols* | | | | | |
|---|---|---|---|---|---|
| $c_{fuel}$ | *Fuel consumpted for the cycle considered* | *[l]* | $\rho_C$ | *Average content of carbon in gasoline [9]* | *[g/l]* |
| $\Delta SOC$ | *Amount of battery energy consumpted from battery for the cycle considered* | *[%]* | $Mm_{CO2}$ | *Molar mass of $CO_2$* | *[g/mol]* |
| $c_{rech}$ | *Litres of fuel necessary to replace 1% of the battery energy by the ICE* | *[l/%]* | $Mm_C$ | *Carbon molar mass* | *[g/mol]* |
| $H_{LHV}$ | *Lower heating value for gasoline* | *[MJ/kg]* | $\phi$ | *Coefficient for incomplete combustion in ICE* | *[ - ]* |
| $E_{batt}$ | *Amount of energy stored into the battery pack* | *[kWh]* | $\varrho_{CO2\text{-}spec}$ | *European average $CO_2$ emission [8]* | *[g/kWh]* |
| $\eta_{ch}$ | *Overall efficiency of charging process* | *[ - ]* | $€_{fuel}$ | *Cost of fuel* | *[€/l]* |
| $f$ | *European percentage of electrical energy produced by fossil sources [8]* | *[ - ]* | $€_{ElEnergy}$ | *Cost of electrical energy* | *[€/kWh]* |
| $\eta_{PS}$ | *Average european thermoelectrical power station efficiency [8]* | *[ - ]* | | | |

Tab. 2. List of symbols.

## 4 Conclusion

The numerical simulation model [6] briefly discussed in this paper has been used in order to analyze the benefits achievable using a Start&Stop energy control strategy on the plug-in hybrid vehicle realized at the Dipartimento di Meccanica of the Politecnico di Milano. The numerical simulation model has been used in order to identify the control parameter set which minimize the fuel consumption and consequently the emission of pollutant. The presented paper shows also the benefit introduced in term of improvement efficiency of ICE motor, letting it to operate only in high efficiency point. At last a set of performance indexes has been established taking into account the pollutant emissions due to the production of electrical energy used to charge the plug-In HEV battery. These indexes has been evaluated considering the average European energy production mix, efficiency and emission. The presented results show that the prototypal vehicle is capable to reduce fuel consumption and pollutant emission also considering that a not negligible part of the energy absorbed by the grid has been produced from fossil sources.

## References

[1] Cheli, F., Viganò, R., Mapelli, F., Tarsitano, D., Realizzazione di un sistema di trazione per un autoveicolo ibrido bimodale, Associazione Italiana per l'Analisi delle Sollecitazioni AIAS, 2007.

[2] Wu, J., Emadi, A., Duoba, M., Bohn, T., Plug-in hybrid electric vehicles: Testing, simulations, and analysis, Proceedings of the 2007 IEEE Vehicle Power and Propulsion Conference, pp. 469-476, 2007.

[3] Emadi, A., Lee, Y., Rajashekara, K., Power electronics and motor drives in electric, hybrid electric, and plug-in hybrid electric vehicles, IEEE Transactions on Industrial Electronics, v. 55, nr. 6, pp. 2237-2245, 2008.

[4] Burke, A., Batteries and ultracapacitors for electric, hybrid, and fuel cell vehicles, Proceedings of the IEEE, v. 95, nr. 4, pp. 806-820, 2007.

[5] http://www.fia.com/en-GB/sport/championships/Pages/alternative.aspx

[6] Cheli, F., Mapelli, F., Manigrasso, R., Tarsitano, D., Full energetic model of a plug-in hybrid electrical vehicle, SPEEDAM 2008 - International Symposium on Power Electronics, Electrical Drives, Automation and Motion, pp. 733-738, 2008.

[7] Mapelli, F., Mauri, M., Tarsitano, D., Energy control strategies comparison for a city car plug-in hev, Conference of the IEEE Industrial Electronics Society, 2009.

[8] Macchi, E., Quale futuro per la generazione di energia elettrica in Italia?, AEIT numero 1/2, 2009.

[9] Office of Transportation and Air Quality, EPA, Average Carbon Dioxide Emissions Resulting from Gasoline and Diesel Fuel, EPA420-F-05-001, 2005.

**Federico Cheli, Ferdinando Mapelli, Roberto Viganò, Davide Tarsitano**
Politecnico di Milano,
Via la Masa, 1
20156 Milan
Italy
federico.cheli@polimi.it
ferdinando.mapell@polimi.it
roberto.vigano@polimi.it
davide.tarsitano@mecc.polimi.it

**Keywords:** plug-in Hybrid HEV, energy control strategy, energetic model, pollutant emission

# Smart Power Control and Architecture for an Efficient Vehicle Alternator - Capacitor - Load System

L. Brabetz, M. Ayeb, D. Tellmann, J. Wang, Universität Kassel

## Abstract

The introduction of additional electrical functionalities will represent a challenge for both the electrical energy and the power balance, which implies the need for good energy efficiency and a low system weight. In particular, safety-critical, power-consuming new systems, e.g. in the area of vehicle dynamics, demand high reliability in the vehicle power supply. The presented concept combines 1) efficient recuperation of the vehicle's kinetic or potential energy even in conventional, IC-engine driven vehicles, 2) the buffering of power peaks and 3) the power management of particular loads. It is based on a modified voltage-controlled alternator and an appropriate charging strategy for a combination of super capacitors, battery and DC/DC converter allowing the use of very compact components. In addition to the theoretical description and modelling, a dedicated hardware-in-the-loop based test set-up was built in order to validate the approach with experimental data and results.

## 1 Introduction

Even in conventionally-driven vehicles, an efficient recuperation by regenerative braking can effectively be used to provide the energy for the increased power consumption of the electrical distribution system. Initial vehicles already in production are using standard or improved alternators and batteries in combination with an energy management which takes into account the current driving condition. The downside of this concept is that the alternator and the battery must be over-dimensioned as long as they operate on a constant voltage level and that the frequency and depth of charge cycles is increased. An approach to address these disadvantages is to add ultra capacitors and to allow a wider range of the alternator voltage. These concepts require DC/DC converters and optional switches and cause negative design impacts on packaging, weight and costs, but provide a higher recuperation rate and allow the downsizing of the battery.

On the other hand, the introduction of additional and safety-critical power-consuming actuators increases the need for voltage stabilization and for high reliability of the power supply. The proposed architecture combines the functionalities of recuperation and power stabilization in order to justify the additional system costs. It is designed to minimize the power requirements to be met by alternator, electrical storage devices and converter, thus reducing the cost impact under the boundary condition of improved energy efficiency. It is based on a voltage-controlled alternator, an appropriate charging strategy for a combination of super capacitors and battery, and takes into account the status of the loads.

## 2 Requirements and Status

The electrical distribution system consists of electrical energy converters, storage and control devices, loads, sensors and electromechanical components. It is defined by its topology, logical schematics, type and number of components and its physical properties. With regard to power and energy, the functional requirements are as follows:

- Sufficient and robust energy supply of all electrical loads with a closely toleranced or even stabilized supply voltage for a defined number of sensitive loads
- Appropriate power characteristics to handle short-term, high-power requirements of specific loads without any functional reduction or EMC impact, in some cases the allocation of a secondary operating voltage, and the implementation of power management as a subfunction of the energy management
- A robust but well-dimensioned energy balance, e.g. providing enhanced starting capabilities even at low SoC, especially in case of a start-stop application, and an efficient predictive energy management
- A high energy efficiency through the reduction of losses and improved component design as well as through the implementation of a high-performance recuperation capability

Furthermore, there are the general automotive technical and economical targets and boundary conditions, which determine the parameters of the applied components such as the type of starter (conventional/start/stop), alternator, battery and voltage control, the use and performance of an additional storage device, DC/DC converters and related switching devices and the power loads and their voltage range.

Consequently, there is a high number of potential architectures and corresponding topologies of which some are already in production or are under investigation, e.g. [1 - 6]. The first approach is to obtain maximum use and functionality by employing regular or improved standard batteries and by applying minimum design changes to a current architecture as shown in Fig. 1a. The solution as in Fig. 1b allows the realization of a low cost start-stop function via a DC/DC converter and the architecture presented in Fig. 1c utilizes the voltage control of the alternator G2 to render possible fast charging of the battery during vehicle braking and thus to allow a moderate recuperation. The optional DC/DC converter stabilizes the voltage supply of sensitive electrical components. The limitations of current batteries can be overcome either by applying a more innovative technology such as LiIon or by adding a second energy storage with complementary technical characteristics as provided by ultra capacitors (Fig. 2a). A solution as shown in Fig. 2b applies a DC/DC converter to the ultra-capacitor and thus makes better use of its energy content due to $E = CU^2/2$. Alternatively, it can be used to provide a power supply with a higher voltage and a low internal resistance. Within these architectures, an efficient recuperation strategy is limited by the maximum power and current of the alternator and the battery. An obvious approach is to let the alternator charge the ultra-capacitor by utilizing an adapted voltage control (Fig. 3a). At higher voltages the possible alternator power output increases considerably, its energy efficiency improves slightly and the low internal resistance of the ultra-capacitors allows the required charging currents. Considering all operating states of the vehicle, this converter must be specified for strong energy flows, which is certainly a drawback. An active cooled linear power switch and a diode as shown Fig. 3b could be used to keep the requirements to the DC/DC converter within an acceptable range.

(a)

(b)

(c)

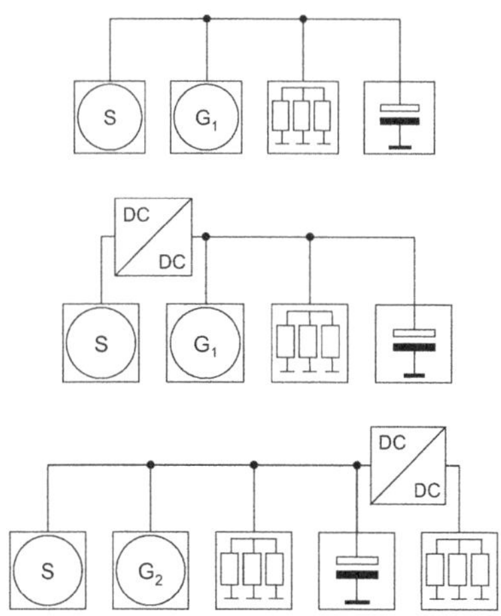


Fig. 1a-c: conventional architectures.

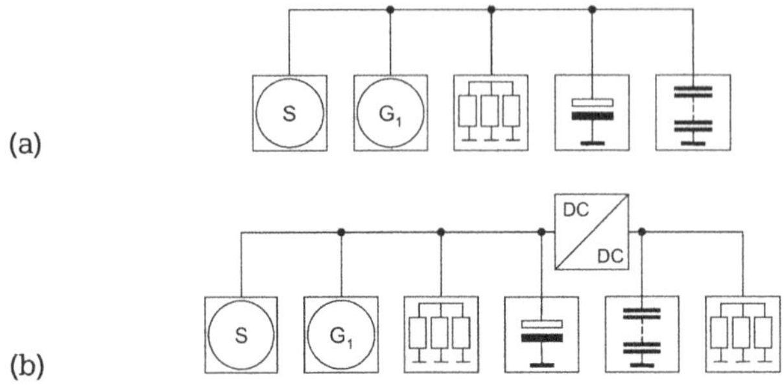

Fig. 2 a,b: use of ultra-capacitors as buffer

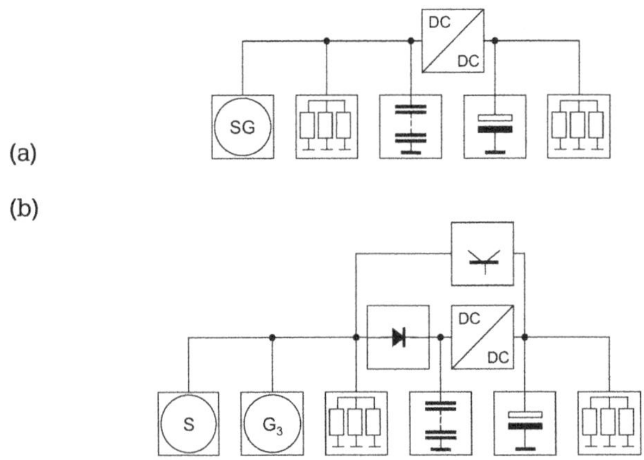

Fig. 3 a,b: use of ultra-capacitors as energy storage

## 3 Architecture

The architecture discussed in this paper (Fig. 4) is based on the type shown in Fig. 3 a and b but uses two relays (or optionally just one, see below) and uses a very different type of control strategy [10]. The basic idea is to switch between two modes of operation in order to minimize the rating of the DC/DC converter. The regular mode of operation, where the alternator is active mainly during negative acceleration or downhill driving, is defined by switch *a* open and switch *b* closed, which corresponds operationally to the architecture shown in Fig. 3a and its functionality. The second operating mode (switch *a* closed and switch *b* open) is applied only if the power flow (bi-directional) exceeds the rating $p_{DC/DC}$ of the DC/DC converter. Thus the converter costs are reduced, however, in this configuration, which corresponds to the architecture shown in

Fig. 2b, recuperation is limited. Without the optional switch *b*, only the second mode is affected and corresponds to the architecture shown in Fig. 2a where the ultra capacitors would be used only partially for energy storage.

In the regular operation mode, the battery is charged and (under certain conditions) discharged via the DC/DC converter which causes considerably less cycling of the battery (the storage of short time power flows is mainly realized by the ultra-capacitor). Furthermore, loads (2) are supplied by the battery and the DC/DC converter. Thus a suppression of transients and, at least partially, a stabilization of their supply voltage is feasible.

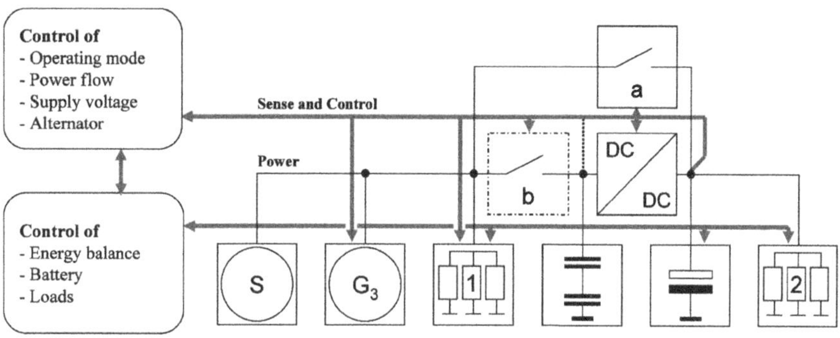


Fig. 4. Proposed architecture

As shown in Fig. 5, a specific control strategy is applied to determine the operating mode. It takes into account the status of the storage devices (SoC and capacitor voltage $U_C$), the power consumption of loads 1 and 2, the power generation of the alternator and the drive cycle (Fig. 5 a-d: regular mode, switch a closed, switch  open, e-f: back up mode, switch a open, switch b closed). A major challenge is a smooth transition between the operating modes without affecting the supply voltage too much while enabling the use of simple switches.

Consequently, the controls of the alternator and the DC/DC converter are sub-functions of the strategy described above. A particular function is the supply-voltage stabilization of loads (2) as it might interfere with the optimum power flow (here the charging/discharging of the battery). One approach is to give priority to the power flow under the boundary conditions of defined minimum and maximum voltages and voltage slopes.

The second group of loads (loads 1, Fig. 4) is in parallel to the alternator and the ultra-capacitors and thus must operate at a wide voltage range ($U_C$, see section 4). The obvious advantage is that the power for these loads is buffered by the ultra-capacitors, which enables short term high power peaks. Furthermore, this part of the power supply needn't be provided by the DC/DC converter. The downside is that only a small group of loads, for instance some heating elements, allow such a voltage swing and that their control and design must be adapted to this new requirement.

As mentioned previously, the second operation mode not only restrains the recuperation capability but inhibits the voltage stabilization function of the DC/DC converter. Therefore, in order to avoid the secondary mode and to apply it merely as a safety back up, the described control unit is to be linked to the vehicle's load management. Thus, its extended control objective is not only to avoid a critical energy balance but also to anticipate a critical power flow situation (Fig. 4).

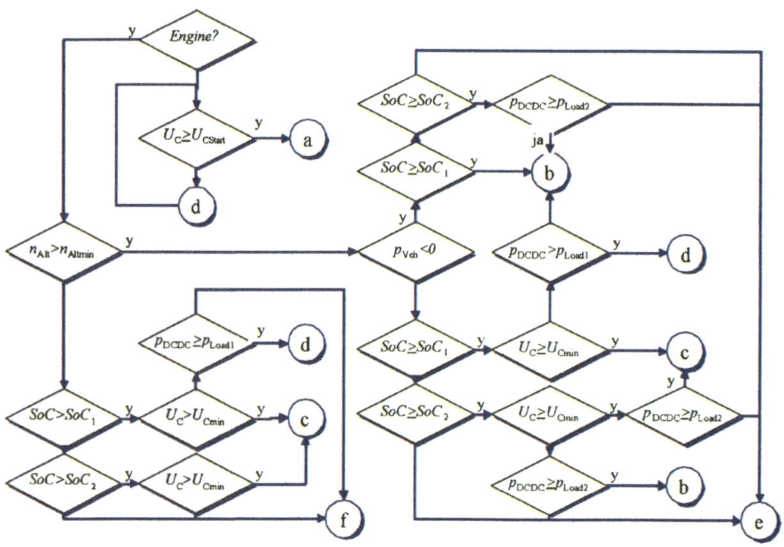


Fig. 5. a-f: $n_{Alt}$, $n_{Altmin}$: rotational alternator speed/minimum, $U_C$, $U_{cmin}$, $U_{CStart}$: capacitor voltages (minimum, minimum.start voltage), $SoC$, $SoC_1$, $SoC_2$: battery state of charge/minimum1/minimum2, $p_{DCDC}$, $p_{Load1}$, $p_{Load2}$, $p_{Veh}$: max. power DC/DC converter, power of loads 1 & 2, recuperation power

## 3 System Characteristics

The energy efficiency of the system improves not only due to the enhanced recuperation capability but also because of the variable and higher operating voltage of the alternator. As to be expected, its efficiency ratio $\eta_{alt}$ improves slightly with the operating voltage as the output power increases at a just marginal rise of losses. This characteristic in conjunction with the considerable thermal inertia allows the use of even a standard alternator at more than twice its regular power rating for a limited time. For a real ultra-capacitor, one obtains an energy efficiency ratio of $\eta_C > 95\ \%$ which exceeds the performance of a battery with only marginal aging effects. For both the alternator and the ultra-capacitor, measurements and simulations [7, 10, 11] have confirmed these results under different operating scenarios, however, appropriate endurance tests have not yet been performed.

Both the system cost and efficiency are to a great extent determined by the capacitance $C$ and its maximum capacitor voltage $U_{Cmax}$. If $C_{cell}$ and $U_{cell}$ are the capacitance and the voltage of a single capacitor, $k$ the total number of capacitors and $n$, $m$ the numbers of capacitors in parallel and in series respectively, one gets:

$$k = mn \quad , \quad U_{C\max} = mU_{cell} \quad , \quad C = C_{cell}\frac{n}{m} = C_{cell}k\frac{U_{cell}^2}{U_{C\max}^2} \tag{1}$$

The maximum energy $E_{rec}$ gained during a recuperation cycle of length $t_{rec}$ is given by:

$$E_{rec} = \int_0^{t_{rec}} U_C I_{alt\_\max} dt = \begin{cases} I_{alt\_\max} U_{C\min} t_{rec} + \dfrac{U_{C\max}^2 I_{alt\_\max}^2 t_{rec}^2}{2U_{cell}^2 C_{cell} k}, & t_{rec} < t_{\max} \\ \dfrac{kC_{cell}U_{cell}^2\left(U_{C\max}^2 - U_{C\min}^2\right)}{2U_{\max\_C}^2}, & t_{rec} \geq t_{\max} \end{cases} \tag{2}$$

$$t_{\max} = \frac{U_{cell}^2 C_{cell} k}{U_{C\max}^2 I_{alt\_\max}}\left(U_{C\max} - U_{C\min}\right)$$

where $U_{Cmin}$ is the capacitor voltage at the beginning of the recuperation and $I_{alt\_max}$ is the maximum alternator current. Hence $U_{Cmax}$ must be as high as possible in order to maximize $E_{rec}$. Taking into account the maximum voltages $U_{Altmax}$ and $U_{Sysmax}$ of the alternator and the system, one gets $U_{Cmax} = min\{U_{Altmax}, U_{Sysmax}\}$. $U_{Cmin}$ should be as high as possible to minimize the range of the supply voltage of the loads *1* and to increase the minimum recuperation power $I_{alt}{*}U_{Cmin}$. On the other hand, the maximum recuperation energy decreases with a higher $U_{Cmin}$. Based on the measured characteristics of the components and the modes of operation, a model [10] has been used to calculate the potential energy saving and to determine the optimum value $k^*C_{cell}$. It takes into account the influence of the drive cycle [8], simplified vehicle controls and dynamics, fuel consumption relationships, electrical power consumption and various system parameters.

## 5 Measurements and Results

While the simulations mentioned above have been used to optimize the architecture, the experimental, hardware-in-the-loop based set-up as described in [11] has been used to verify the results and to parameterize the component models. E.g., Fig. 6 presents the measured and simulated capacitor voltage vs. time, which shows a good correlation.

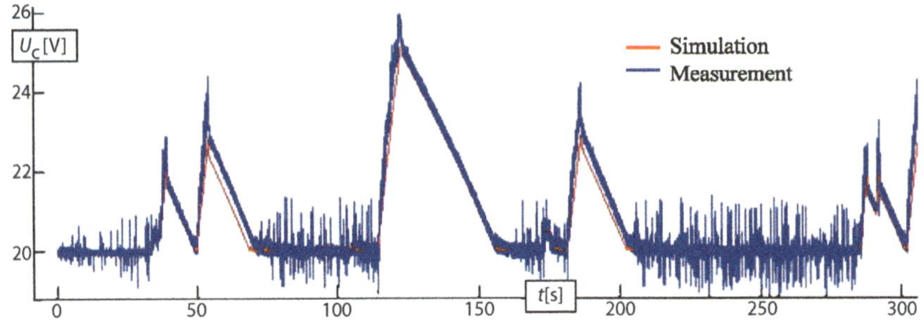


Fig. 6. Simulated and measured capacitor voltage $UC$ vs. time.

Different scenarios of $C$, $U_{Cmin}$ and $I_{load}$ were investigated based on a simulation for a FTP75 drive cycle and a lower middle class vehicle. As expected, the optimum $U_{Cmin}$ depends on $C$ (Fig.7a) and there is no significant gain in fuel savings by increasing $C$ above a given value. Fig. 7b shows that the range $U_{Cmin}$ – $U_{Cmax}$ increases with smaller values of $C$, thus demanding a wider operating voltage range for the loads $I$.

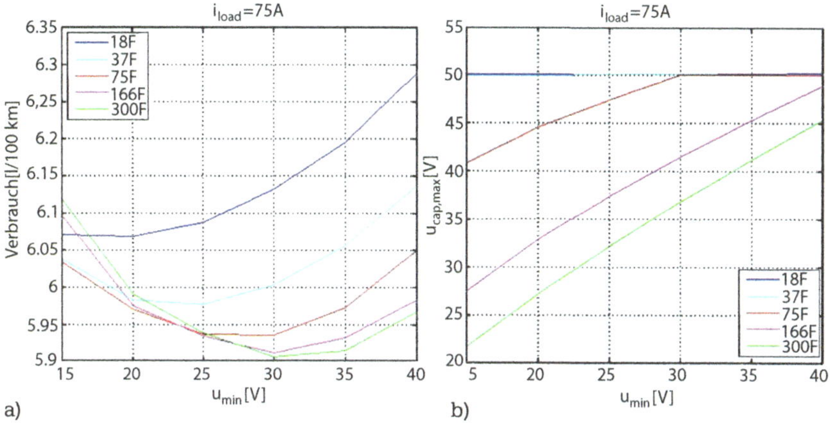


Fig.7. a) Avg. fuel consumption vs. $U_{Cmin}$ and b) max. voltage $U_C$ reached vs. $U_{Cmin}$ for various C over drive cycle.

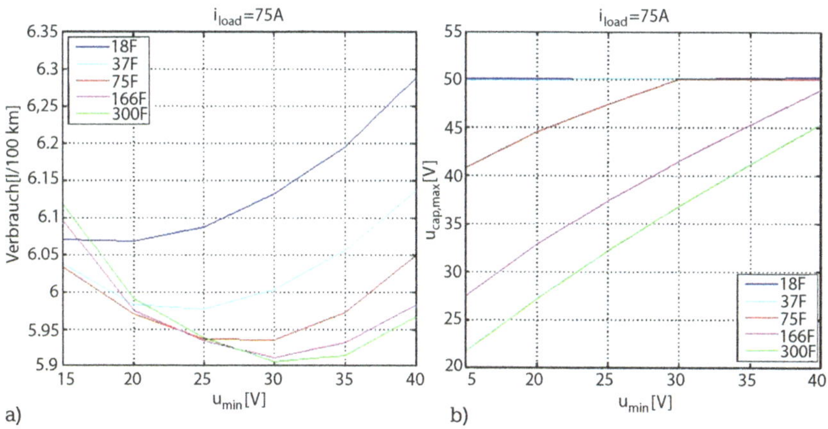


Fig. 8.a) Weight of the installed capacitance vs. $U_{Cmin}$ and b) fuel consumption vs. weight.

Using the installed capacitance and the simulated voltage range, the weight of the required capacitor module was estimated (Fig. 8a). It can be seen that the optimal value of $U_{\mathrm{Cmin}}$ leads to different weights. As shown in Fig. 8b, an installed weight above 2.5 kg has no significant influence on the fuel consumption. In addition, the necessary space required for installation would lead to uneconomical solutions. Due to the fact that the load of the electric consumers is not constant and not directly correlared to the actual driving maneuver, the proposed architecture has further optimization potentials, which will be investigated in future studies.

## References

[1] Richard, D., Dubel, Y., Valeo StARS Technology: A Competitive Solution for Hybridization, 1-4244-0844-X/07, IEEE 2007.

[2] Winkler, J., Esch, S., Mikrohybrid ohne Starter Generator mit den Funktionen Schnellheizung und Rekuperation, Conference Baden Baden, 2005.

[3] Kurz, A., Ultracapacitor alternative solution with starter batteries, Automobil Elektronik, p. 28, 2006.

[4] Schneuwly, A., Prummer, M., Auer, J., Ultra Capacitors: New Energy Storage Concepts in Automobiles, VDI Berichte Nr. 2000, 2007.

[5] Neugebauer, St., El-Dwaik, F., Hockgeiger, E., Mattes, W., Intelligent Contributions for efficient Dynamics of the BMW fleet – the Auto-Start-Stopp Function (ASSF) and Brake-Energy Recovery (BER), VDI Berichte Nr. 2000, 2007.

[6] Brabetz, L., Die Zukunft der Bordnetzarchitektur, Bordnetzarchitekturen, Konsequenzen für das Energiemanagement, Euroforum, München, 2002.

[7] Miller, J.M., McCleer, J.P., Cohen, M., Ultracapacitors as Energy Buffers in a Multiple Zone Electrical Distribution System, Maxwell Technologies, White Paper, 2004.

[8] Ploumen, S., Urhahne, J., Spijker, E., Karden, E., Fricke, B., Regenerative Braking, Possibilities of Brake Energy Recovery, 3rd Int. Automotive EE Systems, 2007.

[9] Rhode-Brandenburge, K., Verfahren zur einfachen und sicheren Abschätzung von Kraftstoffverbrauchspotenzialen, Haus der Technik, 1996.

[10] Brabetz, L., Ayeb, M., Tellmann, D., Efficient vehicle power supply by adaptive energy, charge and heat management of an alternator - super capacitor system, SAE Congress, Detroit, 2009.

[11] Brabetz, L., Ayeb, M., Tellmann, D., Increase of recuperation in vehicles with conventional powertrain, ICSAT, Ingolstadt, 2010.

**Ludwig Brabetz, Mohamed Ayeb, Dirk Tellmann, Jiayi Wang**
Universität Kassel
Wilhelmshöher Allee 73
34121 Kassel
Germany
brabetz@uni-kassel.de
ayeb@uni-kassel.de
tellmann@uni-kassel.de
jiayi.wang@uni-kassel.de

**Keywords:** electrical vehicle architecture, recuperation, powermanagement, energy efficiency, ultra-capacitor

# Advanced Mobile Information and Planning Support for EVs (MIPEV)

P. Bouteiller, Finfortec GmbH
P. Conradi, Steinbeis GmbH

## Abstract

Introducing electric vehicles (EVs) as personal transport vehicles requires new driver information systems. We anticipate new mobile information systems, assisting drivers on their emission-free, but, due to initial restrictions in battery capacity, range-limited journey. These initial limitations will create an opportunity for new, integrated mobility concepts, triggering a new class of mobile, IP-based car driver/traveller information systems. We propose a system environment, based on a combination of a general mobile user interaction system, aggregating individual mobility offers, assisted by co-modality integration. This could be realized by combining private and public transport, new car rental and car sharing/car pooling concepts.

## 1 Introduction

Generally, each system consists of a fixed on-board-unit with electronic lock, charge metering, localising and communication properties in the vehicle, a central server, and a mobile unit, as shown in Fig. 1. The mobile unit is preferably a standard modern smart phone (e.g. iPhone, Nexus, Palm Pre), offering navigation, power management, information and convenience features in combination with IP-based car sharing/rental concepts and public transport timetables, thus maximising the options a user has to reach any destination.

The mobile system part is guided by an algorithm, analysing the route's properties, the user's preferences and his budget and proposing charging or battery swapping stations on the planned route, and, if so required, alternative modes of transport, which will then be passed on to a reservation management system, making sure the driver can enjoy a maximum of comfort and peace-of-mind on his journey.

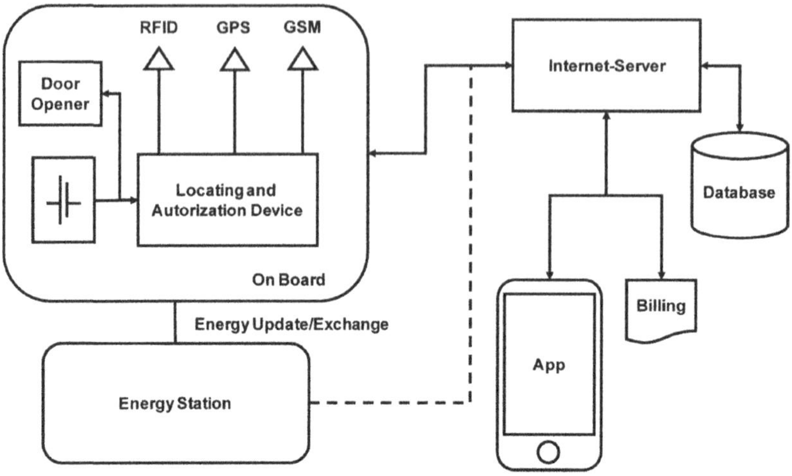


Fig. 1. Architectural view for an individual assistance system for EV

## 2 Social Mobility Networks

We currently witness the transgression of social networks [1, 2] to mobile devices, leading to new, location based community features. At the same time, commercial car rental and car sharing systems are moving towards community-based models, leading to a new breed of socially enabled mobility apps. Travellers will increasingly have a variety of options to organise their trips, be they community based (car sharing, car pooling) or a seamless integration of traditional concepts (car rental, public transport). This could indeed lead to a new class of social mobility networks.

Currently we observe a lot of innovation and dynamic in this market. Some examples are car2go [3] and Zipcar [4] for ad hoc rental, Avego [5], and Autolib [6] for car pooling, Caribo [7] for a mobile agency for arranged lifts. DBRent is offering an iPhone-App to the user for locating pool bicycles nearby [8]. Vélib' offers bicycles in Paris with search and account balance over iPhone [9].

### 2.1 Integrated System Utilising Rental, Pooling, Co-Modality Options

The functions rental, ad hoc rental, pooling and taxi have similar order functions. Thus, the ordering could be combined within a common location based system framework, comparable to a social network that covers a number of different services. This allows for social features like travelling together, group meetings as purpose of a journey, or individual service deployments for network partners. These services are hosted on a cloud-based server architecture and can be administered from anywhere. Users can manage their own information and are taking advantage of the joint service for a variety of applications, comparable to autonomous taxis driven by individual entrepreneurs, who are organised through a central office.

We develop our system from an end-user perspective, taking into account that options and requirements will change in future. The early stage of e-mobility will lead to a generation of cars that are less personal and more expensive compared to traditional ICE cars. They will be taken up by early adopters in megacities [10, 11] who have a higher than average disposition for social networking and innovative concepts. This could lead to new forms of car pooling, rental and sharing, enabled by new apps using the power of social networks.

Thus, a mobile information system using real time information is at the core of our proposed solution, providing users with options and solutions according to their specific requirements. The user's smart phone serves as central and comfortable interface to use and administer all services. It also provides the user with all relevant information concerning the mobility state of his electric car, allowing for a fully automated planning and optimising of co-modality enabled travel. A number of new business models emerge from these initial thoughts.

## 2.2 Planning Activities of the End User

Some activities required for an ad-hoc travel planning with electric cars are depicted below in a state chart, as shown in Fig. 2.

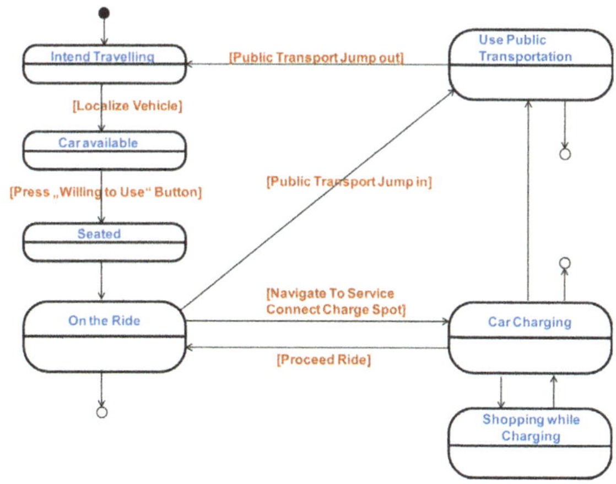


Fig. 2. State Chart/UML of intention and activities of end user

The user in this example has a smart phone and is registered with a social community, where he enters desired destination, time and budget. Driver and rider are matched on the community platform and mutual GPS positioning will facilitate the physical meeting when the trip commences. Administration and billing are fully automated.

The system will be able to calculate an optimal itinerary, including public transport, available rental cars in the vicinity, taxis (which can be ordered by the push of a button), and community rides or any combination of these options depending on the respective user's preferences.

Private offers in the social community system coexist with commercial offers and the system will calculate cost/time profiles for alternative options. Users who take care of recharging electric cars by dropping them at charge spots during or after their journey can benefit from discounts whilst users who prefer to drop cars half or nearly depleted will have to pay a premium.

A number of use cases emerge from these thoughts leading to a new travel experience with social and communication interactions, all coordinated through the interface of a smart phone. These consist of:

- Travel planner, navigation, location based services
- Charging state of battery, remaining range of the car, available charge spots, organisation of battery charge/swap, accounting, billing
- Push-info-service for vehicle swap (co-modality), placement of sharing-offers

The combination of electric cars with personal information systems will lead to exciting life-style options. Navigation support, charge advice, system status data are integrated into electric mobility via multi-touch smart phones. Smart phones and tablets will be the user interface for keeping users and passengers up-to date outside of their car.

## 2.3 System Composition

We propose a general, overarching system design, consisting mainly of an on-board unit, an Internet server, and a smart phone app. The system can be used for a number of use cases, which can be flexibly combined due to the modular system architecture. In principle, the system should work as shown in Tab. 1:

| | | Systems | | | | | |
|---|---|---|---|---|---|---|---|
| | | Vehicle | Driver | Car Passenger | Charge Spot Network | Network | Post Processing / Clearing |
| Tasks | Electronic Assistence | **Board Information System** | **Personal Information System** | **Personal Information System** | **Metering, Communication Devices** | **Server** | **Server, Delivery** |
| | Battery Load / Exchange | old/new Energy Status | Authorize Exchange | | AAA, Charge, Exchange, Recovery | Exchange Management | Billing/km |
| | Transport Operation | AAA, Energy Status | Authorize Navigation Charge | | | | Billing/kWh |
| | Community Features | Travel Offer | Contacts, Travel, Meetings | Contacts, Travel, Meetings | Location Based Services | Social Network | Billing/km |
| | Co-Modality Features | | Transfer | Transfer | | Car Location & Travel Information | Billing/tarif |

Tab. 1. Structural and functional composition of the system

We have to distinguish between driver and passenger. We assume the integration of the car and relevant charge spots into the system and a versatile server architecture for ubiquitous availability.

There are three main tasks: Individual transport with consideration of charging requirements, seamless change of transportation mode (including public transport), and the integration of communities for rental and pooling options, both private and commercial.

From a technical perspective, a common and responsible handling of the limited resources of electric vehicles can be achieved irrespective of the type of ownership, i.e. whether it is a car share, pooling, rental with or without charge spots, or a commercial taxi. The required logistics for driver, passenger or fleet owner can be realised with an intelligent fleet management and utilisation control system, which considers different commercial parameters for each individual agent. Such an intelligent cross-vendor mobility concept will increase customer benefits and satisfaction levels. A single smart phone application can manage all relevant services and even suggest the perfect mode of transportation based on the users current location.

### 2.4 Special Requirements for Electric Vehicles

How can we design the system for future requirements of EV users? Due to the limited range of the car the main difference to ICE cars is the charging requirement. After nearly each use, someone should connect the electric car to a suitable charge point. We provide a draft concept for an information system for use with individual electric cars. It informs the user via smart phone regarding the car's charging state. This information can also be used for an exact range prediction (in conjunction with a navigation system). The required charging procedure after use can be organised in any of the following ways:

- The car is only considered "returned" when connected to an authorised charge spot.
- The user is incentivised to connect the car after use with financial rebates.
- Recharging is in the sole responsibility of the provider.

When selecting a car for rental, the relevant charging state should be indicated, including the computed max. range. The user can then decide whether this range is sufficient for his purpose, whether the range will be sufficient at time of pickup, or whether he should opt for a different car.

## 3 Conclusion

The obvious trend towards the mentioned flexible vehicle usage (e.g. car2go, ZIP Car, Call a Bike flex) pose a challenge when transferred to electric mobility: The quality of service is only possible, when the majority of cars is recharged at regular intervals, preferable after every use. Supported by social network features and economic incentives, as well as an improved charging infrastructure, we believe it will be possible to maintain the best possible level of freedom and spontaneity whilst responsible behaviour is encouraged to the good of the community.

The most relevant aspect will be the connection and administration of various business models via open APIs to the unified infrastructure of social networks, with the provision of mobile accessible real-time information. Based on frequently provided core services for the user, like identifying the current geo-position, provision of real-time information and booking and unlocking of electric vehicles or alternative means of transport, a new breed of applications is evolving, which are required to satisfy future requirements in terms of ecology and comfort for a new era of mobility.

## References


[1] Boyd, D., Ellison, N., Social Network Sites: Definition, History, and Scholarship, http://jcmc.indiana.edu/vol13/issue1/boyd.ellison.html, last visited Feb 14, 2010.

[2] Breiger, R., The Analysis of Social Networks, Handbook of Data Analysis, SAGE Publications, London, 2004.

[3] Geiss, C., car2go – die Zukunft urbaner Mobilität, Daimer Business Innovation, held at IHK Potsdam, Potsdam, 2009.

[4] Keegan, P., Zipcar – The best new Idea in Business, CNNMoney.com, Fortune Magazine, Aug 27, 2009.

[5] Thurn, S., Kelly, H., Surface Transportation 3.0, Feb. 18, 2009, www.cra.org/ccc/docs/init/Surface_Transportation_3.0.pdf, last visited Feb. 14, 2010.

[6] Spinak, A., Chiu, D., et al., SII – Sustainability Innovation Inventory, July, 2008, http://www.connectedurbandevelopment.org/pdf/sust_ii/autolib.pdf, last visited Feb. 14, 2010.

[7] Ander, R., Caribo, Die erste mobile Mitfahr-Community, Sept., 2009, www.save-our-energy.de/attachments/244_-1_5_cariboSept09de.pdf, last visited Feb. 14, 2010.

[8] Stolberg, A., Hoffmann, C., Projektion Call a Bike, Wissenschaftszentrum Berlin für Sozialforschung GmbH (WZB) und DBrent GmbH, 2005, http://www.wzb.eu/gwd/mobi/pdf/hoffmann_stolberg2005_callabikeII.pdf, last visited Feb. 14, 2010.

[9] L'Apur, l'agence d'urbanisme de Paris, Implanter 1451 Stations VELIB' Dans Paris, http://www.apur.org/images/notes4pages/4P27.pdf, last visited Feb. 14, 2010.

[10] Walsh, B., Big-City Buyers Seem Ready for Electric Cars, Time in Partnership with CNN, Jan. 11, 2010, http://www.time.com/time/business/article/0,8599,1953010,00.html, last visited Feb. 14, 2010.

[11] Malorny, C., Krieger, A., Electric Vehicles in Megacities, McKinsey Report, Jan. 11, 2010.



**Philipp Bouteiller**
Finfortec GmbH
Leibnizstr. 55
10629 Berlin
Germany
philipp.bouteiller@finfortec.com

**Peter Conradi**
Steinbeis Innovation GmbH
Lise-Meitner-Str. 10
64293 Darmstadt
Germany
peter.conradi@stw.de

**Keywords:** business model integration, car pooling, car rental, car sharing, co-modality, driver assistance, EV, fleet management, framework, green cars, mobile information systems, mobile travel assistance, public transport, social mobility network

# TU VeLog: A Small Competitive Autarkic GPS Data Logging System

F. Schüppel, S. Marker, P. Waldowski, V. Schindler, Technische Universität Berlin


## Abstract

For the research and development of electric vehicles it is not only important to determine the velocity profile to calculate the energy requirement, but also the car's position in order to establish a charging infrastructure in convenient locations for both user and utilities. For this purpose, the Department of Automotive Engineering of the Technische Universität Berlin has developed a small competitive autarkic GPS data logging system displaying unique features: It does neither require changes to the car nor a special handling by the driver. The battery of the logger lasts for approx. a week. In this paper, we give a validation of the GPS measurement technique as well as an example for an application of the logger.


## 1 Motivation

Today's vehicle usage is normally measured in v-t-diagrams (velocity profiles) such as driving cycles or in mobility surveys like the "Mobility in Germany" project. Driving cycles allow to determine a car's energy consumption and $CO_2$ emissions. Moreover, energy consumption and carbon emissions of different cars can be compared. The driving cycles are often synthesised and as the measurement is carried out under ideal conditions, the actual fuel consumption is usually higher when driving the car on the road. There are several driving cycles among which the most common ones are the 10-15-Mode, FTP 75, and NEDC.

With the help of surveys, the daily driving distances of the respondents can be identified, itemizing the types of commutation (to and from work, shopping, etc.) and urbanization (urban, rural, highway) as well as the number of cars owned by a household. Both types of driving information allow for a statement about the energy consumption of a car and the necessary range of the vehicle. With regard to cars with internal combustion engines (ICE), these data are mostly sufficient. However, the above-mentioned presupposed ideal conditions as well as the individual driving style, which remains unconsidered in the driving cycle analysis, usually result in an increase in fuel consumption under real life conditions. Moreover, there are user profiles that are not represented in the common driving cycles, such as taxis, parcel services, sales representatives or vehicle fleets for on-site service.

With regard to electric vehicles (EVs), however, not only the velocity profile is important, but also the car's position due to the lack of a sufficient charging infrastructure and the limited range of the vehicles. By knowing the car's position, the location of the route, and the charging infrastructure it is possible to predict which user groups will benefit from an EV with today's and tomorrow's infrastructure. Furthermore, it can be established for which distances an EV can be used and when to use a conventional car with an ICE. The reduction in energy consumption can also be calculated. In order to expand the grid and charging infrastructure where it is necessary, i.e. in places with a high operating grade of charging stations, it is also helpful to know where the vehicles are located while parking. In order to have the car fully recharged when it is needed the necessary charging power can be established. For measuring velocity profiles and logging GPS data several measuring instruments are available, the most widely used equipment being the VBOX by Racelogic. However, the VBOX and other commercially available GPS measurement equipment also have certain disadvantages. First of all, the instruments are highly expensive, thus tying-up a lot of capital when using them in high numbers for larger vehicle fleets. In addition, the devices can be stolen. Furthermore, the VBOX relies on external power supply, which necessitates handling by the driver, e.g. by plugging the device into the cigarette lighter.

These facts have motivated the Department of Automotive Engineering of the Technische Universität Berlin to develop an autarkic measurement technique within the framework of the NET-ELAN project, which does neither require changes on the car nor handling by the driver. A small smart competitive data logger called "VeLog" records the GPS position, the current speed and the internal temperature of the vehicle. The device also logs the acceleration values $(x,y,z)$ as well as the humidity and pressure data. In order to keep weight low and size small without losing stamina, the VeLog is only active when a movement is detected. When inactive, the VeLog consumes less than 20mA. Thereby it is possible to log more than 1000km during one week with a 6Ah-battery. The size of the VeLog is approx. 110mm x 95mm x 30mm. It is fixed on the windscreen by suction cups.

## 2 Specifications of the VeLog

The measurement technique that has been developed by the Department of Automotive Engineering of the Technische Universität Berlin uses a data logger for tracking packages. It is distributed by SparkFun Electronics in Boulder (USA). The logger is fixed to a small (60mm x 40mm) printed circuit board (PCB). Sensors for measuring humidity, temperature, and barometric pressure as well as a triple axis accelerometer are included. In order to define the position, a 20 channel SiRF Star III GPS engine module with an integrated antenna is used. The hardware of the logger is based on a 32 bit Arm7(TDMI) platform that is commonly used for automotive applications and consumer electronics. The firmware is built with the free GCC tool chain, which can easily be modified or extended. By updating the firmware it is possible to add or replace sensors. There are many Open Source tools available on the web for the ARM7 platform, by which a new firmware for ARM devices can be programmed. The LPC2148 ARM processor's sleep mode reduces power consumption.

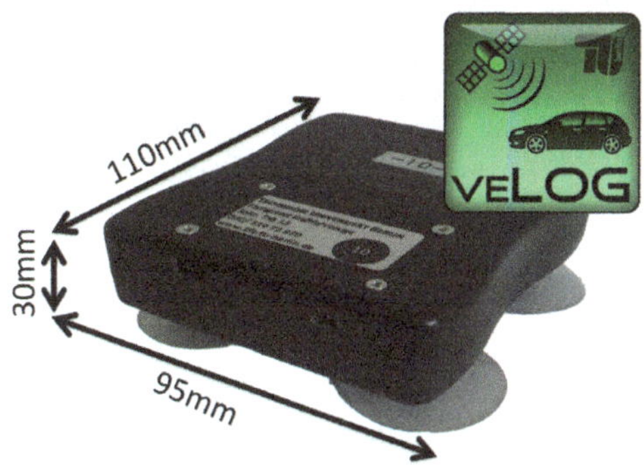


Fig. 1. TU VeLog Dimensions

In order to store the measured data, a micro SD card slot is built into the PCB. Flash cards with a capacity of up to 2GB can be used. Since only a FAT16 file-system is supported, high capacity cards (SD HC) cannot be used. The data volume for one hour vehicle logging is approx. 1.2MB so that SD HC support is not necessary. However, by updating the firmware, a FAT32 file system with flash cards of up to 32GB can be used. For flashing the firmware and downloading the measured data, the VeLog comes with an USB communication port. SparkFun Electronics has created a so-called "boot loader" for flashing through the USB-port, which makes the update process quick and easy without the bottleneck of the RS232 serial port. With the boot loader the compiled firmware is stored on the SD card and flashed on the LPC2148 memory after switching on the logger. When using a lithium-polymer battery it can also be charged via the USB-port.

## 2.1 Sensor Specifications

The four sensors measuring temperature/humidity, barometric pressure, acceleration, and position are described below:

- LIS302DL accelerometer: The LIS 302 is a triple axis accelerometer by STMicroelectronics, which can measure up to 8g with 400Hz. Vehicle data are logged with a 2g scale. With approx. 1mW the power consumption is low. The accelerometer has two interrupts. With the free-fall interrupt it is possible to wake up the logger from power down mode.

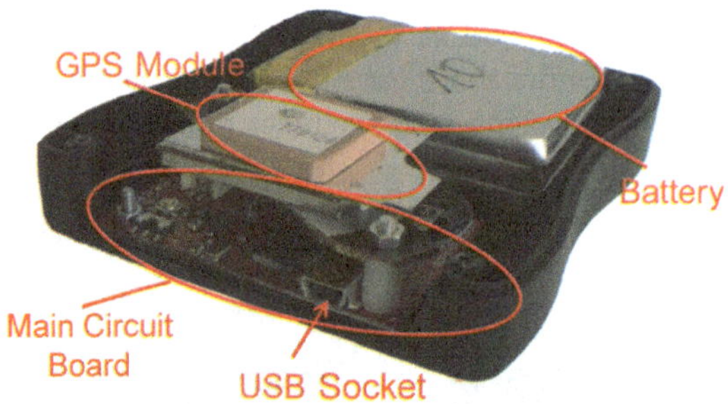


Fig. 2. TU VeLog Assembly

- SHT15 temperature/humidity sensor: The SHT15 is a relative humidity and temperature sensor with a fully calibrated output by Sensirion. Its absolute accuracy for temperature and for humidity amounts to approx. 2%. Power consumption is rate lower than 0.08mW at a 1Hz refresh. The sensor has a power down mode to maximise battery stamina.

- SCP1000 barometric pressure sensor: The pressure sensor by VTI Technologies has a measuring range from 30kPa to 120kPa. Its power consumption amounts to approx. 0,01mA.

- EM-408 SiRF III GPS: The 20-channel EM-408 GPS module by USGlobalSat is based on a high performance SiRF Star III chipset. It has a built-in patch antenna with high sensitivity amounting to -159dBm and a connector for an external antenna. The sat-fix after waking up is very fast because its hot start time is rather short (average of 8 seconds). The positional accuracy amounts to approx. 10m, the velocity accuracy to approx. 0.1m/s. It has a refresh rate of 1Hz, which proves to be a weak point when using the velocity for derivation of the horizontal acceleration. The NMEA output protocol can be chosen by a binary command. The position is sent to the main board at different baud rates through a serial port as an ASCII-string. In order to reduce power consumption by approx. 44mA, the module can be powered down by a 3.3V signal.

- Battery: For the VeLog a lithium-polymer battery triple pack with a capacity of 6000mAh by Union Battery is used. It has a lithium-ion voltage of 3.7V and can be charged by the integrated charging unit.

- Housing: The case is a black plastic box with the size of a standard Eurocard (110mm x 95mm x 30mm) with four suction cups on the top for fixing the VeLog on the windscreen.

## 2.2 Power Management and Data Storage

In order to avoid that the deployment of the VeLog requires changes to the car or handling by the driver, a special energy management is required. Even though logging with a rate of 1Hz, especially the CPU and the GPS-module consume so much energy that the VeLog battery lasts only for two days, an external power supply via the cigarette lighter and a DC/DC converter does not represent an option. The risk of data distortion if the driver forgets to plug in the converter before starting to drive is considered unacceptable. Since the battery with a capacity of 6Ah occupies half of the housing, a larger battery is not feasible, either. The driver's view would be obstructed, since the measuring device needs to be located on the windscreen in order to provide valid GPS signals.

The solution to the problem is an energy management system that powers the logger down when no movement is detected. In a configuration file, it can be chosen how long the VeLog should wait before going to sleep mode. The logger starts up again when the acceleration exceeds the acceleration threshold, which is also defined in the configuration file. The start up procedure is initiated by the already mentioned free fall interrupt of the accelerometer, which forces the CPU, likewise via an interrupt, to wake up. Thereby it is possible to use the sleep mode of the sensors, reducing the power consumption by factor of 10. In sleep mode the battery will last for two weeks. In the case of an average private user driving approx. 300km a week, the battery power will last for more than one week. With special user profiles like taxis, which are driving more than 200km a day, the battery will be depleted much earlier. This will also happen at very low temperatures due to the rising internal resistance of lithium-ion batteries and the freezing of the electrolyte.

Other operating modes are feasible as well, such as continuous or time-controlled logging using the real time clock (RTC) of the CPU. The RTC and the Universal Time Coordinated (UTC) from the GPS module are synced each time the logger goes to sleep or wakes up.

Each sensor has its own sampling rate. The measured data are logged synchronously by using the RTC. With 10Hz the acceleration sensor has the highest sampling rate, whereas the temperature/humidity sensor has, due to its inertia, the lowest rate. The data are stored in two different files: One file contains all measured data except the GPS-signal (date, time, acceleration *X, Y, Z,* battery voltage, ambient pressure, temperature, humidity). This file is stored with a 10Hz sampling rate, although some sensors have a lower rate. For empty channels a placeholder is used. The second file contains the NMEA recommended minimum sentence C GPS-data (message ID, time, fix status, latitude, latitude hemisphere, longitude, longitude hemisphere, speed, bearing, date). It is stored with a sampling rate of 1Hz. Both files have a comma separated ASCII format and can be read with MS Excel or a standard text editor. The tracked route enclosed in the GPS data file with its NMEA format can easily be visualized with Google Maps or GPS map platforms like GPSies.

## 2.3 Optional Functions

The VeLog with its commonly used ARM-platform and Open Source C-code offers several options to add new functions or to modify existing ones. Most of the hardware cannot be changed because it is soldered on the PCB. Only the GPS module, an external antenna and the battery are connected to the PCB with (Japan soldering terminals) connectors and can be replaced without soldering. The source code of the software (=firmware) used for the VeLog as well as the programming environment (WinARM) can be downloaded from the website of SparkFun. Thereby changes to the firmware can be made, the new code can be compiled and afterwards uploaded to the VeLog.

In order to add more sensors, a serial port of ARM-CPU is led through and provided with solder connections. An external sensor, e.g. for measuring the outside temperature, can be connected to it. There are also a number of different GPS modules, which can easily be used as most of them use the NMEA-protocol. They start to work as soon as they have been connected to the five-pin jst-connecter. The firmware needs to be modified if the new module uses a different protocol, baud or refreshing rate, etc. the Department of Automotive Engineering of the Technische Universität Berlin is currently testing the 10Hz GPS module "SkyTraq Venus". With this module, the vertical acceleration can be differentiated from the GPS speed, also when it comes to more dynamic driving maneuvers. More compact or waterproof cases can also be used, provided that a corresponding battery is employed and the new housing is non-metal. Otherwise, the GPS reception could be bad. Without firmware updates it is not possible to maintain power supply with the USB port while measuring because, with the standard firmware, the VeLog stops logging when an USB connector is plugged in. In order to increase battery stamina, bigger or parallel-connected batteries can be used or the power management as mentioned above could be modified. For larger modifications to the firm- and hardware such as installing a cell-phone module, a development board by SparkFun for LPC2148 devices is available. With this board different sensors can also be tested without soldering.

## 3 Sensor Validation

In order to measure the sensor quality of the VeLog, the measured data have been validated with a Racelogic VBOX 3i (fresh calibration certificate). For the designated application solely the GPS-Data is used and validated. The additional sensor will be validated later. The VBOX 3i is a GPS data logging system that logs GPS positions with up to 100Hz (without interpolation). The VBOX has also the advantage of an external antenna. While driving a typical route through Berlin, both GPS loggers were measuring the position and the speed. The measurement was carried out in the inner city with street canyons, where there is a weak GPS signal compared to measurements outside of cities.

### 3.1 GPS-Position Validation

The GPS position was validated via Internet maps or satellite pictures (e.g. OpenStreetMaps, GoogleMaps). Rides at different speeds and with different grades of urbanization (urban, rural and highway) were visualized with the above mentioned map tools. The conformity of the GPS route and the streets were verified. There were no significant variations.

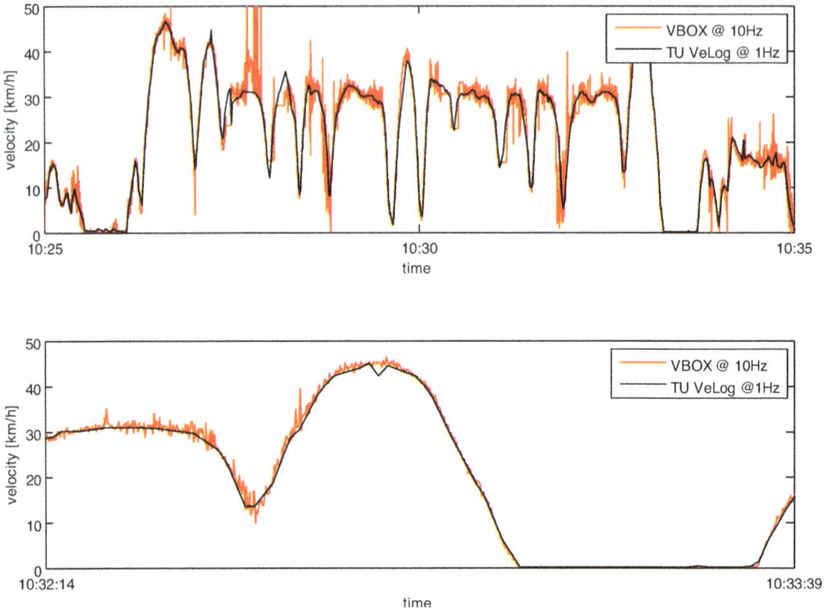


Fig. 3. GPS speed validation

### 3.2 GPS-Speed Validation

The GPS speed signal of the VeLog was validated with the VBOX 3i. The natural sampling rate of the VBOX with 100Hz differs from that of the VeLog (1Hz). In order to reduce the differences caused by the rates, the VBOX data were measured with 10Hz. Due to the higher sampling rate, the VBOX signal is noisier. Apart from this, there are only minor differences between both measurement techniques.

## 4 Application and Results

The VeLog has been developed for use by the NET-ELAN project to allow for an assessment of the technical feasibility of taxis with a battery powered all-electric power train. Due to the daily driving distance of taxis in Berlin, the current battery size and the costs involved, it does not make sense to use a battery the capacity of which is as large as required for the daily driving distance. This requires rethinking our energy saving strategies. The favoured solution consists of recharging the battery while waiting at a taxi stand.

Fig. 4. TU VeLog in the field

Therefore, a measurement with five taxis was performed. The GPS route and speed were logged for four weeks. With the measured data the waiting time at taxi stands could be identified and a typical driving cycle of a taxi was created. Based on the thesis that taxis are recharged while waiting at a taxi stand for longer than 5min and with the help of the driving cycle the battery size was calculated for different scenarios on the basis of a simulation model:

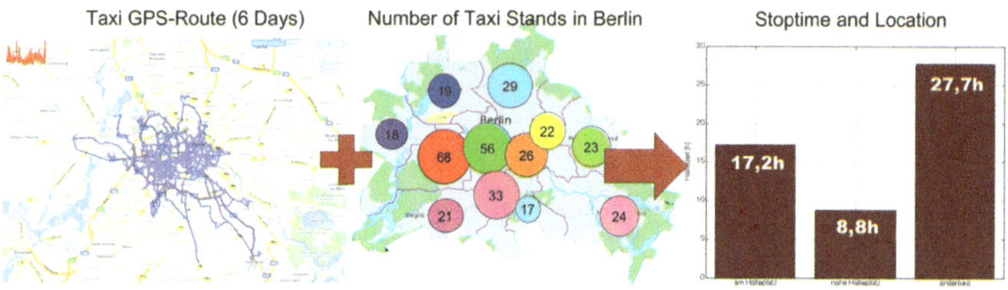


Fig. 5. TU VeLog application

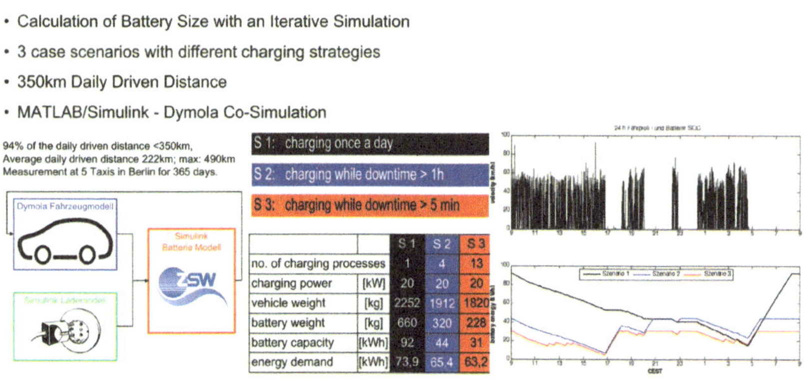


| | | S 1 | S 2 | S 3 |
|---|---|---|---|---|
| no. of charging processes | | 1 | 4 | 13 |
| charging power | [kW] | 20 | 20 | 20 |
| vehicle weight | [kg] | 2252 | 1912 | 1820 |
| battery weight | [kg] | 660 | 320 | 228 |
| battery capacity | [kWh] | 92 | 44 | 31 |
| energy demand | [kWh] | 73,9 | 65,4 | 63,2 |

Fig. 6. TU VeLog application results

## References

[1] Gonder, J., Markel, T., Simpson, A., Thornton, M., Using GPS Travel Data to Assess the Real World Driving Energy Use of Plug-in Hybrid Electric Vehicles (PHEVs), Conference Paper NREL/CP-540-40858, 2007.

[2] Gevatter, H.-J., Grünhaupt, U., Handbuch der Mess- und Automatisierungstechnik im Automobil, Springer, Berlin Heidelberg, 2006.

**Fabian Schueppel, Stefanie Marker, Paul Waldowski, Volker Schindler**
Technische Universität Berlin
Gustav-Meyer-Allee 25 TiB 13
13355 Berlin
Germany
fabian.schueppel@kfz.tu-berlin.de
stefanie.marker@kfz.tu-berlin.de
paul.waldowski@kfz.tu-berlin.de
volker.schindler@kfz.tu-berlin.de

**Keywords:** NET-ELAN, TU VeLog, driving cycle, GPS logger, data logging system, vehicle logging, GPS travel data, electric vehicle

# Monitoring of Batteries and Variable Dynamic Behaviour of an Electric Vehicle

E. Cañibano, J. Romo, M.-I. González, L. De Prada, S. Benito, A. Pisonero,
CIDAUT Foundation

## Abstract

CIDAUT Foundation is a Research and Development Centre for Transport and Energy in Spain. One of CIDAUT´s current lines of work is sustainable mobility, involving the electric vehicle and its infrastructure. In order to advance in this direction CIDAUT has designed and manufactured a technological demonstrator consisting on an electric vehicle. This vehicle is been used to test different powertrain configurations, batteries technologies, electronics, dynamics, etc. The main characteristics of this demonstrator are: the use of two independent motors in the rear wheels; its variable dynamic behaviour; exterior rear mirrors replaced by cameras; exterior lights changed for LED; and the possibility to record all the important parameters from the batteries in real time.

## 1 Introduction

As part of its effort towards Sustainable Mobility, CIDAUT Foundation has designed and manufactured a technological demonstrator consisting on an electric vehicle. This vehicle was developed in CIDAUT's research and development project Minimal Environment Impact Mobility ($M^2$IA, Movilidad de Mínimo Impacto Ambiental). The vehicle was developed to be the testing ground for different type of batteries and motor configurations in order to improve powertrain efficiency in electric vehicles.

## 2 Characteristics

The first relevant characteristic of this vehicle is the use of two independent motors in the rear axle (Fig. 1). Each motor is governed by one controller and both are connected to the PLC (Programmable Logic Controller) of the vehicle.

Fig. 1. Independent electric motors, located in the rear axle

The PLC of the vehicle receives the signals concerning the motors speed as well as the steering wheel position. With this information the PLC calculates the signal to be sent to each motor. Depending on the parameters introduced by the user the dynamic behaviour of the vehicle can be understeering, oversteering or neutral. Figure 2 shows the information displayed in the screen of the vehicle when the vehicle is turning in a curve.

Figure 2 shows, under the red circle, that the steering wheel angle is 430° to the left. This match up with a turning angle of 31° in the left wheel and 25° in the right wheel according to Ackerman relationship. The speed of the centre of gravity of the vehicle is shown in green in the upper part of the screen. To obtain this speed and follow the chosen trajectory, the left motor has been set to run at 17.9 km/h and the right motor has been set at 23.5 km/h. The plot in the middle of the screen shows the evolution of the speed of the vehicle centre of gravity and the speed of both rear wheels.

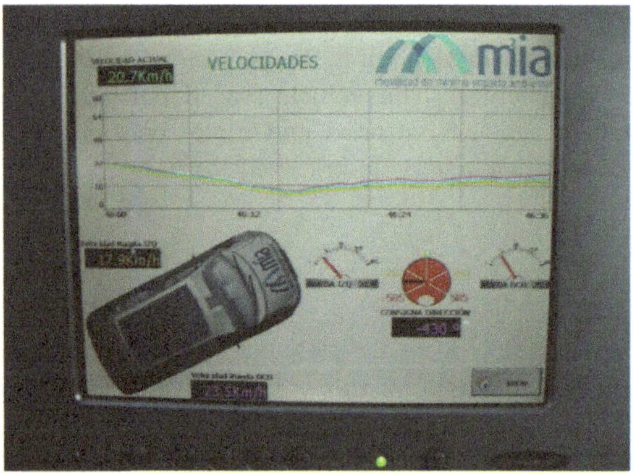


Fig. 2. Information shown in one of the screens

Fig. 3. Micro cameras for rear vision and illumination by LED

The PLC also allows the user to select the kind of driving: snow, economic or dynamic. Most of the driving parameters in the vehicle can be changed using the tactile screen located in front of the passenger seat.

In order to optimize the energy consumption other considerations have been taken into account. To reduce the drag resistance the exterior mirrors have been changed for micro cameras. And to decrease the energy consumption associated to illumination, the incandescent lights have been replaced by LEDs.

The last characteristic to be highlighted of the vehicle is the possibility to collect all the relevant information from the batteries in real time. The vehicle is fed by 18 NiMH batteries, all of them connected to the PLC. The information about the state of health, state of charge, voltage, current and temperature of each of the batteries can be displayed in the screen and recorded for ulterior analysis.

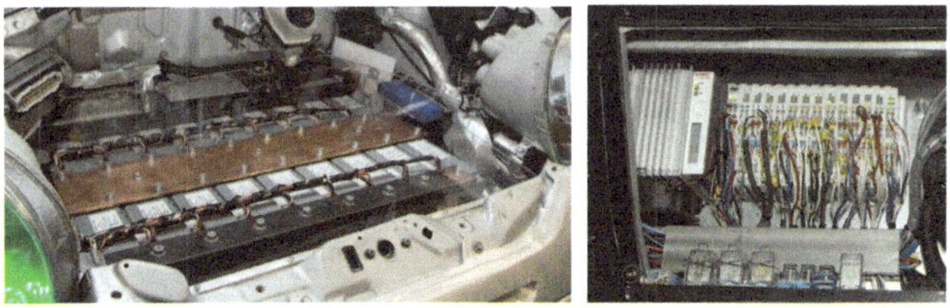

Fig. 4. On the left side the battery pack is shown, as located in the technological demonstrator. On the right a detail of the power electronics is shown.

## 3 Experimental Procedures

In order to evaluate the performance of the electric vehicle a driving test was carried out. The cycle chosen is shown in figure 5, the total test time was 57 minutes, although the total moving time was only 25 minutes. The distance for the whole test was 11.7 km but for the purpose of this paper, only the last 690 seconds are shown.

During the procedure, the ambient temperature was 1.5°C and the relative humidity was 51.5%. In order to register the speed and position of the demonstrator during the procedure, a commercial GPS tracking device was used. The data from this device is shown on the right side of figure 5. The data from the batteries was recorded by the PLC installed within the technological demonstrator. Every 5 seconds, the voltage, the current, the state of charge and the temperature for each battery were recorded.

The driving cycle was performed in February in the Boecillo Technology Park (Valladolid, Spain) in order to simulate an urban driving cycle. The reason to choose this location was to find the closest safe place to CIDAUT's facilities in which interferences with usual traffic condition were minimal.

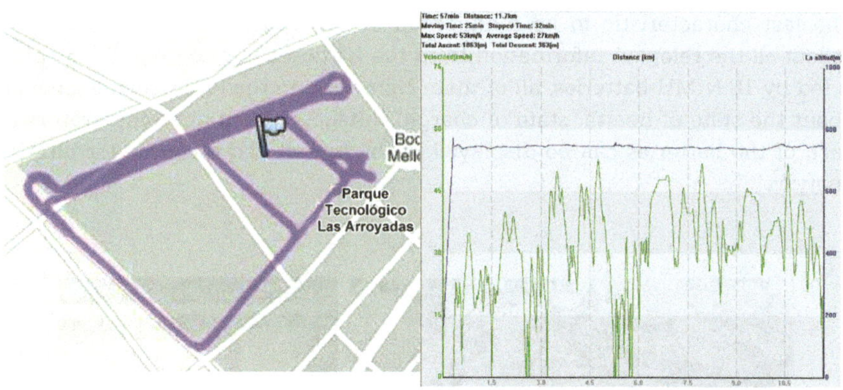


Fig. 5. On the left side the driving cycle is shown, on the right side the speed and time of the cycle is shown, both as recorded by the commercial GPS tracking device

## 4 Results

The data taken during the driving cycle are presented below; the altitude and speed recorded by the GPS tracking devices is presented against time in Fig. 6.

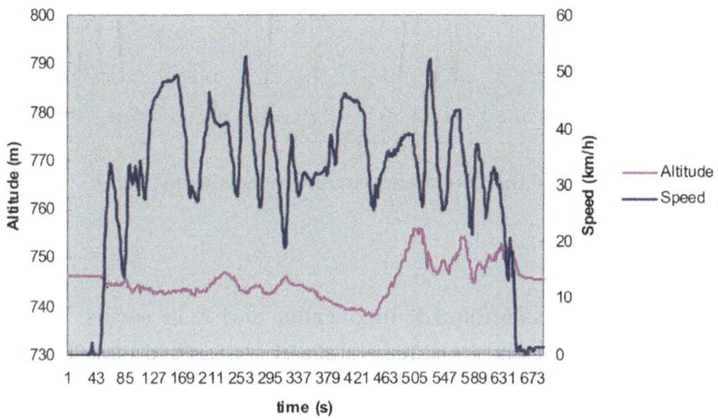


Fig. 6. The evolution of electric vehicle altitude and speed during the driving cycle, as measured from the GPS tracking device

The system is able to measure all the important parameters of the batteries for the 18 elements in the vehicle, so it is possible to detect when one of them is going wrong and to sent and alarm message to stop the vehicle. In figure 7 is shown the voltage evolution of the 18 batteries during the driving cycle.

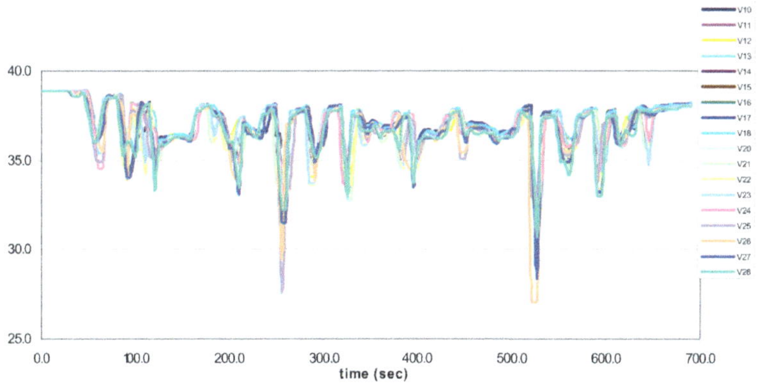


Fig. 7. The evolution of electric vehicle voltage during the driving cycle

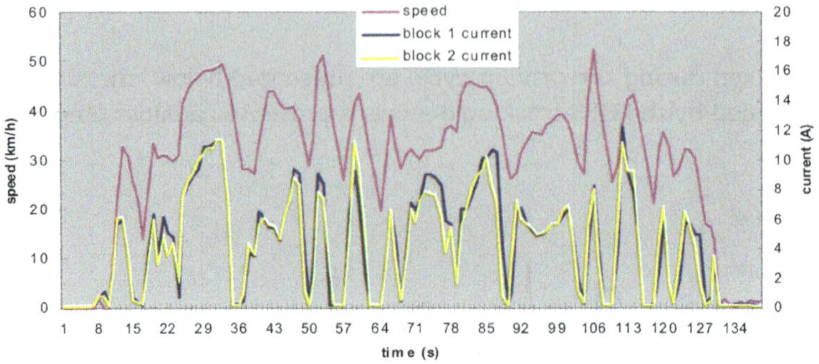


Fig. 8. Evolution of the speed and current versus time

The 18 batteries are assembled 9 in parallel and 2 in series, so in order to obtain stronger conclusions averaged values of the 2 series blocks are going to be analysed.

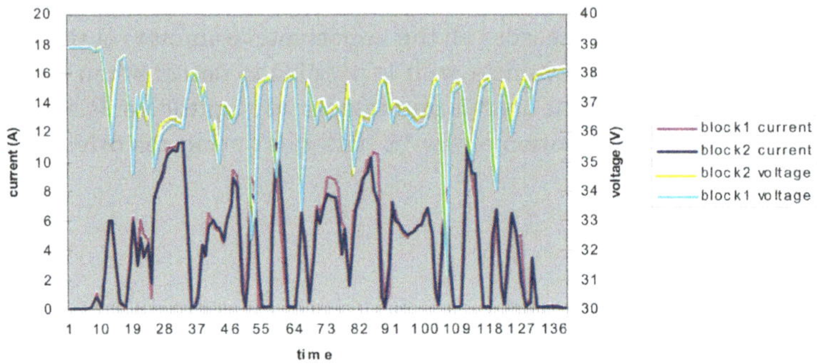


Fig. 9. Evolution of the current and voltage versus time

It is important to notice the direct relationship between the acceleration of the vehicle and the current consumed. There is also a relationship in the decrease in the batteries voltage when a high current is demanded for some seconds. The reception and storage of all the information of the batteries enable the introduction of temperature alarms, to detect a malfunction of the batteries. This storage capability allows showing the driver the real autonomy and the state of the batteries in order to avoid using the vehicle when the level of load is low.

## 5 Conclusions

As a result from M²IA project, CIDAUT has developed a technological demonstrator. This versatile platform allows testing different types of batteries as well as different powertrain configurations in order to improve the efficiency of electric vehicles. CIDAUT has evaluated an array of 18 NiMH batteries. Those batteries were tested in an electric vehicle equipped with two independent motors in the rear axle (see figure 1). This kind of solution makes possible to remove the mechanical differential and thus increase the mechanical capacity. In order to evaluate the efficiency of such powertrain configuration a simple driving cycle was defined (see figure 5). The results from testing the configuration show the relationship among current and voltage with the vehicle speed. The state of charge of the batteries is also monitored in real time via PLC, showing the energy consumption of the chosen configuration.

The versatility of the technology demonstrator allows evaluating the efficiency of other powertrain configurations simply by replacing the current one and then repeat the driving cycle. During this cycle the current, voltage and state of charge in the batteries are recorded in real time, and then compared to assess any change in the overall efficiency.

The capability to display information in real time regarding all the important parameters from the batteries allows:

- To establish alarms when the batteries are malfunctioning.
- To measure the current, voltage, power and consumption depending on the driving conditions.
- To know the state of each of the battery at any time.
- To evaluate the influence of new solutions in powertrain and its associated energy consumption.
- To have an actual measurement of the vehicle's autonomy.

## References

[1] Åhman, M., Primary energy efficiency of alternative powertrains in vehicles, Energy, Volume 26, Issue 11, pp. 973-989, November 2001.

[2] Alonso S., Cañibano E., Peña A. Networked Clean Vehicle 2015 (NCV2015), Spanish research project supported by the Spanish Ministry of Science and Innovation.

[3] Tate E., Harpster M., Savagian P.J. General Motors Corporation, The electrification of the Automobile: From Conventional Hybrid, to Plug-in Hybrids, to Extended-Range Electric Vehicles. SAE 2008-01-0458.

**Esteban Cañibano, J. Romo, M.-I. González,**
**L. De Prada, S. Benito, A. Pisonero**
Fundacion CIDAUT
Parque Tecnologico de Boecillo, p209
Boecillo 47151
Spain
estcan@cidaut.es
javrom@cidaut.es
mangon@cidaut.es
luipra@cidaut.es

**Keywords:** electric vehicle, power train efficiency, sustainable mobility, battery monitoring, green car

# Power Train Efficiency

# Vehicle Energy Management – Energy Related Optimal Operating Strategies for $CO_2$ Reduction

Th. Eymann, Bosch Engineering GmbH
A. Vikas, Robert Bosch GmbH

## Abstract

Further reduction in $CO_2$ will be a major challenge for the automotive industry in the coming years. One of the biggest challenges will be to achieve the proposed EU $CO_2$ emissions limit of 95 g/km in 2020. This will require a 38% reduction in $CO_2$ emissions, based on the average value of new European registrations in 2008 (154 g/km $CO_2$). In addition to combustion engine improvements, vehicle energy management offers a huge potential to reduce $CO_2$ emissions through innovative electronic systems. Starting from a top-down approach, we have developed operating strategies that offer the possibility to optimize the energy usage of the entire system. Energy recuperation is the main element of the operating strategies, and the vehicle systems and components must be adapted accordingly. The energy management strategies described in this paper are based on a "virtual" average 2008 European vehicle. For this vehicle the following driving cycles were investigated in the simulation: New European Driving Cycle (NEDC), Common Artemis Driving Cycle (CADC) and a derivation of the CADC driving cycle. We assumed that combustion engines improvements will reduce $CO_2$ emissions by about 25% in 2020. This paper shows how an appropriate vehicle energy management process can provide the additional $CO_2$ reductions needed to achieve the proposed EU limit of 95 g/km in 2020.

## 1 Motivation and Goals

In view of global warming and the ever increasing consumption of energy worldwide (2009: 12,400 Mtoe, 2030: 17,000 Mtoe), the "$CO_2$ challenge" has gained greater importance in the automobile industry. The transportation industry is projected to be one of the even growing sectors in the future. Hence it is crucial to reduce $CO_2$ emissions through regulations that limit global warming. The $CO_2$ regulations in the EU prescribe an average $CO_2$ emission level of 130 g/km for new registered vehicles in 2012. This is a reduction of

16% compared to the value of 154 g/km achieved in 2008. A limit of 95 g/km is expected for the year 2020. This is a further reduction of 27 % with respect to the limit in 2012, or a total reduction of 38 % with respect to the level archived in 2008.

When we observe the energy flow in a vehicle, it is evident that the greatest opportunities are to minimize the energy loss (e.g. through engine-related measures, avoiding engine idling phases) and to recuperate lost energy (e.g. brake energy, exhaust heat). If we assume a reduction potential of 20 - 25% for the engine-related measures (direct injection, turbo charging, engine downsizing, de-throttling, etc.), then an additional 13 - 18% reduction must be achieved. This paper describes how vehicle energy management can contribute to meeting these additional $CO_2$ reduction requirements.

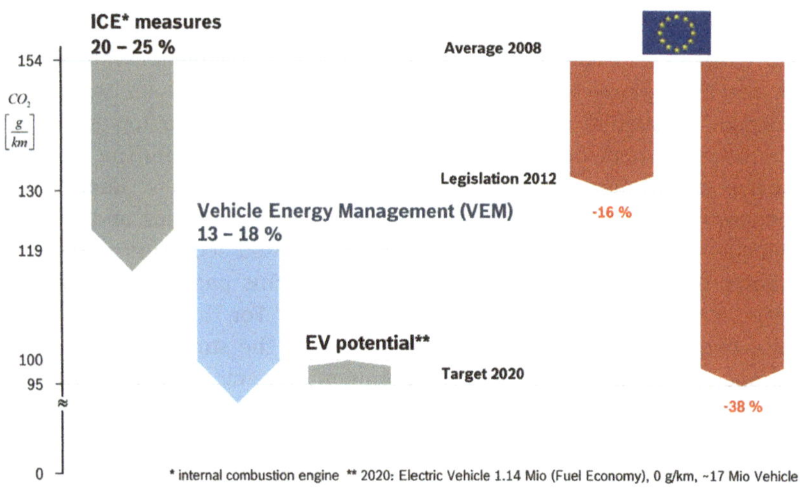


Fig. 1. $CO_2$ reduction potentials and EU legislation

## 2 Energy and Energy Conversion Devices in the Vehicle

Four forms of energy (mechanical (M), chemical (C), thermal (T) and electrical (E)) are required for the operation of a vehicle. Figure 2 illustrates the various forms of energy and their conversion from one form to another. The physical/chemical mechanisms are also specified (e.g. fuel combustion converts chemical energy into thermal energy). In this case, chemical energy is the method of energy storage. Vehicle energy management determines the optimum utilization of this energy flow, the energy conversion devices and the energy storage for $CO_2$ reduction. The complexity of the process also illustrates the potential that is yet to be discovered in energy management with improved and / or additional energy storage and conversion devices.

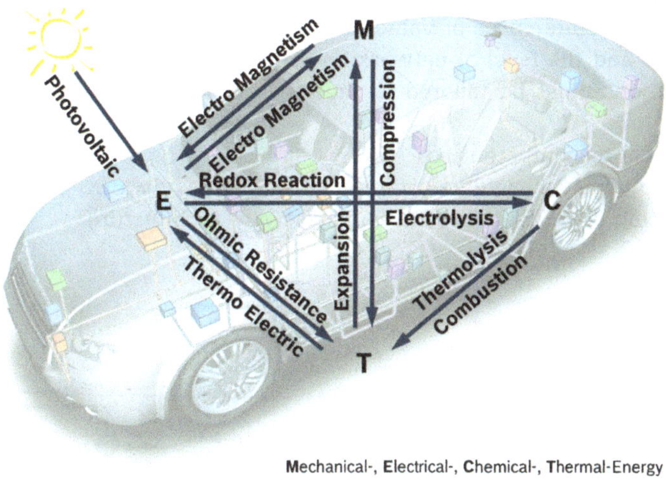


Fig. 2. In-vehicle energy and energy conversions

## 3 Vehicle Energy Management

During the development of an entire system, the components are usually dealt with individually and optimized "locally", without considering the interactions with neighbouring sub-systems and components, as well as the effect on the entire system. Currently the automotive industry already spends effort on the component optimization, e.g. reduction of no-load current of electronic control units (ECU). Also the optimizations of subsystems are done, e.g. stop/ start systems. These measures already exploit a lot of $CO_2$ emission reduction potential. To completely utilize the $CO_2$ reduction potential of vehicle energy management it is essential to derive the requirements using the energy related description of entire systems, and subsequently develop subsystems and components accordingly (top-down-approach).

Hence the vehicle is a very complex system with many subsystems and cross interaction an appropriate simulation model of the entire in-vehicle energy flow must be setup. When the energy flow in the vehicle is represented as a closed loop system, three main functions can be identified: storage, generation and consumption. The vehicle energy management is then reduced to the control/regulation of the energy flow using actuators and energy conversion devices (e.g. fuel cells). In order to accomplish this, the electrical system distributes energy based on subsystem requirements, as illustrated in figure 3.

The determined solutions must be embedded into existing E/E-architectures and software architectures of vehicles. Over the different segments the E/E-architecture and the level of networking of the components deferrers highly. These solutions must be tailored in the mean of cost. Three key points are necessary:

- Simulation competencies
- Reasonable vehicle data communication and bandwidth
- Existing of a E/E-architecture and software architecture

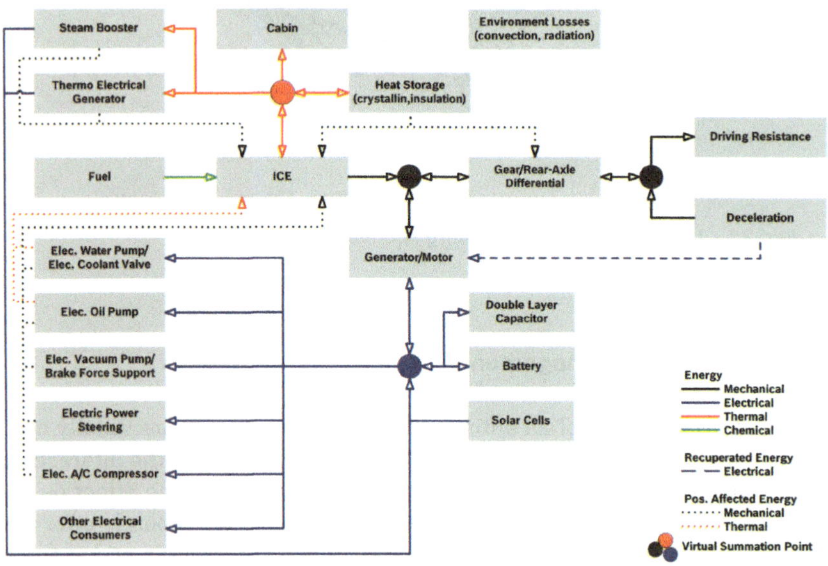


Fig. 3. In-vehicle Energy flow e.g. electrified ancillary components

## 4 Simulation Results

For the system consideration, we defined a vehicle that has the average values of $CO_2$ emissions and vehicle weight for EU27 in 2008. This vehicle is a compact car with a mass of 1320 kg, powered by a gasoline engine with $CO_2$ emissions of 154 g/km. The simulation was carried out for the following driving cycles:

- New European Driving Cycle (NEDC); total distance approximately 11 km
- Common Artemis Driving Cycle (CADC); total distance approximately 170 km
- Combined cycle; total distance approximately 42 km

The combined cycle represents the driving behavior of an average German driver, and it includes city and country roads as CADC cycle components. In the simulation, the fuel consumption was determined using the following parameters:

- Without recuperation
- With recuperation, voltage levels: 14 V, 28 V and 42 V
- Maximum current during recuperation phase 250 A

In order to use off-the-shelf technologies and thus maintain the costs for the power net, the maximum current was limited to 250 A. The minimum power net wattage was 350 W, which corresponds to the power required to meet the electrical energy consumption requirement in NEDC. Since a typical real-world vehicle has additional electrical loads that are not required in NEDC (mainly comfort feature loads), fuel consumption was calculated using a power net wattage of 1250 W. In addition to this, the fuel consumption in the theoretical case was calculated such that the entire recuperated energy could be used as electrical energy. The simulation results are summarized into the table. For the 350 W power net scenario (i.e. the minimum requirement for NEDC), recuperation provided a 7.8% reduction in fuel consumption regardless of the voltage level. However, a much greater reduction can be achieved for 1250 W power net scenario, particularly at higher voltage levels.

| | NEDC [l/100km] | | CADC [l/100km] | | Combined [l/100km] | |
|---|---|---|---|---|---|---|
| w/o Recuperation, 350W | 6.51 | - | 9.04 | - | 8.82 | - |
| 14V@250A, 350W | 5.99 | -7.8% | 8.41 | -6.7% | 8.13 | -7.6% |
| 28V@250A, 350W | 5.99 | -7.8% | 8.41 | -6.7% | 8.13 | -7.6% |
| 42V@250A, 350W | 5.99 | -7.8% | 8.41 | -6.7% | 8.13 | -7.6% |
| $I_{Theo.}$ > 250A, 350W | 5.45 | -16.3% | 6.89 | -23.8% | 6.46 | -27.0% |
| w/o Recuperation, 1250W | 7.83 | - | 10.65 | - | 10.59 | - |
| 14V@250A, 1250W | 7.31 | -6.6% | 9.66 | -9.3% | 9.49 | -10.4% |
| 28V@250A, 1250W | 7.01 | -10.5% | 9,16 | -13.9% | 8.94 | -15.6% |
| 42V@250A, 1250W | 6.89 | -12.1% | 8.88 | -16.5% | 8.63 | -18.5% |
| $I_{Theo.}$ > 250A, 1250W | 6.77 | -13.5% | 8.50 | -20.2% | 8.23 | -22.3% |

Tab. 1. Simulation results

Figure 4 shows the simulation results of the electrical energy consumption by the power net (1250 W) and the brake energy recuperation (14 V, 28 V and 42 V with max. 250 A) in NEDC. The 1250 W power net consumption corresponds to 3.78 kWh/100km of electrical energy. If the generator and the powertrain components are considered, the required powertrain energy would be $E_{\mathrm{M}}$ = 16 kWh/100km and the fuel consumption would be 1.83 l/100km. This is equivalent to 43.6 g/km of $CO_2$ emissions. With 28 V recuperation, the same power net consumption would require an electrical energy of only 2.1 kWh/100km ($E_{\mathrm{M}}$ = 8.89 kWh/100km or 1.01 l/100km) and would reduce the $CO_2$ emissions by 19.36 g/km. This corresponds to a 10.5% reduction in fuel consumption in NEDC. See table 1, third row from the bottom.

If we consider CADC and the combined cycle in a similar manner, we see a clear trend that brake energy recuperation can provide significant reductions in fuel consumption at higher power net wattages enabled by higher voltage levels. Furthermore, since the brake proportion, the acceleration proportion and the speed levels are higher in the CADC and combined cycle, the fuel saving potential is also greater in these cycles, see table.

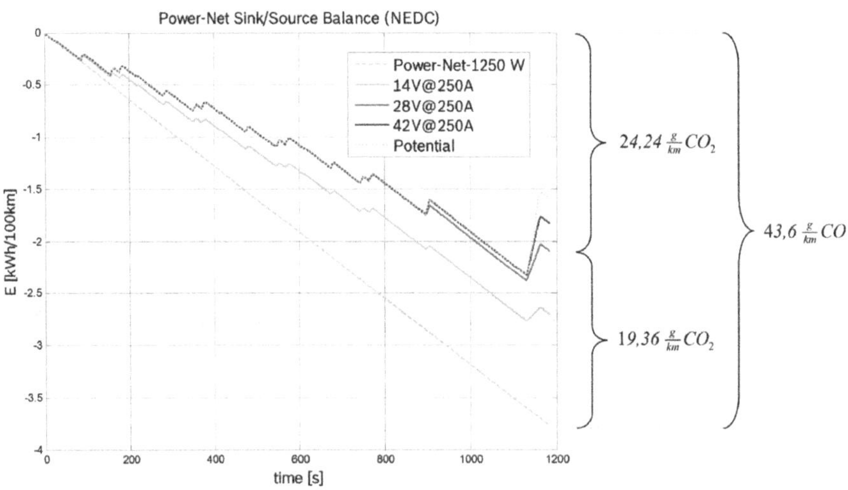


Fig. 4. Simulation result of the power net consumption in the NEDC

## 5 Operating Strategies

The analysis depicted in the energy control loop structure (figure 3) results in five operation modes, which are defined as follows:

- Driving: The state of the energy storage device is above an upper threshold. The mechanical energy produced by the combustion engine is used exclusively to move the vehicle.
- Charging: The state of the energy storage device is below a lower threshold. The energy storage device is charged.
- Discharging: The state of the energy storage device is within the upper and lower thresholds. The energy storage device is discharged according to a function of the current and predicted future efficiency of the combustion engine and power net loads.
- Generating: The state of the energy storage device is within the upper and lower thresholds. The energy storage device is charged according to a function of the current and predicted future efficiency of the combustion engine and power net loads.
- Recuperation: Energy is recuperated (e.g. during coasting phases).

The recuperation mode is always active in parallel to the other four modes. If three states are defined (e.g. high, medium and low), which can be assumed for the devices of the classes "storage", "generator" and "load", respectively, then the logic structure is as shown in figure 5. As an example for the three possible states, the energy storage device "battery" can be:

- high: The state of charge is above the upper threshold.
- medium: The state of charge is between the upper and lower thresholds.
- low: The state of charge is below the lower threshold.

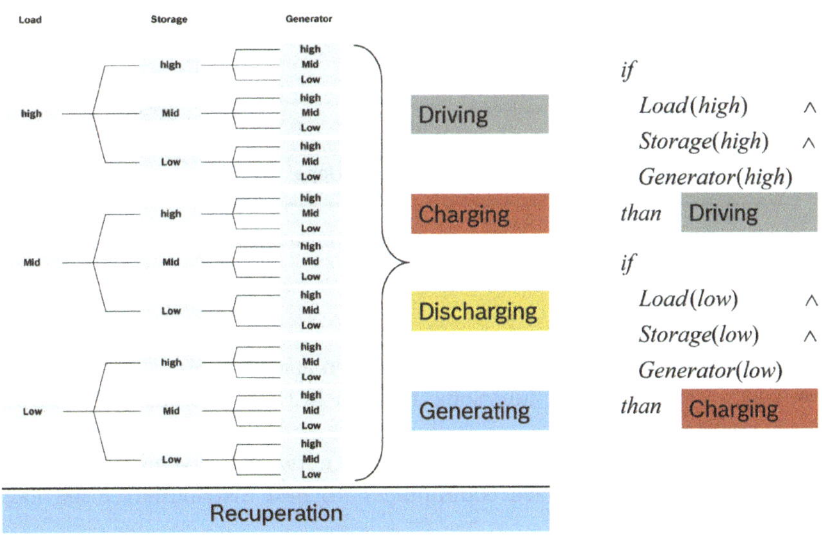


Fig. 5. Operating strategies

Similarly the states can also be selected for the loads. For the generator "combustion engine", the efficiency coefficient for conversion from chemical to mechanical energy is selected as the state variable. The operating strategy can then be determined using logical "If-Then" statements, as shown in the example on the right side of the figure 5.

## 6 E/E-Architecture

Usually the vehicle fleet is classified in six segments: "mini", "small", "lower medium", "medium", "upper medium" and "large". The $CO_2$ emissions in EU are regulated based on the curb weight. For the vehicle segments individual $CO_2$ targets can be calculate. We assumed that the improvement of the combustion engines will reduce $CO_2$ emissions by about 25% in 2020. Figure 6 shows the gap of the different vehicle segments to the $CO_2$ target in 2020 after the ICE improvement (from "mini" approximately 10% to "large" approximately 25%).

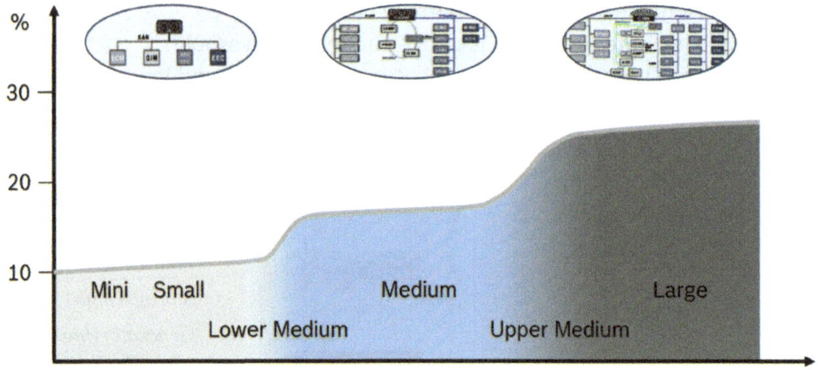


Fig. 6. Gap to $CO_2$ target 2020 after ICE optimization

There is no silver bullet but architecture pattern are available:

- Level 1: independent functions; isolated ECU for customised solutions e.g. easy Stop/Start
- Level 2: Information exchange over data network between the ECU
- Level 3: Master – Slave communication; first distributed closed loop functions
- Level 4: Vehicle Control Unit; high complex closed loop control and high cross functionality

In the "mini" and "small" segment is it feasible to close a gap of 10 – 12 %. Considered measures are: Stop/Start, thermal management, electrical power steering and high efficient generator including active rectifier. The appropriate architecture could be Level 2.

In addition to the previous one the "medium" vehicle segment needs strong recuperation combined with the cross domain function "Sailing" (moving vehicle with engine off) to close a gap of 12 – 17 %. In some cases the integration of new energy conversion devices could be necessary, e.g. thermal electric generator. The appropriate architecture would be at least Level 3.

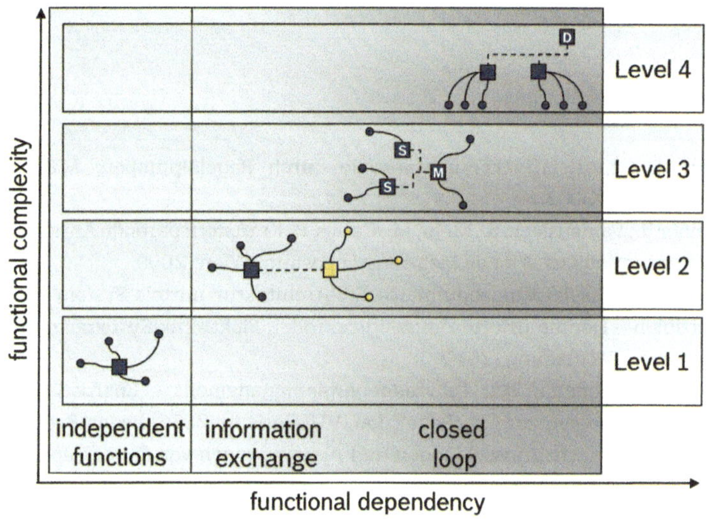


Fig. 7. Architecture level of functional complexity versus functional interdependency

The high end segment needs in addition to the previous measures strong hybridization (mild hybrid or full hybrid). Due to the high system complexity architecture Level 4 would be appropriate, see figure 7. Which architecture pattern is appropriate for the intermediate phases depends on a case by case decision – mainly based on costs.

## 7 Summary

EU regulations will require a $CO_2$ reduction of approximately 38 % in the EU27 in 2020 as compared to the average value in 2008. Assuming that engine improvements will reduce fuel consumption by 25 %, an additional 15 % must be achieved by other methods. The vehicle energy management strategies that are described in this paper can contribute to meeting these additional $CO_2$ reduction requirements. Key components are E/E-architecture, brake energy recuperation and flexible operation strategies. A holistic approach to managing the energy efficiency of the complete system can help to optimize the energy storage devices, generators and loads. The prerequisite, however, is the electrification of ancillary components.

## References

[1] Voigt, D., Kraftstoffverbrauchsvorteile durch Regelölpumpen, MTZ, 12/2003, Jahrgang 64, 2003.

[2] Greiner, J., Vahlensieck, B., Mohr, M., Casals, P., Kraftstoffsparende Antriebssysteme, 30. Internationales Wiener Motorensymposium, Wien, 2009.

[3] Stolz, W., Preis-/leistungsoptimierte E/E-Architektur mittels System Engineering Prinzipien – Der integrierte Fahrzeugcontroller, Elektronik-Systeme im Automobil, Euroforum Fachtagung, 2010.

[4] Vikas, A., Eymann, Th., Fahrzeugenergiemanagement – Energetisch optimale Betriebs-strategien zur $CO_2$ Reduktion, VDI-Berichte 2075, Baden-Baden 2009.

[5] Fischle, H., Neuste Entwicklungen und Anwendungen von Super-Caps als Energie und Leistungsspeicher, Elektrische Energiespeicher, VDI-Berichte 2058, Fulda, 2009.

[6] Rauchfuss, L., Hindorf, K., Einfluss und Potenzial von Nebenaggregarten auf den Kraftstoffverbrauch, Energieeinsparung durch Elektronik im Fahrzeug, Baden-Baden Spezial, 2008.

[7] Gross, J., Hartmann, S., Merkle, M., Entwicklungstrends und zukünftige Lösungen für Start/Stopp Systeme, VDI-Berichte 2075, Baden-Baden, 2009.

[8] Robert Bosch GmbH Kraftfahrtechnisches Taschenbuch, 25. Auflage, Vieweg, 2003.

**Thomas Eymann**
Bosch Engineering GmbH
Postfach 13 50
74003 Heilbronn
Germany
thomas.eymann@de.bosch.com

**Athanasios Vikas**
Robert Bosch GmbH
Postfach 13 55
74003 Heilbronn
Germany
athanasios.vikas@de.bosch.com

**Keywords:** vehicle energy management, holistic top-down approach, recuperation, 2 voltage power net, optimized operating strategies, E/E-architecture

# Active Engine Management Sensors for Power Train Efficiency

P. Slama, Infineon Technologies Austria AG

## Abstract

"Clean performance"- the challenge with regard to improved driving performance and at the same time lower fuel consumption goes on. In addition to this lower exhaust limits are set by law. This bears the need for even more advanced engine management systems in future. These challenging targets result in higher requirements to engine management sensors with respect to accuracy, electromagnetic compatibility (EMC) and optimum operation with an increasing variety of applications. The development of a brand-new active engine management sensor generation based on giant magnetoresistance (GMR) has just been finished. The unique features of this new technology are high accuracy (repeatability), fast start-up calibration and high EMC robustness.

## 1 Advanced Sensors in Engine Management Systems

Active magnetic sensors in advanced engine management systems provide a digital switching signal, which maps the mechanical teeth profiles or the magnetic domains of a passing pole wheel. Subsequent processing in the microprocessor determines the current speed or angle position of the target wheel from this switching signal. This data is further processed for accurate ignition control and misfire detection [1]. The sensor need to provide a phase accurate output signal for magnetic input frequencies from 0 to 10 kHz and over an amplitude range of approximately 1 to 100. The required temperature range is from -40°C to 150°C with an air gap range of 0 to 3.5 mm [1]. Optimization of engine management systems shall improve $CO_2$ emissions. Target is minus 30% for gasoline and minus 10% for diesel engines. In order to meet this challenging goal the sensing system has to provide more accurate information and provide additional features for start-stop, hybrid and electric vehicles. The task of the sensor is therefore to provide the switching flanks with good reproducibility and high angle accuracy relative to the target wheel. Nevertheless the mounting tolerance of the sensor should be relatively large to keep the system costs low.

### 1.1 More Precise Engine Control and Misfire Detection – Higher Accuracy Requirements

The goal of future crankshaft sensors is delivering even more accurate digital signals with respect to phase jitter and phase accuracy. The challenging target is < +/- |0.025| [° crank] phase jitter (repeatability) and < +/- |0.3| [° crank] absolute phase accuracy.

The GMR effect compared to other technologies has higher sensitivity and a superior signal to noise ratio. Due to these characteristics the new active Infineon integrated (one chip solution) iGMR sensor can cope with these upcoming ambitious demands.

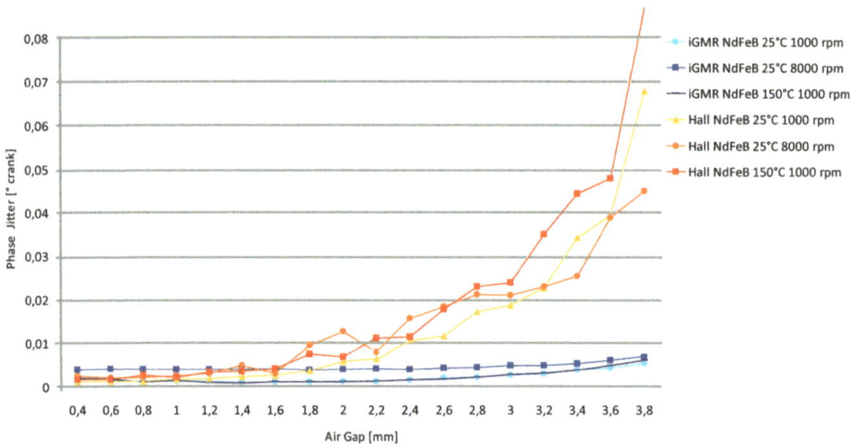


Fig. 1. Phase jitter (repeatability) performance comparison Hall vs. GMR technology sensor versus revolutions per minute (rpm), air gap and temperature in back bias application

Fig. 1 shows a phase jitter (repeatability) performance comparison between Hall and GMR technology sensor in back bias applications. The same NdFeB magnet and dimensions were used for both types. The GMR technology sensor proves stable phase jitter performance versus both temperature and air gap. In contradiction to that the phase jitter performance of the Hall sensor is strongly dependent on input signal size (air gap) and temperature. For the Hall cell the thermal noise is getting the dominant factor at elevated temperatures.

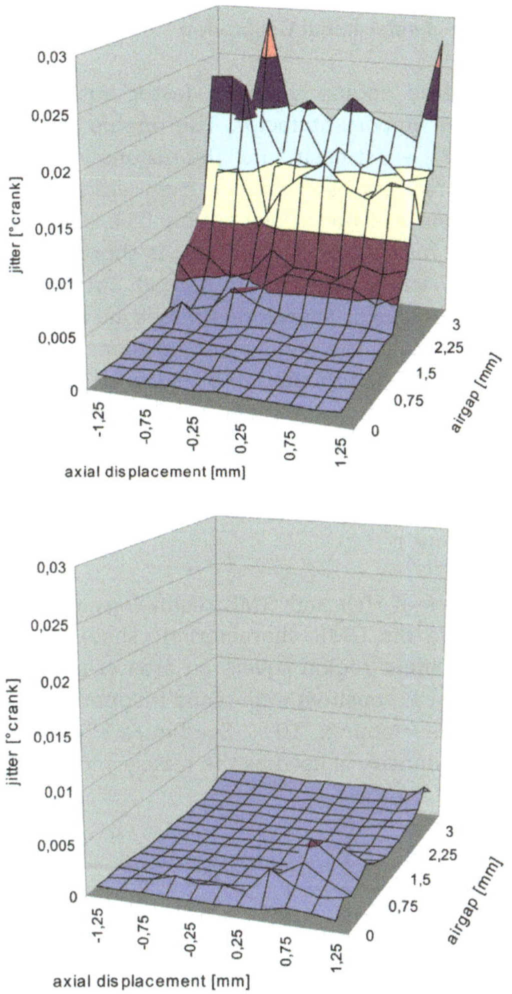


a)

b)

Fig. 2. Phase jitter (repeatability) performance comparison Hall (a) vs. GMR (b) technology sensor versus air gap and axial displacement in magnetic encoder application

Figure 2 shows a phase jitter (repeatability) performance comparison between Hall and GMR technology sensor in magnetic encoder application. The same magnetic encoder wheel was used for both types. The GMR technology sensor proves stable phase jitter performance versus both air gap and axial displacement. In contradiction to that the phase jitter performance of the Hall sensor is strongly dependent on input signal size (air gap). The new active Infineon GMR technology crankshaft sensor meets the accuracy requirements of future engine management systems.

## 1.2 Start-Up Behavior – Faster Initial Calibration

With respect to optimized engine control a faster initial calibration of the system at start-up is an important factor. The engine management system can earlier determine the exact position of the pistons. As a consequence the combustion cycle can precisely be controlled from earlier phase on. This will lower both $CO_2$ emissions and fuel consumption. The main advantage of the GMR sensor is that the initial calibration phase is shorter compared to other sensing technologies. Prior to issuing the second output pulse the sensor calibration phase was finished. Therefore the second pulse delivers highly accurate phase information to the engine management system. Compared to this a Hall based sensor needs (maximum) 6 edges for calibration. Therefore the calibration behavior of GMR based sensors is superior. The actual rotational direction information is transferred via a pulse width modulated (PWM) sensor output protocol. An advantage of the GMR sensor is that from the first output pulse onwards the correct rotational direction information will be transferred (at first start-up per default).

The characteristic curves of Hall and GMR technology sensors are different. Besides the linear range the GMR characteristic shows an area where the signal is clamped. The linear region covers an area of approx. – 5 mT up to + 5 mT. The GMR sensor is sensitive to in-plane magnetic flux density, $B$. The typical resistance change is approx. 10 %. See figure 3. The resistance dependence on magnetic flux density is used as the measuring principle.

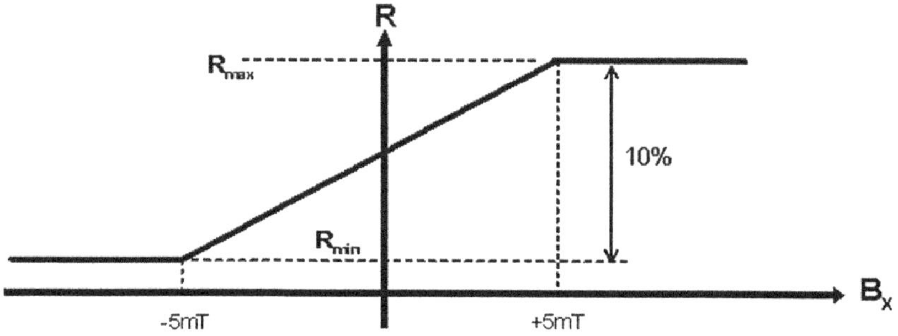


Fig. 3. GMR characteristic curve

Because of the dedicated GMR characteristic curve the electrical input signal size can perfectly adapted to the analog-to-digital converter (ADC). Any kind of gain control resp. PGA (programmable gain amplifier) as it is needed for Hall sensors is not required. The consequence out of that is that the sensor calibrated mode can be entered at the third edge of the differential input signal $B_{x,diff}$. See figure 4.

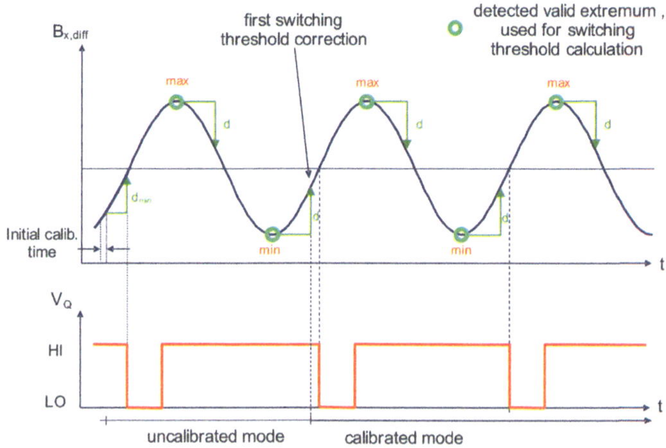


Fig. 4. Sensor start-up at magnetic rising edge

In order to calculate the calibrated mode switching threshold the input signal has to be tracked and the extreme values have to be determined. After an initial calibration time $t_{on}$ the magnetic signal is tracked by an ADC and monitored within the digital circuit. For detection the signal needs to exceed an internal threshold (digital noise constant $d_{min}$). After power on, when the signal slope is identified as a rising edge and the signal change exceeds $d_{min}$, the first pulse is issued at the output. The first output pulse includes the right direction information. As soon as the second extrema is detected, the switching threshold can be calculated. All following output switchings are done at zero crossing of the signal on the magnetic rising edge. The sensor is in calibrated mode from that point. From 2nd output pulse on the advanced iGMR sensor provides accurate position information of the crankshaft.

In case the start-up of the sensor occurs at the magnetic falling edge, the first output pulse will be issued on the first following magnetic rising edge. A " virtual zero" switching point has to be calculated as only one extreme value is determined so far. Depending on start-up position this switching point is close to the calibrated switching point. See figure 5.

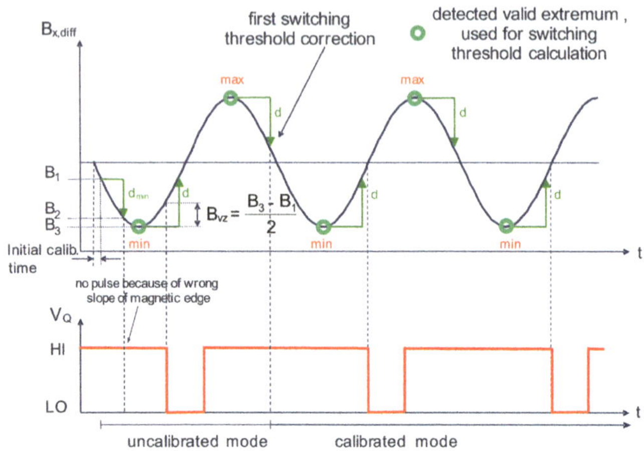


Fig. 5. Sensor start-up at magnetic falling edge

### 1.3 Hybrid and Start/Stop Engine Requirements

Hybrid vehicles and start/stop engines significantly reduce $CO_2$ emissions compared to normal combustion engines. The worldwide trend towards start/stop engines is ongoing. Start/stop engine implies that the combustion engine is stopped at vehicle stand still. The engine is re-started when the driver intends to continue driving. This concept is considered to be most efficient in city traffic ("stop and go mode") for emission reduction and fuel consumption saving. A start/stop engine can reduce $CO_2$ emissions up to 10%. Depending on the version hybrid vehicles can lower $CO_2$ emissions even more up to 30%.

These two concepts require besides higher accurate sensing systems a crankshaft sensor with rotational direction detection feature. During engine stop mode it happens, that the crankshaft rotates backwards because of mass moment of inertia. With this new feature the sensor detects a backward rotation of the crankshaft and the engine management system does not lose the exact position of the pistons. After standstill of the target wheel (no differential magnetic input signal), there is no timer present which resets the sensor. As soon as the target wheel moves again (a differential signal appears) the output pulses are triggered by the last calculated threshold (zero crossing calculated with last detected offset). At re-start of the engine the engine management system is still in calibrated operation mode, wrong injections can be avoided.

For rotational direction determination Infineon uses an additional sensing element "iGMR5" (see figure 6.). This sensor in the center of the IC is used to provide a typically 90° phase shifted signal. Depending upon the rotation direction of the target wheel, the signal of the center probe anticipates or lags behind for 90°. This phase relationship can be evaluated and converted into rotation direction information by sampling the signal of the center probe in the proximity of the zero crossing of the "speed" bridge signal. The evaluation of the rotation direction is interesting only at low rotation speed, since a direction of rotation reversal can take place only there. Hence at low rotation the already existing ADC can be used in time multiplexing to convert also the "direction" signal of the center probe without losses of phase accuracy in the output signal. After a change of the rotational direction a 90 µs output pulse is being issued at the magnetic falling edge. In this way the reference to the mechanical edge of the gear tooth is still consistent [1].

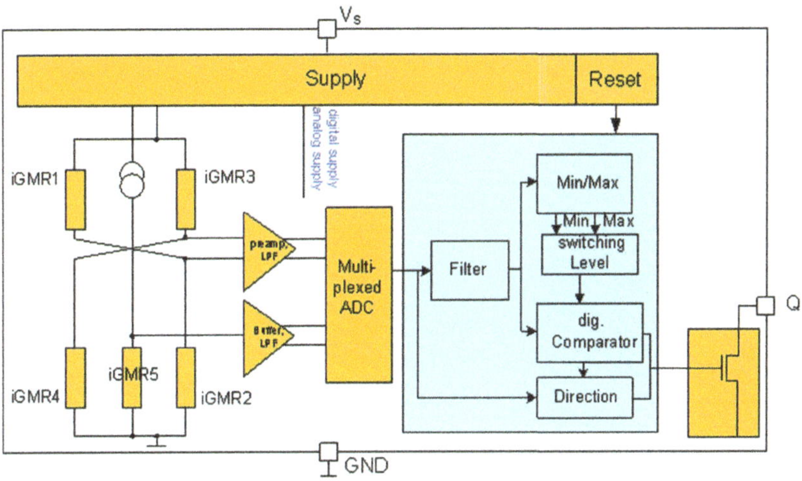


Fig. 6. iGMR chip block diagram

## 2 Summary

Infineon presents a family of new speeds sensor for engine management applications based on an integrated GMR technology with the sensing element and signal conditioning circuit on one IC. Sophisticated calibration and an offset compensation algorithm based on digital signal processing, together with high sensor sensitivity, accounts for superior performance compared to Hall sensors. Investigations on GMR stability prove that the technology meets the requirements of the automotive industry regarding reliability and lifetime.

## References

[1] Kapser, K., Zaruba, S., Slama, P., Katzmaier, E., Speed Sensors for Automotive Applications based on Integrated GMR Technology, In Valldorf, J., Gessner, W., Advanced Microsystems for Automotive Applications 2008, Springer, Berlin 2008.

**Peter Slama**
Infineon Technologies Austria AG
Siemensstr. 2
9500 Villach
Austria
peter.slama@infineon.com

**Keywords:** powertrain efficiency, engine management sensors, GMR sensor, giant magnetoresistance

# Thermal Systems Integration For Fuel Economy - TIFFE

C. Malvicino, R. Seccardini, Centro Ricerche Fiat


## Abstract

The TIFFE project is devoted to the development of an innovative integrated vehicle thermal system to improve the on board thermal management and the energy efficiency. The project main contents are a dual loop air conditioning where one loop transfers the cooling power and one loop rejects the heat, a two-levels temperature heat rejection system to dissipate the high temperature heat (e.g. engine waste heat) and to cool locally the vehicle auxiliary systems, a new generation of compact fluid-to-fluid heat exchangers and use of innovative coolants (e.g. nanofluids): to improve the heat rejection and redesign the heat exchangers. The approach benefits can be summarized in a cost reduction, due to resize and integration of the thermal systems, and fuel economy increase of 15% on real use.


## 1 The TIFFE Project

The major contents of TIFFE are:

- an integrated thermal system based on two cooling loops: one at high temperature (i.e. 90 deg) devoted mainly to reject the heat of the thermal engine and one with coolant flowing at a temperature 5 deg higher the ambient and dedicated to the cooling of all the engine auxiliary system (intake air, power steering, ...)
- a new coolant (nanofluid) to improve the heat transfer to enable new heat exchangers geometries and sizing.
- a compact refrigeration unit placed in the engine bay able to maintain the temperature of a cold and hot circuit.
- new coolant-to-air and coolant-to-fluid heat exchangers with enhanced heat transfer features.
- control strategies to minimize the energy consumption and optimize the systems operation.

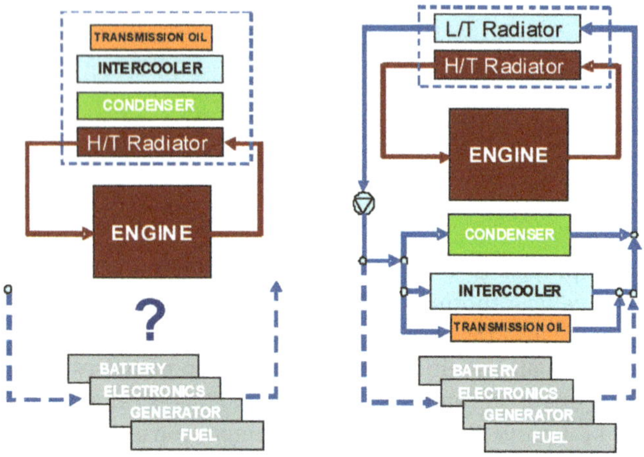


Fig. 1. a) Standard system; b) TIFFE system: two cooling loops one for the engine and one for all the other vehicle auxiliary systems and a loop dedicated to cooling power transfer

The project involves two major car manufactures (Fiat represented by the Centro Ricerche Fiat and Ford) together with two leading automotive suppliers (Denso Thermal Systems and Maflow) and two important and acknowledged research institutions (SINTEF and University of Braunschweig) originating from two EU countries and one associated state. Within the project two vehicle demonstrators will be realised and validated:

- a gasoline passenger car with Stop & Start function
- a diesel light commercial vehicle with hybrid power train

Fig. 2. The TIFFE (Thermal systems integration for fuel economy) consortium

Both vehicles will undergo a complete series of road and climatic chamber tests and a long range road test to verify the reliability and effectiveness of the system.

The major impact of the TIFFE approach are highlighted in figure 3 and here below summarised:

- **Energy management** (also and especially in case of hybrid powertrain or Stop & Start function): based on the enhanced control of the electrical storage unit, the reduced demands of power to reject the heat and the opportunity to keep easily the generator and, when present of the power electronics, temperature never over 5°C higher than the environment. In addition to that the system will increase the efficiency of the air conditioning system, contributing to increase the vehicle fuel economy in real use.
- **Engine and Combustion efficiency**: the fine control of the engine, engine fluids, engine auxiliary systems and of the intake air temperatures (charge air cooler) allows to increase the engine overall efficiency.
- **Aerodynamics:** the re-design of the vehicle front end and of the engine bay reduces the design constraints allowing to reduce the vehicle Cx and the cooling drag.

A series of preliminary activities have been successfully completed to assess the compact refrigeration unit feasibility and advantages and criticisms.

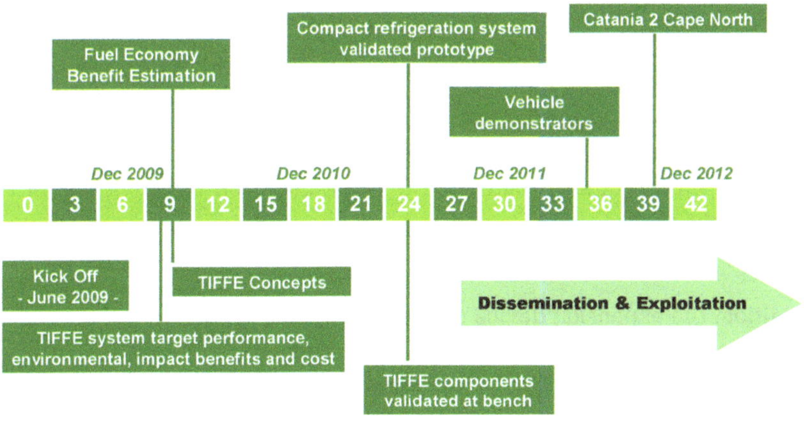


Fig. 3. The TIFFE project scheduling

## 2 Compact Refrigeration Unit

Starting from 2011, January 1st, all new passenger cars produced in Europe cannot have a HVAC system using a refrigerant with a GWP (Global Warning Potential) higher than 150 and that implies that the preset refrigerant R134a is more allowed. This rule, at present regarding only the passenger cars, probably will be soon extended to all the vehicles.

The refrigerant alternatives: $CO_2$ (R-744) or flammable gas (e.g. R-152a, R-1234yf); both of them show critical aspects:

- **Carbon Dioxide ($CO_2$)** R-744 when used as refrigerant, which has a minimal global warming potential, when applied as a refrigerant, operates at very high working pressures (30–140 bar) requiring pipes and corresponding gaskets are capable to withstand the operating pressures. Furthermore, the system additional costs are still very high [1].
- **Flammable refrigerants** - the known R-152a and R-290 (propane) - and the recent HFO-1234yf have low GWP compliant with the limits of the aforementioned standard, but are, however, flammable. The HFO-1234yf has the lower flammability characteristics and has been selected to replace the R134a even if the cost are still quite high and there are open issues to be solved regarding the system efficiency [2].

To cope with these issues the compact refrigeration unit (figure 4) concept has been conceived based on available technology [3, 4] and allowing the pipes length reduction and the refrigeration system confinement in the engine bay. The unit allows also the implementation of the heat pump function without any major modification of the CRU but requires only a change of the coolant circuitry.

The CRU is constituted of three parts:

- Evaporator: where coolant fluid subtracts heat from the intermediate fluid that is used to cool the blown in the cabin by the HVAC module.
- Condenser (gas cooler, when $CO_2$ is used), where the coolant transfers heat via the intermediate fluid to the radiator on the front end of the vehicle.
- Internal heat eXchanger (IHX).

In addition, the expansion device can be easily integrated.

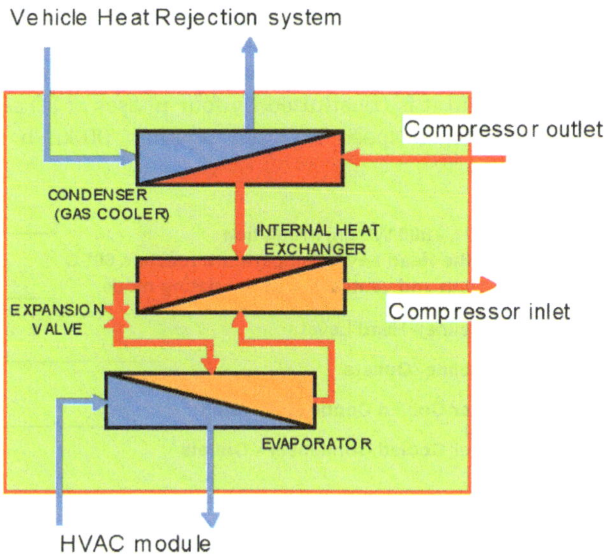


Fig. 4. Functional scheme of the refrigeration system: the unit constitutes a module installed in the engine compartment able to produce the required cooling power. The compressor can be mechanically or electrically driven.

## 3 Water Cooled Condenser

A system were only the condenser is water-cooled has been realised and tested on two vehicle demonstrators: a passenger car and a light commercial vehicle. In both cases R-134a refrigerant has been used. The condenser positioning in the engine compartment allows to reduce the length of the lines, to improve the refrigerant condensation increasing the homogeneity and limiting the cost of repairing in case of small accidents (no refrigerant losses, radiator substitution less expensive than the condenser one...).

### 3.1 Thermal Performances

The system has been evaluated measuring the thermal performance (cool down) and fuel consumption on NEDC cycle at 28°C – 50% R.H. and the 35°C – 60% R.H. [5]. The cool down performance is equivalent to the standard production system with air cooled condenser, as shown in figure 5, despite the use of the intermediate heat exchanger. The cool down test is performed in a climatic wind tunnel at 43°C, 30% R.H. and 800 W/sqm of irradiation. The A/C is regulated in max cooling and ventilation and in recirculation mode. The test lasts 120 minutes and it is constituted by four phases of 30 seconds each performed at different vehicle speed: 30 km/h, 60 km/h, 90 km/h, idle.

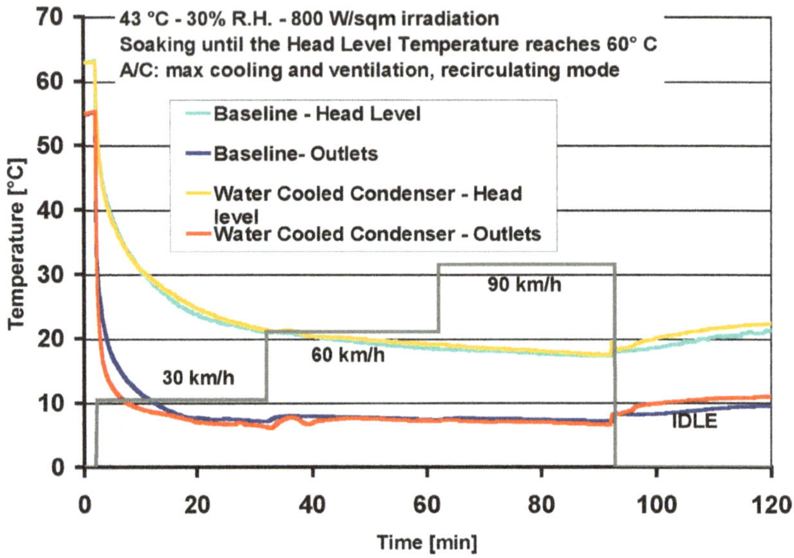


Fig. 5. Cool down performance with water cooled condenser

This confirms the fact that the water cooled condenser improves the refrigerant condensation at least to compensate the inefficiencies due to the intermediate thermal level heat exchange level.

The results of the fuel consumption measurement are reported in figure 6. The water cooled condenser allows to reduce the fuel consumption due to the air conditioning system even in case of severe climate conditions. The improvement is due to the better condensation allowing a reduction of the refrigerant high pressure of about:

- 2 bar in the first phase of the cool down
- 4 bar in the second part of the cool down
- 1 bar in the NEDC test at 28°C
- 2 bar in the NEDC test at 35°C

| | | Air Cooled | Water Cooled | Max $CO_2$ saving |
|---|---|---|---|---|
| | Cycle | A/C Fuel Consumption [l/100 km] | | g/km |
| A/C ON<br>15 °C - 70 % R.H.<br>Set Point 20 °C | EUDC | 0.7 | 0.5 | -4.8 |
| | ECE | 0.3 | 0.3 | 0.3 |
| | NEDC | **0.43** | **0.37** | **-1.6** |
| A/C ON<br>28 °C - 50 % R.H.<br>Set Point 20 °C | EUDC | 2.1 | 1.9 | -5.8 |
| | ECE | 0.6 | 0.6 | -0.1 |
| | NEDC | **1.15** | **1.06** | **-2.2** |
| A/C ON<br>35 °C - 60 % R.H.<br>Set Point 23 °C | EUDC | 3.5 | 3.5 | 0.1 |
| | ECE | 0.6 | 0.6 | -1.1 |
| | NEDC | **1.65** | **1.63** | **-0.7** |

Fig. 6. Fuel overconsumption due to the air conditioning system. The water cooled condenser allows to increase the overall system efficiency even in case of severe climate conditions, despite the intermediate temperature and heat exchange level.

The improvement of the condensation process is very important in case of the adoption of the HFO-1234yf that requires an improvement of the condensation to achieve the same efficiency of the R-134a system [3].

## 4 Water Cooled Evaporator (Secondary Loop A/C)

A system with an indirect expansion lay-out (figure 7), where the cooling power is delivered to the cabin by means of an intermediate fluid (e.g. water-glycol), has been then realized and tested. This approach has the following advantages:

- the refrigerant is entirely contained in the engine bay, allowing a safe use of flammable refrigerant;
- it allows shorter lines on low pressure side, implying a lower charge, reduced leak and reduced pressure drop during operation (high efficiency).

On the other hand, the presence of an intermediate fluid introduces an additional heat resistance and time constant to the system but the intermediate fluid can operate as "cold storage". This plays a relevant role in case of Stop & Start or mild hybrid powertrains where it is important to maximize the benefit of the powertrain architecture limiting as much as possible the engine re-start for thermal comfort reasons. In addition the thermal buffer presence allows to generate the cooling power when is more convenient from energy point of view (e.g. cut off condition partially recovering the engine kinetic energy).

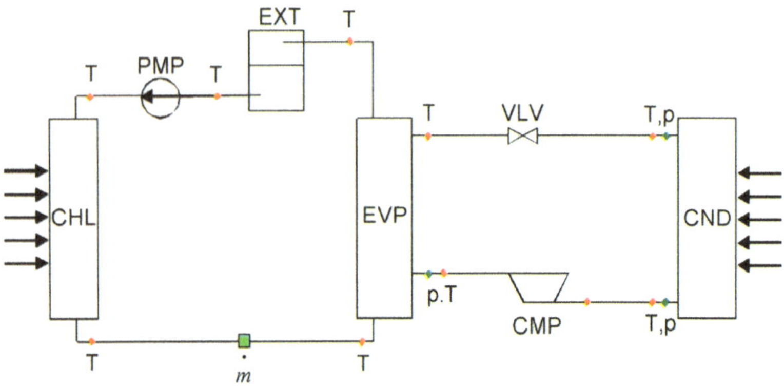


Fig. 7. Block diagram of the system with indirect expansion Legend: *CHL:* chiller, *EXT*: expansion tank; *PMP*: pump, *EVP*: evaporator, *VLV*: thermostatic valve, *CMP*: compressor, *CND*: condenser *T*: temperature measure point, *p*: pressure measure point *m* coolant mass flow.

Designing and controlling in an appropriate way "cold storage", by means of dedicated control logics within the frame of this project, we may have significant improvement in the thermal performance in those conditions. The used chiller is an aluminium plate heat exchanger realized on purpose, the electrical coolant pump is normally used as additional pump for engine cooling and cabin heating.

## 4.1 Thermal Performances

As previously indicated the vehicle with the indirect expansion system has been evaluated measuring its cooling performance during the cool down test and its overall efficiency during NEDC cycles (Fig. 10) at different temperatures. The original scroll compressor (60 cc) has been replaced with a 80 cc piston compressor with fixed displacement. The indirect expansion system allows a slight reduction of the fuel consumption at low and mid thermal loads but implies higher energy demand in case of high thermal load. It must be underlined that the indirect expansion assures a good thermal comfort in case of Stop & Start powertrain without requiring the engine re-start so allowing to take the maximum benefit from Stop & Start or similar powertrains. Figure 8 shows the results acquired comparing the same car equipped with standard A/C and Stop & Start and then with an indirect evaporator A/C system. The curves show that the indirect expansion system assures lower outlets temperature fluctuations and more stationary temperatures at head level.

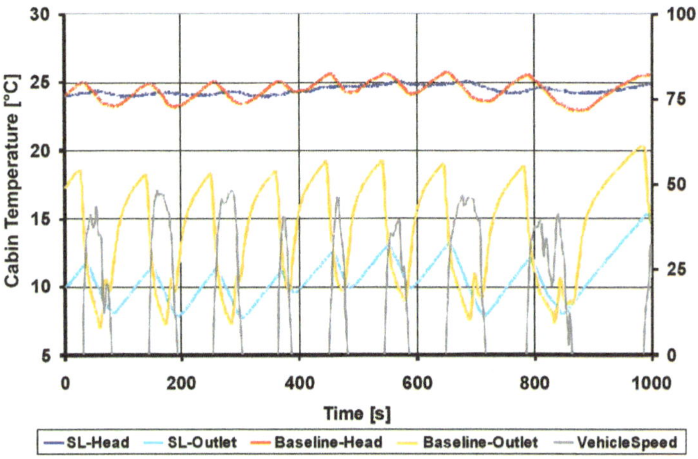


Fig. 8. Comparison of a car with standard A7V and Stop & Start and the same vehicle with indirect expansion A/C and Stop & Start

The indirect evaporation system has been evaluated also in terms of cooling performance (cool down), as shown in figure 8. The performance is slightly worse than the baseline system especially in the first minutes where the outlets temperatures are slightly higher than the baseline due to the increased system thermal inertia. The cool down test conditions are the same described in the previous paragraph.

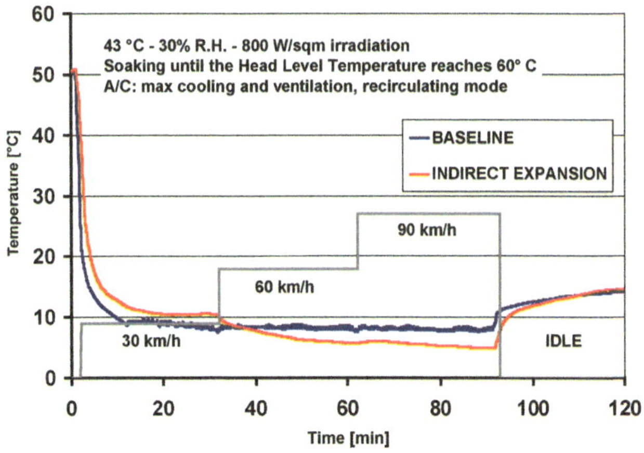


Fig. 9. Indirect expansion cool down compared with the baseline cooling performance

In addition to these characteristics it must be underlined that the decoupling of the evaporator by means of a secondary loop eliminates also all the NVH criticisms of the air conditioning in the cabin (expansion valve whistling, compressor pulsatios etc.).

| | | Air Cooled | Water Cooled | Improved Water Cooled | Max $CO_2$ saving |
|---|---|---|---|---|---|
| | Cycle | A/C Fuel Consumption [l/100 km] | | | g/km |
| A/C ON 15 °C - 70 % R.H. Set Point 20 °C | EUDC | 0.7 | 1.0 | 0.6 | -2.1 |
| | ECE | 0.3 | 0.4 | 0.3 | -0.8 |
| | NEDC | **0.43** | **0.62** | **0.38** | **-1.3** |
| A/C ON 28 °C - 50 % R.H. Set Point 20 °C | EUDC | 2.1 | 2.6 | 2.0 | -2.7 |
| | ECE | 0.6 | 0.6 | 0.6 | -1.3 |
| | NEDC | **1.15** | **1.33** | **1.08** | **-1.8** |
| A/C ON 35 °C - 60 % R.H. Set Point 23 °C | EUDC | 3.5 | 3.6 | 3.5 | 0.0 |
| | ECE | 0.6 | 1.0 | 1.0 | 10.6 |
| | NEDC | **1.65** | **1.95** | **1.91** | **6.7** |

Fig. 10. A/C fuel overconsumption, two version of the system have been developed and tested. The second and improved version, where optimised control strategies have been adopted, shows lower fuel consumption at mild summer conditions and still is responsible of higher consumption in severe summer conditions.

## 5 Conclusions

The idea to adopt a double loop to decouple the refrigeration circuit will be tested on a vehicle demonstrator in the next future within the TIFFE project. Nevertheless the preliminary activities carried out mainly by Centro Ricerche Fiat to assess the advantages and drawbacks of water cooled condenser and water cooled evaporator gave interesting and positive results. The major advantages can be summarised as follows:

Water cooled condenser:

- Safer use of flammable refrigerants.
- Reduced fuel consumption thanks to the improved and homogeneous condensation.
- Reduced lines and rubber hoses length.
- Reduced leak rate and refrigerant charge.

Water cooled evaporator (indirect expansion):
- Safer use of flammable refrigerants.
- Thermal buffer allowing to guarantee a good thermal comfort maximising the benefits of Stop & Start and mild hybrid powertrain.
- Reduced lines and rubber hoses length.
- Reduced leak rate and refrigerant charge.

In the next years the open issues (cost, efficiency improvement, system integration and packaging) will be solved taking benefit of the European project TIFFE started in June 2009 and ending at the end of 2012.

## References

[1] Malvicino, C., The 4 Fiat Pandas Experiment and other considerations on refrigerants, SAE Alternate Refrigerant Systems Symposium, Phoenix (AZ), USA, 2008.

[2] Zilio, C., Monforte, R., A small vehicle MAC system exploited with R-1234yf: bench results, 3rd Mobile Air Conditioning, Vehicle Thermal Systems and Auxiliaries, Turin, 2009.

[3] UltimateCooling™ System Application for R134a and R744 Refrigerant, N.S. AP, SAE AARS, Phoenix (AZ), USA, 2007.

[4] Baker, J., Ghodbane, M., Rugh, J., Hill, W., Alternative Refrigerant Demonstration Vehicles, SAE AARS Phoenix (AZ), USA, 2007.

[5] Malvicino, C., Mola, S., Clodic, D., Mobile air conditioning fuel consumption and thermal comfort assessment procedure, 7th IIR Gustav Lorentzen Conference on Natural Working Fluids, Trondheim, Norway, 2006.

**Carloandrea Malvicino, Riccardo Seccardini**
Centro Ricerche Fiat
Strada Torino 50
Orbassano, 10043
Italy
carloandrea.malvicino@crf.it
riccardo.seccardini@crf.it

# Estimation of the Driver-Style Economy and Safety via Inertial Measurements

S. M. Savaresi, V. Manzoni, A. Corti, Politecnico di Milano
P. De Luca, Teleparking Srl

### Abstract

This work proposes a method to quantify the driver-style economy and safety via inertial measurements. It presents a low-cost and vehicle-independent system for the acquisition of the variables related to the dynamics of a bus. Then, it describes an algorithm which calculates the over-consumption of energy with respect to different reference profiles. A second algorithm counts the number of unsafe behaviours, quantifying the safety. Finally, experimental results shows the algorithm applied to real data.

## 1 Introduction

A massive usage of public transport systems gives a big contribution in term of pollution reduction thanks to the smaller ratio of fuel per passenger per kilometer compared to the private mobility [1]. An example of public transport mean widely used in urban mobility is the bus. It is characterized by a big mass and, in the urban area, its average speed is usually low. In these conditions, the aerodynamic power is almost an order of magnitude less than the inertial power. So, how the driver accelerate and brake – the so-called driving style – has a direct impact on the fuel consumption, which corresponds to the emissions of $CO_2$ and $NO_x$. However, it affects the safety and the comfort of the passengers. This paper presents a method for the quantification of the driver-style economy and safety via the measurements of the longitudinal speed and the longitudinal and lateral acceleration.

This paper is organized as follow: in section 2, the experimental vehicle, the sensors and the data acquisition campaign are presented. Section 3 describes concisely the driver-style estimation algorithm. The result of the application of the algorithm to real data and the discussion of the results are presented in section 4. Finally, section 5 summarized the presented content and gives suggestion for further work.

## 2 Experimental Setup

In this section, the vehicles, the sensors and the design of the data acquisition campaign will be described.

### 2.1 Vehicle and Sensors

The technical characteristics of the bus are listed in table 1 (see also [2]).

| *Property* | *Value* |
|---|---|
| *Mass* | *11,545 kg* |
| *Frontal surface (w × h)* | *2.55 m × 2.96 m = 7.55 $m^2$* |
| *Length* | *11.98 m* |
| *Drag coefficient (Cx)* | *0.33 (estimated)* |

Tab. 1. Technical data of the bus used for the data acquisition campaign.

The weight and the frontal surface will be parameters of the estimation algorithm presented in section 3. The bus is equipped with:

- a commercial GPS (Global Positioning System) receiver connected via serial line RS-232 to a real-time ECU (Electronic Control Unit) for the logging of the data. GPS aims to provide the position and the speed along the path with a sampling frequency of 1 Hz;
- an automotive IMU (Inertial Measurement Unit), connected to the ECU through a CAN bus. It provides the longitudinal and the lateral accelerations with a sampling frequency of 25 Hz.

The GPS is mounted on the lateral window, in order to see the open sky to maximize the receiver accuracy. The IMU is also mounted on the lateral window, assuring its alignment with the longitudinal and the lateral axis of the vehicle. The real-time unit hosts a 40 MIPS microchip microprocessor that runs the software for data processing and logging. The power for the full systems is provided by a 12-Volt battery which guarantee an autonomy of one day. The architecture of the instrumented vehicle is shown in figure 1.

The described experimental system does not require to have the access to the vehicle data buses and signals. In such way, it is a low-cost and vehicle-independent way for the acquisition of the variables related to the dynamics. This system has also been successfully tested on other buses with different technical characteristics.

Fig. 1. The urban bus equipped with a commercial GPS receiver, an inertial platform and a controller unit.

### 2.2 Data Acquisition Campaign

The robustness of the estimation is an important property of the algorithm. The campaign has been designed to get enough data to develop an algorithm which manage different drivers, road characteristics and traffic conditions. The data has been collected for five days along two different urban lines by 11 different drivers.

## 3 The Driver-Style Estimation

The algorithm returns a bi-dimensional index:

- the driver style economy with respect to two different references;
- the driver style safety for the passengers in term of the number of unsafe behaviours.

The signal between two next stops is called start-and-stop (hereafter abbreviated as SAS) and it is the input signal for the estimation algorithm. We decide to split the track into SASs because we want to considers the average speed of a SAS as truth. This hypothesis seems quite strong, but leaves the driver free to manage the traffic conditions and the schedule of the bus stop. The first step of the algorithm is to identify the beginning end the end of a SAS. While the precision of the position estimated by the GPS is about 5 m, sufficient to identify the beginning and the end of the selected line, the precision of the speed is highly affected by the number of available satellites, especially when the vehicle travel at speeds close to 0 m/s. This step is implemented using a finite-state machine with tuned thresholds. The thresholds have been set to work correctly in the 95% of the cases. Once the single start-and-stop has been identified, the algorithm takes it as input and calculates the two indexes.

### 3.1 The Driver-Style Energy Estimation

The block diagram of the calculation of the over-consumption is represented in figure 2.

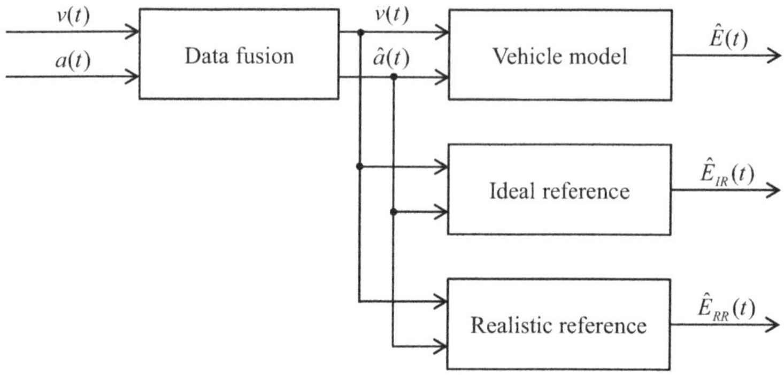


Fig. 2. The block diagram of the driver-style economy estimation algorithm.

The data fusion block integrates the low frequency of the GPS speed signal with the high frequency of the longitudinal accelerometer through the frequency split method. The output of the block is then passed to another entity which calculates the energy spent by the vehicle using a simplified model of the vehicle dynamics which takes into account only the inertial power needed for the acceleration and the inertial power to win the aerodynamic drag. Two terms are not considered in our model, i.e. the inertial power for braking (which is not recoverable since there is not electric engine on board) and the gravitational force due to slopes. In our case it is supposed equal to zero for the flatness of the city's morphology where we did the data acquisition campaign.

The energy calculated is then compared with the one obtained from two reference profiles.

- The ideal reference. This is the less amount of energy needed to move a vehicle for the same distance in the same time. This profile is called ideal profile since it is not usually feasible due to traffic conditions (see figure 3). It is the solution of a minimization problem.
- The realistic reference. This reference is computed from the real speed profile, output of the data fusion (see figure 3). This reference is designed to smooth the high acceleration.

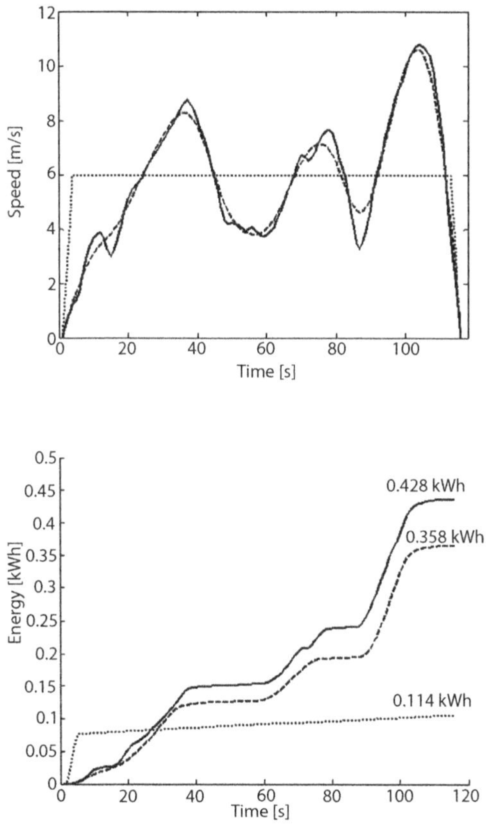


Fig. 3. On the top part of the figure, the comparison among the real speed (solidline), the ideal reference (dotted line) and the realistic reference (dashed line). On the bottom part, the same comparison in term of the energy (same convention).

### 3.2 The Driver-Style Safety Estimation

While for the seated passengers a good driving mainly affects the comfort, the driver-style can be determinant for the safety of the people who stand. We identified two main causes which affect the passengers' safety on buses: high acceleration and high variations of acceleration, the so-called *jerk* [3]. For each sample we define an acceleration-related index $f(a)$ and a jerk-related index $g(j)$. Then we calculate the norm of the vector which has the two indexes as components. Finally, we count how many samples have a norm greater or equal to a threshold with respect to the full-length of the SAS. The threshold is a tuning parameter; it was chosen experimentally, according to the sensations of an expert driver. Figure 4 shows the application of the algorithm to a start-and-stop; if we consider the dotted threshold, the percentage of the unsafe behaviors is the 8.0%. On the other way round, if we choose the solid threshold, there are not any unsafe behaviors.

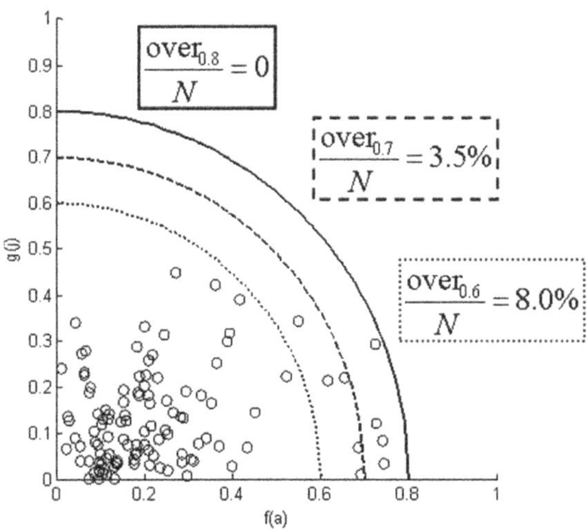


Fig. 4. The figure shows a start-and-stop represented on the plane jerk-acceleration. Three hypothetical thresholds are represented with different line styles.

## 4 Experimental Results

The algorithm has been applied to different drivers who drove along different urban lines in different time-slots during a day. Figure 5 shows a full lap along the urban line no. 1. The lap is composed by 33 start-and-stop. The white bar indicates the ratio between the real consumption and the one of the ideal reference, the gray bar represents the ratio between the real consumption and the realistic reference. The larger the bar is, the higher is the absolute consumption of energy. The solid line indicates the realistic reference, the dashed one the weighted average. On the right part of the figure examples of high and low over-consumption start-and-stop are represented.

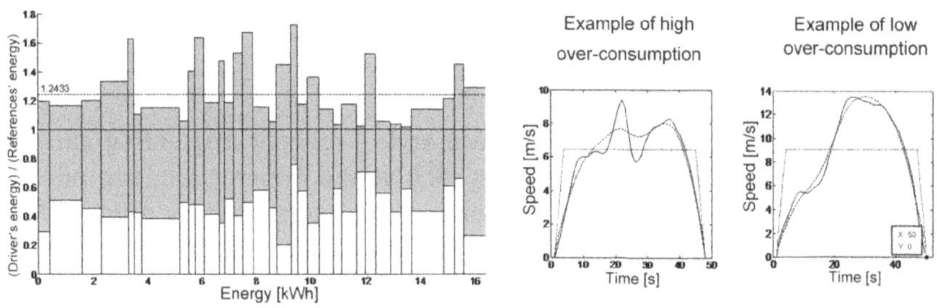


Fig. 5. On the left, the 33 start-and-stop which compose a lap along the urban line. On the right, examples of high and low over-consumption start-and-stop. For the line convention, see the label of Fig. 3.

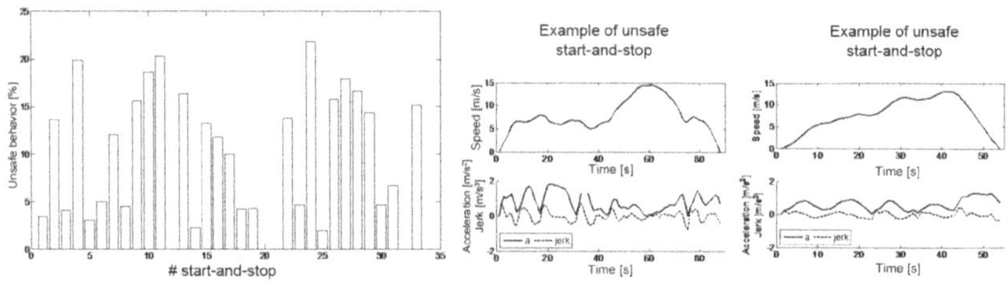


Fig. 6. On the left, the safety index of 33 start-and-stop. On the right, the plots of the signals of speed, acceleration and jerk of a safe and a unsafe safe start-and-stop.

Figure 6 shows the driver-style safety index for the same lap as above. Each bar represents a start-and-stop, the height of the bar the number of the unsafe behaviours. If we have a look to an example of safe start-and-stop, we can note that both the absolute value of the acceleration and the jerk are low due to the smoothness of the speed signal. Conversely the start-and-stop with a low level of safety has high acceleration and jerk.

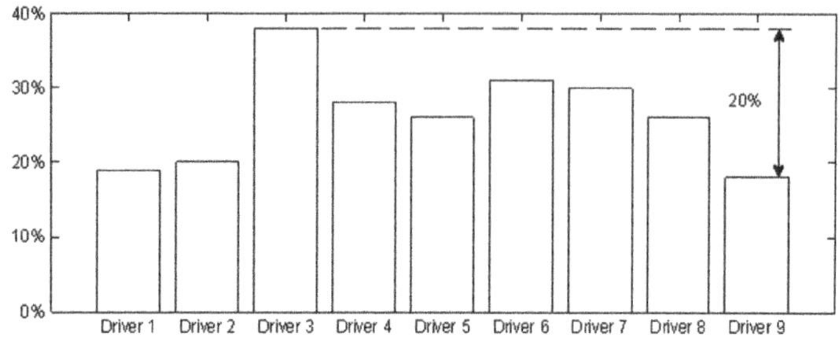


Fig. 7. Driver-style economy index aggregated by driver.

For the sake of data analysis, the driver-style economy and safety can be aggregated along different dimensions, such as the driver, the urban line or the time-slot. Figure 7 shows the economy index aggregated by drivers. The evidence is that there could be significant differences in term of driver-style economy among different drivers. This kind of chart could be useful for a rewarding campaign to motivate drivers to drive well.

Finally, the aggregation of the economy index by daily time slot is proposed. The bar plot in figure 8 shows a correlation between the over-consumption and the morning and the evening hours. This phenomena is predictable since those time-slot correspond to the rush hours, or rather when the traffic is high.

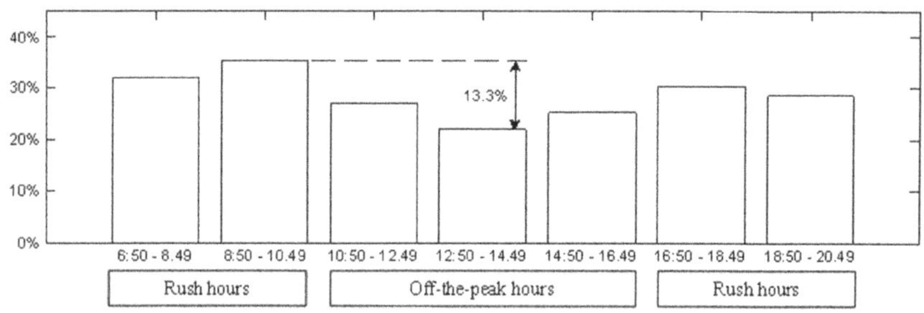


Fig. 8. Driver style economy index aggregated by daily time-slots. The plot shows that traffic of rush hours increment the fuel consumption.

## 5 Conclusion and Outlook

In this paper we present a low-cost and system independent hardware and method to estimate the variable related to the vehicle dynamics. The system is used to collect data for the development of an algorithm which estimate the driver-style in term of economy and safety. The algorithm is then applied to real data, showing the existence of differences among different drivers and a dependence of the fuel consumption with respect to the daily time-slot. Future works will extend the signal pre-processing to take into account slopes. Moreover, a broader data acquisition campaign can statistical validate the data.

## Acknowledgments

This work has been partially supported by the National Research Project "New algorithms and applications of identification and adaptive control" funded by the Department for Education, University and Research of Italy.

## References

[1] Bouwman, M.E., Moll, H.C., Environmental analyses of land transportation systems in The Netherlands, Transportation Research Part D: Transport and Environment, vol. 7, no. 5, pp. 331-345, 2002.

[2] Krajnovic, S., Davidson, L., Numerical Study of the Flow Around a Bus-Shaped Body, Journal of Fluid Engineering, vol. 125, issue 3, pp. 500-509, 2003.

[3] Cook, P.A., Stable Control of Vehicle Convoys for Safety and Comfort, IEEE Transactions on Automatic Control, vol. 52, no. 3, 2007.

**Sergio M. Savaresi, Vincenzo Manzoni, Andrea Corti**
Politecnico di Milano
Piazza L. Da Vinci, 32
20133 Milan
Italy
savaresi@elet.polimi.it
manzoni@elet.polimi.it
corti@elet.polimi.it

**Pietro De Luca**
Teleparking Srl
Via Giovan Battista Pergolesi, 27
20124 Milan
Italy
pietro.deluca@teleparking.it

**Keywords:** driving-style, public transport, data fusion, signal processing, fuel economy, safety

# Safety & Driver Assistance

# Stable Road Lane Model Based on Clothoids

C. Gackstatter, P. Heinemann, S. Thomas, Audi Electronics Venture GmbH
G. Klinker, Technische Universität München

### Abstract

In the following, the main concepts of a road lane model that keeps track of an arbitrary number of lane borders are presented. Information from an existing lane detection device, a gyrosensor, and map data are merged and filtered to create a road model with a desired number of road lines. The model is based on clothoids and continuously provides positions, angles, and curvatures of the border lines of the vehicle's own lane as well as of several neighboring lanes. Particularly on urban roads, in situations with upcoming turning lanes, or when the lane detection system fails to detect road lines, the model can still provide plausible information. This information significantly simplifies the situation analysis in further algorithms that rely on a lane detection system and require detailed information on current road lanes.

## 1 Introduction

Various advanced driver assistance systems are based on vision-based lane detection systems to determine characteristics of the current road and its lanes. Examples are lane departure warning or the lane keeping assistant that use the lane detection to support the driver to keep the lane. If the driver is distracted or careless, the system warns the driver by vibration in the steering wheel or by directing him back to the middle of the lane.

Difficulties arise when the lane detection system fails to recognize road lines or interprets objects as road lines that are not actually part of the current road. Especially in urban regions systems reach their limits. Vision-based approaches like in [1] use edge-oriented methods for a robust detection of road lines and approximate the course of the lanes by clothoids. These methods still fail if line markings are completely missing or covered by other vehicles. Subsequent systems that are based on lane detection have to deal with these inaccuracies and react properly.

Current driver assistance systems that rely on a lane detection method are mainly focused on highway situations. If the system needs only information on the own lane or if the vehicle drives on a highway with clear road markings, approaches as described in [2] can be used. It presents a multi sensor approach that fuses image measurements and map data to improve the tracking of road boarders in highway scenarios. Another method as shown in [3] uses six cues and a particle filter to achieve robust lane tracking. This system is robust against dramatic lane changes and discontinuous changes in road characteristics. By using the Distillation Algorithm to track the vehicles pose relative to the road and their width, the system seems to be stable in situations that are critical for vision-based lane trackers. A comparison of lane position detection and tracking techniques is presented in [4]. One of the methods listed is described in detail in [5]. This approach uses a road model based on clothoids and tracks them by a linear vehicle dynamics model. This road model, however, encounters problems when edge detection fails in complex situations.

If detailed information on the current road is needed and when driving on urban roads, it becomes challenging to guess the road geometry in situations when line detection fails. This happens for example if road lines are occluded by other vehicles, or if the curvatures are very high. Therefore, algorithms are needed to decide which detected road lines contain useful information, which lines should be ignored, and to determine the position of road lines that were not detected at all. The advantage of our system is that additional information on the current road is provided continuously, in particular with regard to urban roads, at intersections, and exit ramps.

The road lane model presented in this paper provides a stable estimate of the current road geometry at any time. It selects suitable road lines from a vision-based lane detection system and guesses the characteristics of missing road lines by neighboring lines and a gyrosensor as illustrated in Fig. 1. For each line, a Kalman filter is used to keep track of the lateral distance to the vehicle, the heading direction, and the clothoid parameters.

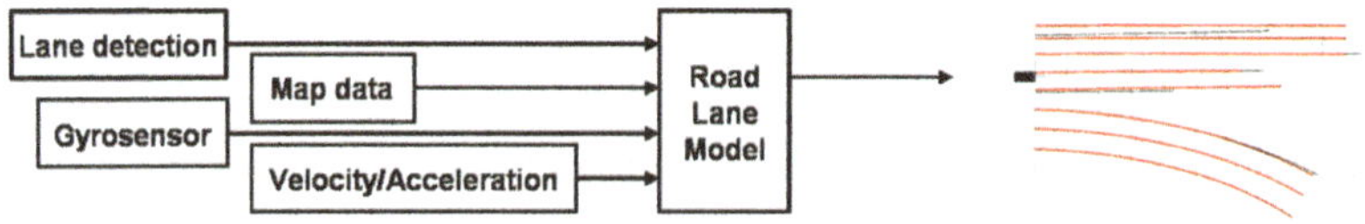


Fig. 1. Road lane model system overview

## 2 Road Lane Model

The road model in its current implementation keeps track of eight road lines, four on the right side and four on the left side of the vehicle. This allows providing information on at least four lanes at any time. More lines could be modeled easily, but due to their high lateral offset, they are unlikely to be recognized by lane detection systems. The benefit of modeling parallel road lines individually becomes obvious on urban streets, close to exit ramps, or at the beginning or ending of turning lanes as depicted in Fig. 2.

Fig. 2. Road lane model in selected situations

Our implementation includes Kalman filtering of the line parameters as well as a sensor fusion of the data from lane detection with the yaw rate of the gyrosensor (section 2.2). At each time step, suitable road lines from the lane detection system are selected and integrated into the current model (section 2.3). If no line is found at a certain position, we estimate the line parameters by the inner neighboring line. Since a parallel curve is needed at this step, we perform an optimization of clothoid parameters to obtain a nearly parallel clothoid even at high curvatures (section 2.4). In addition, the yaw rate sensor combined with the current map data continuously provides information on the relative movement between the vehicle and the road (section 2.5).

### 2.1 Basics of Clothoids

Clothoids are a special type of curve that is used for road construction to avoid abrupt changes of the steering angle when driving from straight to circular road section and vice versa. They are defined by their begin curvature $c_0$, a constant curvature change rate $c_1$ and their total length $l$. The current curvature of a clothoid after length $l_c$ equals

$$c(l_c) = c_0 + c_1 * l_c \tag{1}$$

The tangent angle $\tau$ at a length $l_c$ describes the change in orientation and is obtained by integration over $l_c$:

$$\tau(l_c) = c_0 * l_c + 0.5 * c_1 * l_c^2 = \frac{c_0 + c(l_c)}{2} * l_c \tag{2}$$

## 2.2 Application of the Extended Kalman Filter

The Kalman filter is a set of mathematical equations [6]. It offers a convenient way to estimate the exact state of a technical system by drawing conclusions from defective observations. For the lane model, we use a distinct Kalman filter for each potential road line to estimate eight different parameters. The clothoid parameters $c_0$, $c_1$, and $l$, the lateral offset $d_y$, and the heading angle $\psi$ are determined by the lane detection system. The heading angle represents the angle between the vehicle axis to the road. The gyrosensor combined with map data provides additional information on the yaw rate $\gamma$ and the lateral acceleration of the line $a_y$. Therefore, the state vector for each line that we want to keep track of comprises the following parameters:

$$\hat{x}_k = \begin{pmatrix} d_y & v_y & a_y & \psi & \gamma & c_0 & c_1 & l \end{pmatrix}^T \tag{3}$$

All parameters (except the clothoid parameter) are given in relation of the own vehicle. Estimations on the current acceleration provide additional information for the velocity $v_y$ at the next time step and also the current velocity helps to estimate the next lateral offset $d_y$. Similarly, the yaw rate provides information for the next yaw angle $\psi$ and the curvature change $c_1$ of the clothoid is crucial to determine the curvature $c_0$ at the beginning of the clothoid. These connections are considered in the process function of the discrete Kalman filter that estimates the next state after a time step $\Delta t$ by the parameter estimates of the current state. Due to the heading angle that we have to consider when calculating the next clothoid parameters $c_0$ and $l$, the process step of the Kalman filter is not linear and an extended Kalman filter has to be applied. In comparison to a standard Kalman filter, the extended Kalman filter provides estimates even on non-linear processes. The process function used for the road lane model relates the current state to the next state as follows:

$$\tilde{c}_0 = \frac{1}{\frac{1}{c_0} + \Delta r}, \tilde{c}_2 = \frac{1}{\frac{1}{c_2} + \Delta r}, \tilde{l} = l + \Delta r * \tau, \tilde{\tau} = \tau \tag{5}$$

All four conditions should be considered to obtain a suitable clothoid, particularly in case of high curvatures. If we neglect one of the parameters, the resulting clothoid may considerably differ from the desired parallel curve as demonstrated in Fig. 3. Since a clothoid has only three parameters, we used a least-squares method to determine an optimal set of clothoid parameters. After performing only a single step of the Gauss-Newton algorithm, the values converge and we observed that optimal parameters imply only minimal changes in length $l$ and angle $\tau$. This result corresponds to the observation made in [8], that different curvature parameters can lead to similar curve shapes.

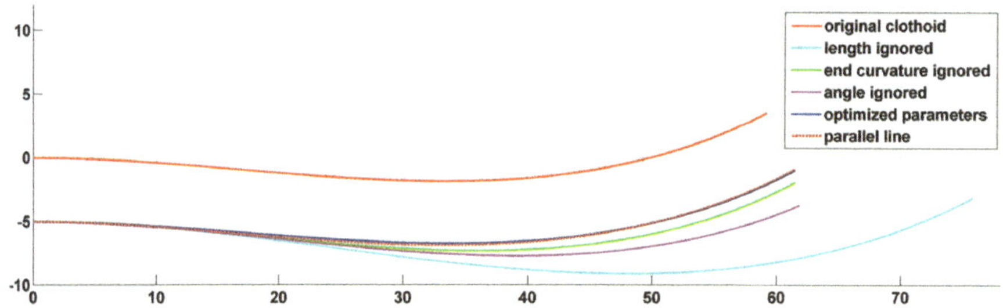


Fig. 3. Clothoid parameter estimation for a parallel line to a clothoid with $c_0$ = -0.01, $c_1$ = 0.0006, $l$ = 60

Therefore, the results of the numerical optimization can be approximated by leaving the expected length $l$ and change in orientation $\tau$ unchanged and by adjusting the values $c_0$ and $c_2$. To ensure the expected change in orientation at the expected length, the sum of the begin and end curvature has to be

$$c_{0,optimized} + c_{2,optimized} = \frac{2\tilde{\tau}}{\tilde{l}} \tag{6}$$

Since this differs from the sum of the expected curvature values, we add half of the resulting difference to $c_0$ and $c_2$:

$$\begin{aligned} c_{0,optimized} &= \tilde{c}_0 + \frac{1}{2} * \left( \frac{2\tilde{\tau}}{\tilde{l}} - (\tilde{c}_0 + \tilde{c}_2) \right), \\ c_{2,optimized} &= \tilde{c}_2 + \frac{1}{2} * \left( \frac{2\tilde{\tau}}{\tilde{l}} - (\tilde{c}_0 + \tilde{c}_2) \right) \end{aligned} \tag{7}$$

An example of a clothoid with optimized curvature parameters is shown in Fig. 3. In particular for clothoids that change between positive and negative curvature or with high curvature in general, the optimization improves the resulting shape of the clothoid significantly.

### 2.5 Parameter Estimation by Gyrosensor

The data fusion with the gyrosensor becomes important if not a single line is identified by the lane detection system. In combination with the current speed $v$, acceleration $a$, and the expected road curvature $c_r$, provided by map data, information on the yaw rate $\gamma$ and the lateral acceleration $a_y$ relative to the road are available:

$$\begin{aligned} \gamma &= \gamma_s - c_r * v * \cos\psi, \\ a_y &= (\gamma_s - c_r * v * \cos\psi) * v * \cos\psi + a * \sin\psi \end{aligned} \tag{8}$$

In both equations, the difference between the measured yaw rate $\gamma_s$ and the expected yaw rate $c_r v * cos\psi$ due to the current road curvature is needed. The result is the yaw rate relative to the current road geometry. For the lateral acceleration, the relative yaw rate needs to be multiplied with the current velocity and the heading angle $\psi$ should be considered to get the acceleration in the direction perpendicular to the street. The effects on the road model when only the lateral acceleration is available and road lines are not detected are analyzed in detail in the following section.

## 3 Results

We tested the road lane model in different situations on urban roads, rural streets, and highways to ensure the stability of the model in situations when vision-based lane detection fails. Each of the evaluations depicted in Fig. 4 to Fig. 7 shows the distances of road lines to the vehicle and a comparison of the current yaw rate values. The red graphs in Fig. 4 - 7 show the estimated lateral offsets $d_y$ of the inner four road lines, while the black graphs represent the lateral offsets measured by the vision-based lane detection system. If a line is not detected or contains no useful information for the current road model, its black graph switches to zero. If all graphs of one color shift up or down as in Fig. 5 at $t = 40s$, this indicates a lane change to the right or left respectively.

In the figures 5 to 7 we purposely ignore all detected lines for a few seconds to observe the process of the road model in comparison to the actual road lines that are still depicted in grey color. That means the model estimates the current positions of road lines solely based on the measured yaw rate and map data. Furthermore, the lower graphs show a comparison of the measured yaw rate in red color and the expected yaw rate due to the current road curvature in black. Negative values are due to a right turn, positive values follow from a turn to the left. It becomes obvious that differences between these two values affect the lateral offsets in the road lane model.

In Fig. 4 we have a common inner city situation that demonstrates the difficulties of vision-based lane detection on urban roads and particularly in tight curves. Suitable lines are not constantly detected, so that the model can only rely on single, occasionally observed lines or on the measured yaw rate and map data. In the case shown, we have a tight right curve with turn rates up to -10 deg/s. Still, the model estimates the lateral offsets qualitatively correct by continuously observing the yaw rate and selecting temporarily detected lines at $t = 64s$ and $t = 67s$. When several road lines are detected again after $t = 70s$, all lateral offsets will be adjusted.

The next example evaluated in Fig. 5 represents a highway scenario. Due to the good visibility of road markings, the lane detection system will usually not fail. Also the map data on highways is based on the more precise Advanced Driver Assistance Systems (ADAS) quality and provides appropriate values for the current road curvature. So even if lane detection fails as simulated from $t = 32s$ to $t = 48s$, we can keep track of the actual distances between $t = 32s$ and $t = 41s$ and recognize the lane change to the right around $t = 40s$. However, due to the high speed (approximately 130 km/h), the gyrosensor is sensitive to slight steering corrections as can be seen in the yaw rate image at $t = 40.5s$. This causes a positive change of the lateral offsets in the road model to the extent that it erroneously assumes another lane change to the left.

On urban roads, when driving at an easy rate (approximately 50 km/h), this problem becomes less eminent. As illustrated in Fig. 6, we can detect a lane change without the mentioned effect. The model follows the actual positions of road lines for 10 seconds with a final deviation of not even 1m. This result demonstrates the potential qualities of the model to track road markings independently from measured lines for a certain time period. To achieve this performance, the quality of the map data and the gyrosensor is crucial. Fig. 7 presents the effect of poor digitalized map data. Although the vehicle did not perform any lane change, the model wrongly estimates lateral offsets that imply several lane changes.

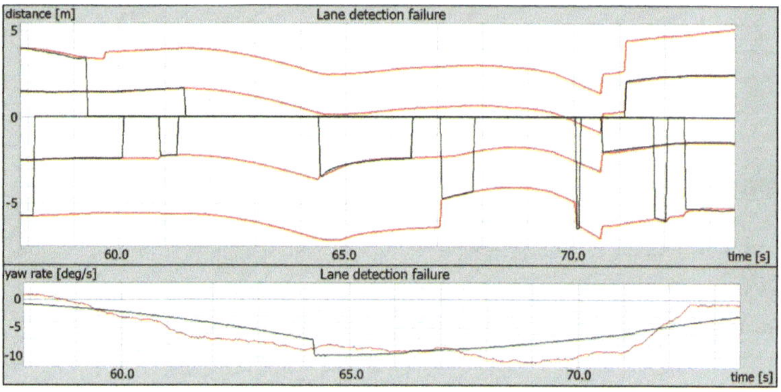


Fig. 4. Line positions and yaw rates (lane detection failure)

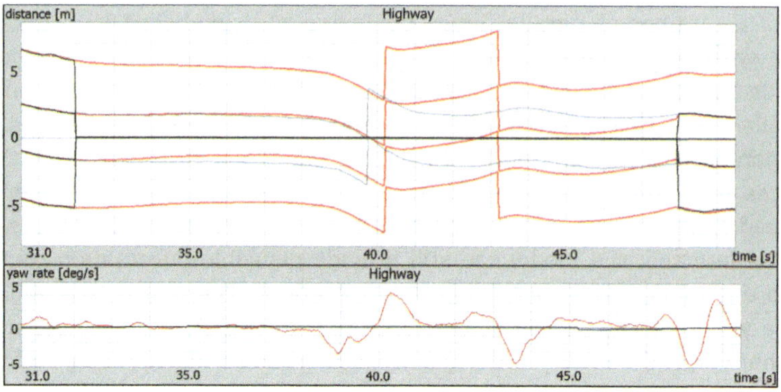


Fig. 5. Line positions and yaw rates

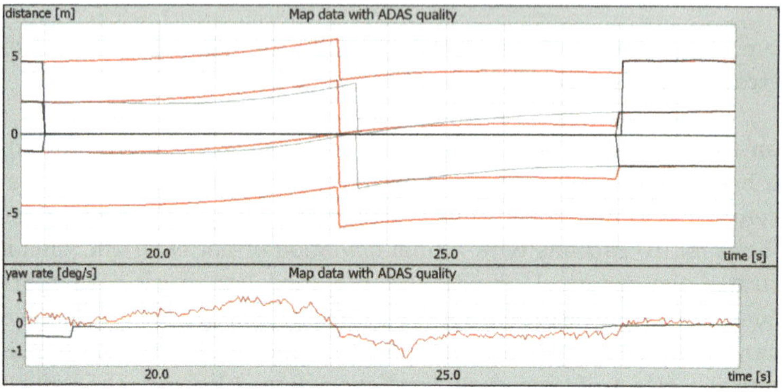


Fig. 6. Line positions and yaw rates (ADAS quality)

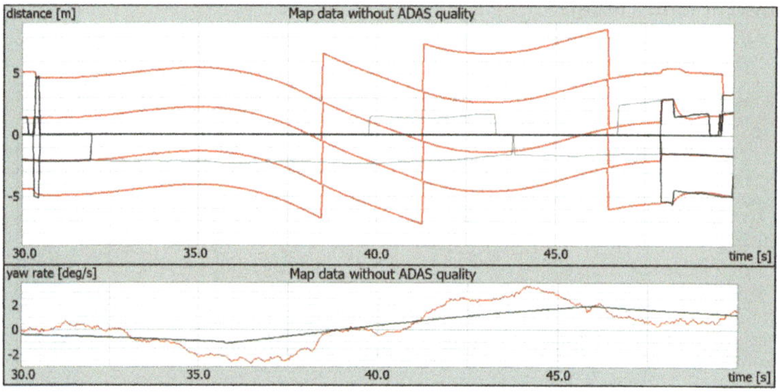


Fig. 7. Line positions and yaw rates (non-ADAS quality)

## 4 Conclusion

In this paper we presented the main concepts of a road lane model that keeps track of a number of parallel lanes by fusing a vision-based lane detection system with a gyrosensor, velocity, and map data. The model is based on clothoids and delivers positions, angles, and curvatures for a desirable number of parallel lines. It is robust against missing lines or erroneous detections by the vision-based system. Suitable road lines are selected by their length and distances to neighboring lines. Missing lines are estimated by numerically optimized clothoids that describe parallel curves even at high curvatures. If no lines are detected, the model can still provide plausible information by tracking the ego motion and comparing it to map data. The better this map data and the yaw rate values, the longer the model delivers plausible information. In further steps, the road geometry of map data can be improved by additional foresighted sensors. Also the yaw rate parameters may be adapted on highway scenarios to be able to deal with high velocities.

## References

[1] Goldbeck, J., Huertgen, B., Lane detection and tracking by video sensors, Proceedings of IEEE Intelligent Transp. Systems, pp. 74-99, 1999.

[2] Cramer, H., et. al, A new approach for tracking lanes by fusing image measurements with map data, IEEE Intelligent Vehicles Symposium, pp. 607-612, 2004.

[3] Apostoloff, N., Zelinsky, A., Robust vision based lane tracking using multiple cues and particle filtering, IEEE Intelligent Vehicles Symposium, pp. 558-563, 2003.

[4] McCall, J., Trivedi, M., Video-based lane estimation and tracking for driver assistance: survey, system, and evaluation, IEEE Intelligent Transp. Systems, pp. 20-37, 2006.

[5] Dickmanns, E., Mysliwets, B., Recursive 3-D Road and Relative Ego-State Recognition, IEEE Trans. on Pattern Analysis and Machine Intell., vol. 14, no. 2, pp. 199-213, 1992.

[6] Welch, G., Bishop, G., An Introduction to the Kalman Filter, SIGGRAPH 2001, Annual Conference on Computer Graphics, course 8, 2001.

[7] Bäumker, M., Rechenverfahren der Ingenieurvermessung, 2002.

[8] Swartz, D., Clothoid Road Geometry Unsuitable for Sensor Fusion, IEEE Intelligent Vehicles Symposium, pp. 484-488, 2003.

**Christina Gackstatter, Patrick Heinemann**
Audi Electronics Venture GmbH
Sachsstr. 18
85080 Gaimersheim
Germany
christina.gackstatter@audi.de
patrick.heinemann@audi.de

**Sven Thomas**
Institut für Informationsverarbeitung, Fakultät für Elektrotechnik und Informatik
Leibniz Universität Hannover
Appelstr. 9a, 30167 Hannover
Germany
thomas-sven@gmx.de

**Gudrun Klinker**
Technische Universität München
Boltzmannstr. 3
85748 Garching
Germany
klinker@in.tum.de

**Keywords:** driver assistance, road model, lane model, clothoids, Kalman filter, sensor data fusion

# Utilizing Near-Infrared and Colour Information in an Automotive Wide-Dynamic-Range Night Vision System for Driver Assistance

D. Hertel, C. De Locht, Melexis

## Abstract

Vision-based driver assistance systems utilizing both colour and near-infrared (NIR) can significantly reduce the risk of traffic accidents by improving night visibility of the roadway and obstacles. The night vision system described here combines within a single imager wide dynamic range (WDR) technology with sensitivity extended into the NIR, and colour capability. The multiple-slope CMOS pixel covers a dynamic range of more than 120 dB. Adaptive control algorithms ensure detection reliability by maintaining the incremental signal-to-noise ratio (iSNR). Colour sensing is based on a modified colour filter array (CFA) with a 4th channel to capture monochrome information from the imager's entire sensitivity range including the NIR component of the headlight spectrum that is invisible to the driver. Virtual NIR filter algorithms extract relevant colour information that is then fused to the luminance information, resulting in up to 70% higher sensitivity than conventional colour cameras.

## 1 Introduction

Reduced visibility at night substantially increases the risk of traffic accidents caused by misjudgement of dark roadways and failure to recognize the presence of pedestrians and obstacles. If an accident does occur, passive safety systems such as seat belts and airbags can reduce the percentage of fatalities, but only to a limited degree [1]. Any further reduction of traffic fatalities requires the deployment of active driver assistance systems. Video-based driver assistance systems provide additional image information in situations where visibility is poor (night vision) or obstructed (side or rear vision). Applications range from driver assistance with forward and rear view images displayed on a dashboard screen to active safety systems where the video footage is analyzed by software that issues warnings or triggers vehicle controls. On the front-end of these systems are video cameras that need to capture scene details (such as the location, shape, size, brightness, and colour [2]) reliably under all circumstances.

Night vision traffic scenes are very challenging not only for the driver's eye but similarly for cameras. The available illumination is often limited to the reach of the headlights; high-beam headlights increase that range but cannot be used in the face of approaching traffic. The intrascene dynamic range – the intensity ratio between light sources and illuminated objects – can be extreme, far exceeding that of linear CMOS imagers. For example, approaching traffic's headlights cause blinding that severely limits the ability of eye or camera to detect darker details of roadway and obstacles. If the dynamic range of a camera is too narrow to accommodate the intrascene dynamic range, important object details will be missing from the corresponding image. Once lost, these details are impossible to recover, even through image post-processing [3]. Since the design of automotive video cameras is governed by performance vs. cost tradeoffs, it is important to base that design on a clear understanding of the non-negotiable performance thresholds such as sensitivity, signal-to-noise ratio, and dynamic range. Therefore the model-based design approach that was used to quantitatively define and meet the performance requirements of automotive night vision included models of night vision environment, detection of scene detail, and human colour night vision.

### 1.1 Wide Dynamic Range of Night Vision Scenes

A quantitative model of the image capture environment was used to determine dynamic range requirements. Scene photometry measures the luminance of individual scene elements such light sources (headlights, taillights, traffic lights) and illuminated object surfaces (traffic signs, road markings, pedestrians, and road surfaces) [4]. Scene photometry was applied to a range of typical night traffic scenes to determine the luminance ranges for light sources and illuminated object surfaces [5]. The example in figure 1 (left) shows that the luminance ratio between bright car headlights and unlit pedestrians can range from 100 to more than 500 thousand.

Photospace describes the environment as a statistical distribution that is populated by measuring average scene light levels for a great number of application-typical scenes. We extended the photospace model by building distributions based not on scene averages but rather on the probability that a scene element of a given luminance will occur [3, 5]. The extended photospace distribution for automotive vision by day and night show clusters representing light sources at high luminance levels and illuminated surfaces at lower luminance levels. The luminance ratio within night scenes ($10^6$) is anticipated to be higher than that of daylight scenes ($10^4$ to $10^5$), due to the much lower luminance of illuminated objects at night. For the camera to successfully detect image detail simultaneously in both the brightest (differentiation of vehicle lights) and darkest objects (pedestrians) at night, a dynamic range of $10^6$:1 (120 dB) is required.

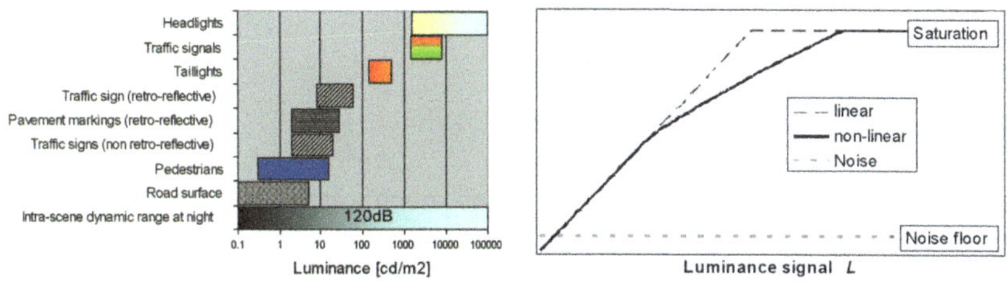


Fig. 1. Luminance ranges [Cd/m$^2$] for light sources and illuminated surfaces in typical night traffic scenes (left), extending dynamic range with a non-linear imager (right).

## 1.2 Reliable Wide Dynamic Range Detail Detection

Dynamic range specifies the range of input intensity signals $L$ where an imager can reliably capture information about both dark and bright details in the same scene at the same exposure settings. Information about scene detail is carried by small variations of the input intensity signal, for example the signal difference between an unlit pedestrian and the darker background. Detail detection by a camera requires it to have a minimum level of incremental signal-to-noise ratio (iSNR),

$$iSNR = \frac{g(L) \cdot L}{\sigma_D(L)} > threshold \tag{1}$$

with input intensity signal $L$, output signal $D(L)$, incremental gain $g(L) = dD(L)/dL$, and image noise $\sigma D(L)$ [6]. If the signal is lowered towards the imager's noise floor, iSNR decreases until the noise floor makes it impossible to detect shadow detail. If the signal is increased towards the imager's saturation level, the onset of saturation will clip further signal increases and thus cause the incremental gain to go to zero.

The dynamic range can be extended both by lowering the noise floor and by shifting saturation signal to a higher input signal level. This can be achieved by a nonlinear response curve. At a so-called breakpoint, the response curve's incremental gain is suddenly lowered, so that saturation is delayed to higher intensity levels than for a linear imager (figure 1, right). However, this extension comes at the cost of a sudden decrease of iSNR at the breakpoint, which sets a firm limit to the dynamic range extension that can be achieved by nonlinear WDR imagers. The threshold criterion for iSNR has to be maintained at every intensity level within the dynamic range.

### 1.3 Challenges of Night Colour Vision

Human night vision is characterized by two mechanisms: Increasing activation of the monochrome receptors ("rods"), and restriction of colour vision to brighter colours ("unrelated colours"). Colour vision in very dark scenes is concentrated in the bright coloured highlights. These are relatively small and isolated and therefore called "unrelated colours" [7]. For example, relevant traffic lights are perceived as colours, but poorly illuminated objects (foliage, pedestrian's clothes) are perceived as monochrome. The concentration of colour information in the bright lights represents a unique WDR challenge. For a monochrome system it is sufficient to render the brightest lights near the maximum signal level of the pixel in order to minimize halos and flare, thus maximizing differentiation between different lights. In the case of colour this may not be sufficient because a coloured light near the maximum signal level will be too de-saturated to be recognizable. For example, if a green colour is captured at 95% signal level then the difference between the R-, G-, and B- responses are so small that the resulting colour information is barely distinguishable from 'white'. Therefore for night vision the dynamic range of a colour system has to be even wider than that of a monochrome system.

Fig. 2. Same scene showing related colour at day time, and unrelated colour at night time.

Night vision colour capture and processing has to balance the lower SNR of the imager at low signal levels with the higher signal amplification necessary to reconstruct a full colour image. Night vision images should be such that the illuminated dark details are rendered as a sufficiently bright monochrome image while "unrelated colour" objects such as traffic lights are rendered in their correct hue and with sufficient saturation so that they can be distinguished and recognized by the human observer (see figure 2). In other words, the best compromise will be to reconstruct a good luminance image then add those unrelated colours relevant to the traffic scene. In display-based driver assistance applications it is important to provide "natural"-looking images that meet the expectations of the viewer so that the image content can be recognized and interpreted easily.

## 2 VIS-NIR True-Colour Night Vision System Based on WDR Colour Imager

Night vision sensitivity can be increased by extending the range of spectral energy that is utilized for image sensing. CMOS imagers are sensitive to NIR up to 1000nm, limited by the cut-off of the silicon. This makes it possible to utilize the large invisible portion of the spectrum emitted by incandescent headlamps. The range of visibility can be extended even further without blinding oncoming traffic if dedicated NIR high-beam headlights are used. However, conventional colour cameras cannot utilize the NIR spectrum because the filter materials used in standard CMOS and CCD technology are highly transparent to NIR. If the illumination contains high amounts of NIR, for example in coloured lights with incandescent lamps, the signal contributions to RGB from the NIR are much higher than those from the actual colours so that colours become extremely de-saturated. Therefore colour cameras need to suppress the NIR signal by an IR-cut filter, practically removing up to 80% of the total radiation emitted by a tungsten headlight. Solutions have been proposed to improve the sensitivity of colour imagers in the presence of the NIR cut filter by adding a large proportion of filter-less 'white' pixels to the RGB filter pattern [8]. These filter-less pixels have about three times higher sensitivity, but do not utilize NIR. Single-chip true-colour night vision cameras with NIR sensitivity have to utilize different technologies to separate the NIR from the colour signal. Examples are fast switching liquid crystal filters, or on-chip interference filter arrays requiring the deposition of 17 thin layers [9]. The colour information is then fused with the luminance information from a wider range of the spectrum, for example NIR or FIR (far infrared), to provide either true-colour or false-colour images. However, the high cost and larger camera dimensions make these solutions less suitable for low-cost automotive applications.

## 2.1 NightBrite™ – WDR Imager Technology with Extended NIR Sensitivity and Colour Capability

The Melexis night vision system combines the following four key technologies within a single image sensor: Firstly Melexis "Avocet" multiple-slope CMOS pixel technology extends the dynamic range to that required by night vision traffic scenes. Secondly Autobrite® adaptive control algorithms for exposure and wide dynamic range (WDR) compression ensure there is iSNR-based reliability of detail detection at each intensity level within the dynamic range. Then the AutoView™ tonemapping algorithm compresses WDR image information into the lower bit depth of image processing pipelines and image displays without any loss of detail. Finally NightBrite™ technology combines WDR with increased spectral sensitivity and colour capability. The NightBrite™ version of the imager has a modified colour filter array (CFA) with a 4th channel to capture monochrome information from the imager's entire sensitivity range that includes the near infrared (NIR). A virtual NIR filter reconstructs the chrominance components from the mixed colour plus NIR signals. An adaptive fusion algorithm adds 'relevant' colour information to the bright monochrome image.

The "Avocet" WDR imager is based on a standard CMOS pixel, making it attractive for low-cost automotive video applications. The 3T-architecture allows high fill factor and thus good dark sensitivity. The pixel is optimized so that in comparison with standard CMOS pixels the NIR sensitivity edge is shifted by more than 100 nm towards the IR, with cut-off at about 1100 nm.

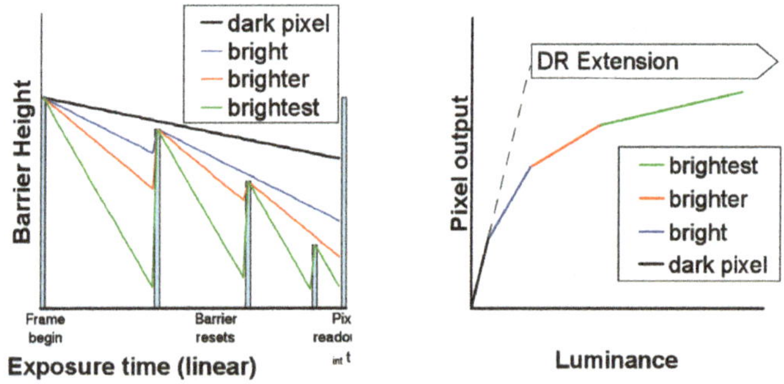


Fig. 3. Time-voltage diagram (left) and response curve (right) for multiple partial reset operation [3].

Dynamic range is extended by multiple partial resets of the CMOS pixel, also referred to as multi-slope or charge spilling. Photoelectrons generated during exposure will discharge the parasitic capacitance of the pixel. The rate of discharge increases with light intensity: dark pixels will only be partially discharged; very bright pixels will be completely discharged (saturated) even before the end of exposure time. This can be prevented by partially recharging the pixel before the end of exposure. As the recharge is only partial, it does not affect the signal of dark pixels but overwrites (resets) the signal of bright pixels that would saturate at or before the end of exposure. The reset pixels resume registering photons, but with the shorter remaining exposure time. The corresponding response curve will be piecewise linear, but with a lower incremental gain after the breakpoint that corresponds to the reset. This shifts the saturation signal to a higher input signal level, extends the dynamic range to match that of the natural scene, and compresses the imager output into the fixed dynamic range of the image display. The partial reset can be repeated up to 6 times, thus extending the response curve by linear sections of progressively decreased slope after each reset point. The 6 reset barriers are controlled by AutoBrite®, an algorithm that adapts exposure time and WDR extension to the intrascene dynamic range while maintaining a minimum level of iSNR throughout the extended dynamic range. Assuming an increased iSNR threshold of 5, which is based on the SNR requirements of colour processing, a linear imager will have a dynamic range of only 36 dB, compression with a single reset will extend the dynamic range to 62 dB, and two resets will reach 98 dB. It requires six resets to reach 138 dB, a level that matches and even exceeds the dynamic range requirements of automotive night vision [6].

The AutoView™ tonemapping algorithm is based on histogram equalization techniques and is implemented directly on the imager. It improves visibility of image details throughout the entire output range, and optimizes image contrast to improve the "naturalness" of the scene when reproduced on a display.

Fig. 4. NightBrite pixel arrangement, and raw images of colour chart: P'-channel, R'G'B' before and after virtual NIR filtering.

The NightBrite™ version of the WDR imager is fitted with a customized colour filter array where a filterless pixel P' replaces the second G'-pixel of the Bayer pattern (figure 4, left). The benefits of P' are two-fold: Firstly, it has a 'panchromatic' spectral response that includes the NIR and is equivalent to that of a monochrome imager. The panchromatic P'-pixel is more than 3 times as sensitive as a colour imager, capable of delivering a much brighter luminance image (figure 4). Secondly the P'-pixel delivers the reference signal necessary to estimate the NIR signal contribution to the R'G'B' pixel signals. The P'-signal is used in a 'virtual NIR cut filter', a customized colour calibration and correction method that effectively removes unwanted NIR contributions to the colour signal. The virtual NIR cut filter shows reasonably good performance in reconstructing colours even under illumination conditions with high amounts of NIR, for example incandescent light at a colour temperature of 3200 K having 20% signal in the visual, and 80% in the NIR range (figure 4). The filter array relies on standard RGB filter materials and fabrication processes so that it can be produced at costs comparable to conventional RGGB imagers, but without the cost of a physical NIR cut filter. Image fusion combines VIS and NIR image information, and represents them it in a way that can be easily interpreted by the human observer viewing video on a dashboard screen. True-colour image fusion is best suited for this application because it uses the extra information from the NIR to enhance the spatial and luminance information while preserving the colour information from the visible range. Since NightBrite™ utilizes the large NIR component of tungsten headlights that is invisible to the driver, it is 70 % more sensitive than conventional colour cameras. Furthermore, high-beam NIR headlights can be used to increase visibility beyond the distance range of headlights at low beam. Even when NightBrite™ is used with an NIR-cut filter, for example in applications where highest colour fidelity is required, it is still 25 % more sensitive than conventional colour cameras.

## 2.2 NightBrite Examples

The example shown in figure 5 demonstrates the NightBrite™ technology for a typical urban night traffic scene.

Fig. 5. Urban night traffic scene: panchromatic P' image (top left), RGB color image after virtual NIR filtering (top right), and final image after image fusion (bottom).

The monochrome image at the top left of figure 5 shows that the wide spectral sensitivity of the panchromatic P'-pixels delivers a sufficiently bright monochrome image even at a frame rate of 30 fps. The colour image on the top right is comparatively dark, consisting mainly of the "unrelated colours" of traffic lights and vehicle taillights against an almost black background. The spectral energy in the visible band is small compared to the total energy within the imager's sensitivity range so that the luminance component of the filtered colours is much lower than the luminance from the P'-pixels. WDR compression with adaptive Autobrite® control ensures that the bright highlight colours are preserved. Image fusion (bottom) then combines the bright monochrome background with the unrelated colours, leading to a more natural rendering of the scene that contains all traffic-relevant colour information from traffic lights and car signal lights.

Fig. 6. Urban night traffic scene taken with NightBrite™ (left), and linear colour camera (right).

The effectiveness of NightBrite™ compared to that of a conventional linear CMOS colour camera is shown in the example of an urban night traffic scene (figure 6). The image on the left taken with NightBrite™ has a brighter background than the linear image (right) due to higher spectral sensitivity, but lower contrast because AutoBrite® compressed a wider luminance range into the same display range. WDR compression and virtual NIR filtering enabled capturing red colour in the bright centres of the brake lights (figure 6, left). Without compression, bright taillights turn white due to signal saturation so that they become indistinguishable from headlights (figure 6, right).

## 4 Conclusions

A VIS-NIR colour night vision system based on a WDR CMOS imager with a modified colour filter array has been demonstrated. A virtual NIR filter successfully removes the NIR from the colour signals, and image fusion utilizes the extra NIR sensitivity to increase the brightness of the luminance channel. First prototypes show the concept to be viable.

## References


[1] Traffic Accident Causation in Europe (TRACE), Review of crash effectiveness of Intelligent Transport Systems, Project No. 027763, Deliverable D4.1.1 – D6.2, 55, 2007.

[2] Frieser, H., Photographische Informationsaufzeichnung, Focal Press, New York, 1975.

[3] Hertel, D., Betts, A., Hicks, H., ten Brinke, M., An adaptive multiple-reset CMOS wide dynamic range imager for automotive vision applications, Proc. 2008 IEEE Intell. Vehicles Symp., 614, IEEE Press, Piscataway, NJ, 2008.

[4] Carlson, P. J., Urbanik, T., Validation of photometric modelling techniques for retroreflective traffic signs, Transp. Res. Rec. 1862, 109, 2004.

[5] Hertel, D., Marechal, H., Tefera, D., Fan, W., Hicks, R., A low-cost VIS-NIR true color night vision video system based on a wide dynamic range CMOS imager, Proc. 2009 IEEE Intell. Vehicles Symp., 273, IEEE Press, Piscataway, NJ, 2009.

[6] Hertel, D., Extended use of ISO 15739 incremental signal-to-noise ratio as reliability criterion for multiple-slope wide dynamic range image capture, SPIE 7242, 724209, 2009.

[7] Fu, C., Li, C., Luo, M. R., Hunt, R. W. G., Pointer, M. R., Quantifying colour appearance for unrelated colour under photopic and mesopic vision, Proc. IS&T 15th Color Imag. Conf., 319, 2007.

[8] Honda, H., Iida, Y., Itoh, G., Egawa, Y., Seki, H., A novel Bayer-like WRGB color filter array for CMOS image sensors, SPIE 6492, 64921J.1, 2007.

[9] Koyama, S., et. al., A day and night MOS imager spectrally adjusted for a wide range of color temperatures, SPIE 7249, 72490S, 2009.



**Dirk Hertel**
Melexis, Inc.
10 Wilson Road
Cambridge, MA 02138
USA
dhe@melexis.com

**Cliff De Locht**
Melexis
Transportstraat 1
3980 Tessenderlo
Belgium
cde@melexis.com


**Keywords:** road and passenger safety, driver assistance, night vision, pedestrian recognition, automotive colour night vision, near-infrared night vision

# ADOSE – Bio-Inspired In-Vehicle Sensor Technology for Active Safety

J. Kogler, Ch. Sulzbachner, E. Schoitsch, W. Kubinger, M. Litzenberger,
AIT Austrian Institute of Technology GmbH

## Abstract

Reliable Advanced Driver Assistance Systems (ADAS) are intended to assist the driver under various traffic, weather and other environment conditions. The growing traffic requires sensors and systems which handle difficult urban and non-urban scenarios. For such systems new cost-efficient sensor technologies are developed and evaluated in the EU-FP7 project ADOSE (reliable Application-specific Detection of road users with vehicle On-board Sensors, providing the vehicle with a virtual safety belt by addressing complementary safety functions.

## 1 Introduction

The EU-funded project ADOSE is evaluating new sensor technologies and sensor systems. Such sensors are necessary for ADAS like lane departure warning, collision warning, high-beam assist or side impact detection. Fig. 1 illustrates the various sensors from different project members and additionally, the operating distance of each sensor is depicted.

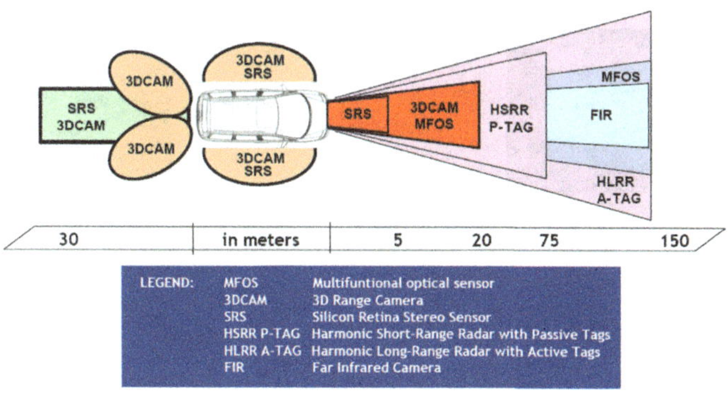


Fig. 1. Overview of all sensor technologies evaluated and considered in ADOSE

The approach of the Austrian Institute of Technology (AIT) towards reduction of the ADAS costs is to use a "Silicon Retina Stereo Sensor" (SRS). The SRS is specifically tailored to serve as a pre-crash warning and preparation sensor for side impacts. Pre-crash applications must reliably react in real time to prepare the vehicle (e.g. activate the pretensioner, preparation of a side airbag) for the imminent impact (which, in case of side impact, cannot be avoided by a reasonable reaction of the impacted vehicle). For the pre-crash sensor, it is necessary to take distance measurements of objects approaching the sensor. Two silicon retinas have therefore been coupled to a stereo vision unit, allowing distance information to be extracted from moving objects in the viewed scenery.

## 2 Silicon Retina Technology

Derived from the human vision system, the bio-inspired silicon retina sensor is a new type of imager. Conventional optical sensors capture images at a fixed frame-rate. A silicon retina optical sensor provides only timed event-triggered information, which means the sensor delivers information about the illumination changes ('events') in the visual field. The sensor detects intensity changes in positive (ON-event) and negative (OFF-event) direction in an observed scene, with each pixel delivering its address and event data separately and independently. The so-called "address event representation" (AER) was proposed in 1991 by Sivilotti [1] for transferring the state of an array of neurons from one chip to another. An early implementation of an artificial retina has been carried out by Fukushima et. al [2] in 1970 and the first retina imager on silicon basis is described in the work from Mead and Mahowald [3], which have also established the term "Silicon Retina". This type of sensor is intended to overcome certain obstacles in classical vision systems, which are depicted in Fig. 2.

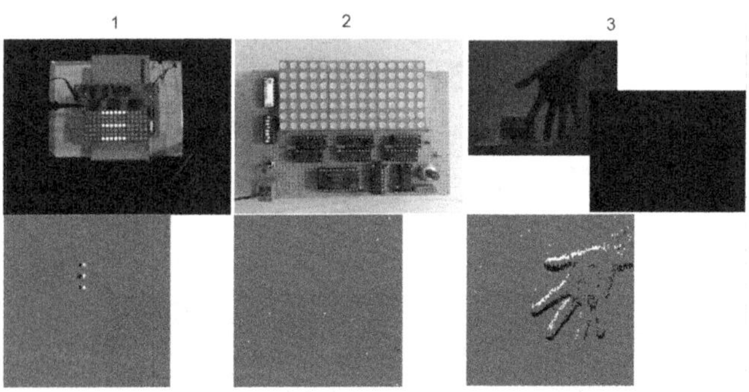


Fig. 2. Advantages of the silicon retina sensor technology, (1) high temporal resolution, (2) data transmission efficiency, (3) wide dynamic range

First, the high temporal resolution allows quick reactions to fast motion in the visual field. Due to the low resolution and the asynchronous transmission of address events (AEs) from pixels where an intensity change has been occurred, a temporal resolution up to 1ms is reached. In Fig. 2 (1) the speed of a silicon retina imager compared to a monochrome camera (Basler A601f@60fps) is shown. The top image in column (1) shows a running LED pattern with a frequency of 450Hz. The silicon retina can capture the LED hopping sequence, but the monochrome camera can not capture the fast moving pattern and therefore, more than one LED column is visible in a single image.

The second advantage is the on-sensor pre-processing because it reduces significantly both memory requirements and processing power. In Fig. 2 (2) the efficiency of the transmission is illustrated. The monochrome camera at top in the column (2) has no new information over time, but the unchanged image must be transferred after an image has been captured. In case of silicon retina imagers underneath no information has to be transferred with exception of a few noise events are visible in the field of view which must be transferred.

The third benefit of the silicon retina is the wide dynamic range up to 120dB, which helps to handle difficult lighting situations, encountered in real-world traffic and is demonstrated in Fig. 2 (3). The top image pair shows a moving hand in an average illuminated room with an illumination of ~1000 $lm/m^2$ and captured with a conventional monochrome camera. The second image of this pair below shows also a moved hand captured with a monochrome camera and an illumination of ~5 $lm/m^2$. In case of the monochrome sensors only the hand in the well illuminated environment is visible, but the silicon retina sensor covers both situations, what is depicted in the image below in Fig. 2 (3).

## 3 Stereo Vision Sensor

The silicon retina (SR) stereo sensor for processing of 3D stereo information uses two silicon retina sensors and Lichtsteiner et. al [4] describes in his work the silicon retina sensor used for the stereo system. Derived from the side impact detection use case defined in the project the stereo vision system must fulfil different requirements. In Fig. 3 the detection area of the camera system is shown. The vehicle (at least 0.5m wide) is approaching the camera system from the side with a maximum speed of 60km/h and the reaction time of the car equipment is assumed to be 350ms. These parameters are responsible for the chosen stereo vision system parameters, which are:

- Detection Range: objects must be detected with high confidence well before activation of countermeasures. Due to the assumption of a system reaction time of 350ms and the maximum speed of 60km/h, a detection range of 6 meters is required. Between 6m and 5m is the main operating distance.
- Field of view (FOV): for the given sensor resolution of 128x128 (pixel-pitch 40μm) pixels and a baseline of 0.45m, lenses with a field of view of 30° and a focal length of 8.5mm are chosen. The large baseline is necessary to reach the required depth resolution of three consecutive detections during one meter of movement at a distance of 6 meters.

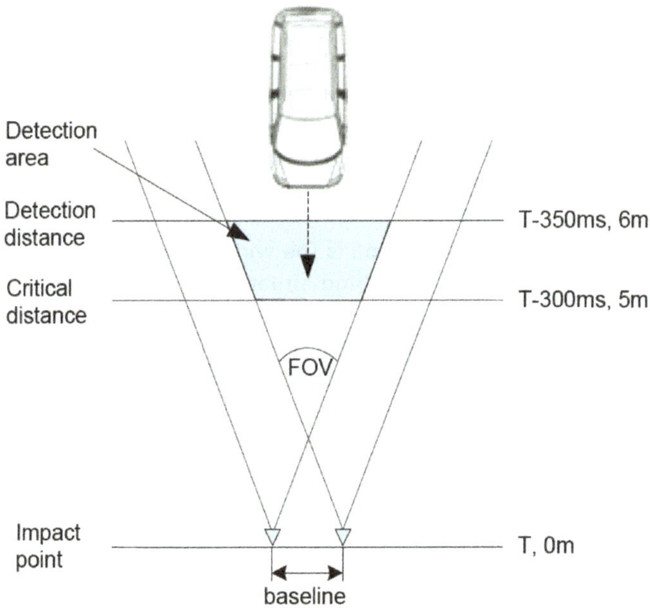


Fig. 3. Overview about the stereo vision system configuration

## 4 Stereo Vision Algorithm

Conventional area-based stereo algorithms that use monochrome or colour images aim for calculating dense depth images, i.e. providing depth information for every pixel of the input images. The task of stereo vision processing is the matching of corresponding pixels between the left and the right image, the so called "stereo matching". For this problem, a large variety of different approaches can be found in the literature [5]. In Section 4.2 state-of-the-art algorithms for silicon retina cameras are tested. Aside from several pre-processing steps, which are required by some stereo matching approaches, it is important for stereo vision algorithms to use calibrated cameras and rectified images which are described in section 4.1.

### 4.1 Calibration and Rectification

The data acquired from the cameras are not prepared for line-by-line matching respectively event-by-event matching, because the epipolar lines are not in parallel. Therefore, a rectification of the camera data is carried out which is described in detail by Scheer [6]. Before this rectification can be done, the cameras have to be calibrated. With conventional cameras the calibration pattern (Fig. 4 on the top) is captured in different views from the right and left camera and the crossings of the pattern are used for the calculation of the camera parameters.

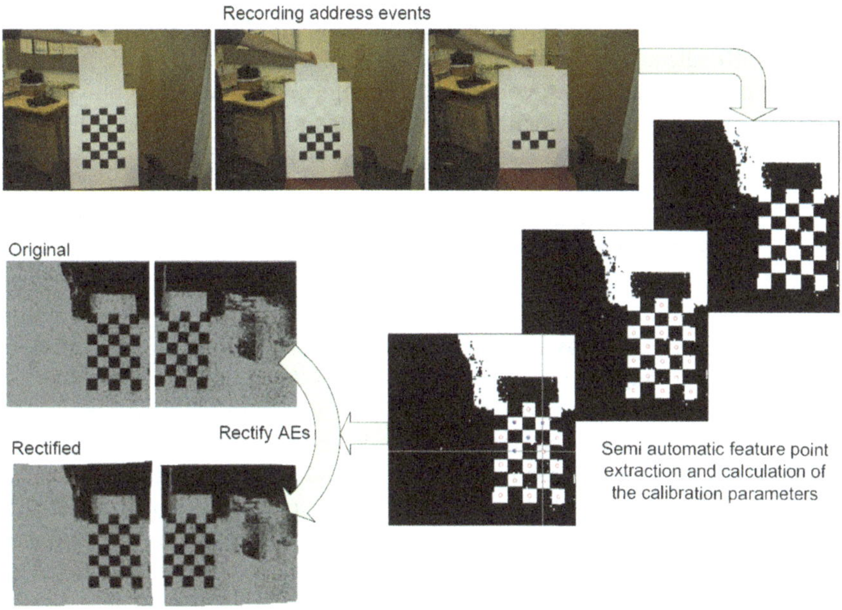


Fig. 4. Calibration and rectification of silicon retina cameras

For silicon retina imagers, it is not possible to capture the calibration pattern if there is no movement more precisely no change in the intensity. In case of silicon retina sensors an alternative approach is necessary. In Fig. 4 on the top the calibration pattern is visible in a stable position and only a white paper is moved up and down in front of the calibration pattern. During this time all address events are collected and written into one output file. The collected address event data are converted into a binary image which is used for the extraction of feature points. Instead of the crossings from the calibration pattern the centres of the squares for extraction of corresponding features are used. The right side in Fig. 4 shows the semi-automatic extraction of the feature points, because not all centres found are supporting the calibration process. For the calibration itself the calibration from Zhang [7] in combination with the calibration toolbox from Caltech for Matlab [8] is used. All data extracted from the binary images are loaded via the external interface into the calibration engine and the results are applied on silicon retina data for the calibration and rectification step. The left side of Fig. 4 shows an example of rectified silicon retina data from the left and right camera.

### 4.2 Evaluation of State of the Art Algorithm with Silicon Retina Cameras

In case of silicon retina cameras it is a challenging task to handle the asynchronously incoming address events (AEs) for the stereo matching process. Hess [9] worked with silicon retina data and used a global disparity filter in his work to find a main disparity and combined this information with the outcome of a general disparity filter which evaluated the confidence and possibility of a match.

In our work [13] we evaluate the option to use standard stereo vision algorithms for AE data from silicon retina imagers. There are the area-based approaches, which are described and compared in the work of Scharnstein and Szeliski [10], where we use a "Sum of Absolute Differences" (SAD) algorithm with different window sizes as representative of this category. The second class are feature-based approaches, where in the work from Shi and Tomasi [11] a description of features in general can be found and a more detailed description of an implementation for feature-matching is presented in the work of Tang et. al [12]. For the feature-based approach with silicon retina data a "Centre-of-Gravity" matching is applied. In our work we have compared both state-of-the-art approaches and measured the accuracy of the distance estimation, and also the pre-processing step of the data conversion is described in detail. The accuracy of the distance estimation had at least an average relative disparity error of 7% respectively 17%. These results are not satisfying and therefore some improvements took place. In Fig. 5 new results for both algorithm are shown, but with the improvement of calibrated and rectified image data as outlined in Section 4.1. The left side of Fig.5 shows the results of the area-based approach, using 500 image pairs for each result and evaluating three different distances with four different window sizes.

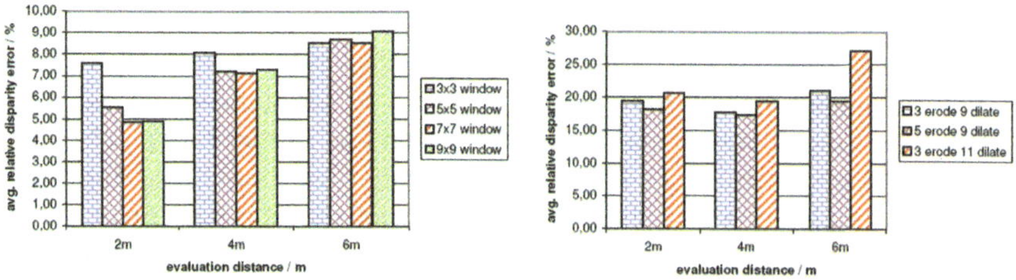


Fig. 5. Results of an area-based (SAD) (left) and of a feature-based (Centre-of-Gravity) (right) algorithm with calibrated and rectified camera data

Due to the used calibrated and rectified data at least an error of at least 5% can be measured. The right side shows the results of the feature-based algorithm results also in three different distances and with a variety of morphological operations. For each result are 500 image pairs used and the perceivable improvements with the feature-based approach and rectified camera data are minor. The improved results are still not satisfying and therefore a new approach for the stereo matching of silicon retina has to be implemented. A first outlook of this algorithm is given in section 4.3.

### 4.3 Stereo Algorithm Approach Specialized for Silicon Retina Data

Conventional block-based and feature-based stereo algorithms have been shown to reduce the advantage of the asynchronous data interface, throttle the performance and do not reach the necessary accuracy for the estimation of the distances. Therefore a novel algorithm approach based on locality and timely correlation of the asynchronous data event streams of both imagers was applied. An overview of the algorithm with all blocks is depicted in Fig. 6.

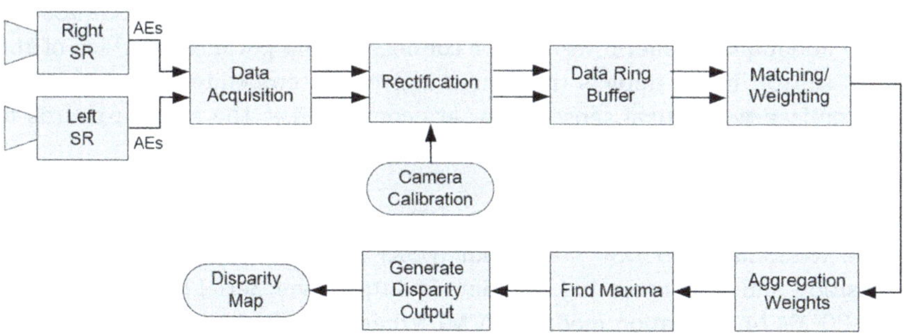


Fig. 6. Blockdiagram of the novel stereo algorithm approach for silicon retina cameras

First, the sensor is calibrated to affect the distortion coefficients and the camera parameters as mentioned in section 4.1. Then, the events received by the embedded system (described in section 5) are undistorted and rectified to obtain matchable events lying on parallel and horizontal epipolar lines. The rectified events from the left and right imager are stored in a ring buffer structure to exploit the asynchronous behaviour of the camera. The algorithm reads from the ring buffer and starts processing the stereo data. At first, all events are matched and each matched event gets a weight corresponding to their time difference and is stored into a buffer afterwards. The time correlation and weighting of the matches found is followed by an aggregation, used for the improvement of confidence of the matched events. After the aggregation the search of the maxima starts and the corresponding disparity of the highest weight is written into a result buffer, the so-called disparity map. This approach does not use collecting and image building mechanisms as the approaches described in [13], because the data is directly used from the buffer and matched against the opposite side. Early prototypes showed that by using this algorithm the advantage of the SR technology could be fully exploited at very high processing rates.

## 5 Embedded System

The embedded system used for the described stereo vision approach to perform data acquisition, and pre-processing is based on a TMS320C6455 single-core fixed-point "Digital Signal Processor" (DSP) from Texas Instruments. Due to the high performance requirements of the stereo vision algorithms, the TMS320C6474 is dedicated to data processing. Both cores are based on a C64x+ DSP core from Texas Instruments. The most significant difference between both DSPs is that the TMS320C6474 consists of three C64x+ DSP cores rather than one. This has a noticeable effect on the Peak MMACS of each DSP. Another difference in terms of data acquisition is that the TMS320C6474 has no adequate parallel interface for connecting the parallel interface of the optical sensor. Fig. 7 depicts the schematics of the embedded system, which compromises two optical sensors that are connected to the adapter-boards to handle the data traffic.

The adapter-board is connected to the "External Memory Interface" (EMIF) of the TMS320C6455 DSP starter kit (DSK) for data acquisition. After the acquisition and pre-processing, the data is output over Serial RapidIO™ to the TMS320C6474 evaluation module (EVM), where it is further processed by the stereo algorithm.

First, the sensor is calibrated to affect the distortion coefficients and the camera parameters as mentioned in section 4.1. Then, the events received by the embedded system (described in section 5) are undistorted and rectified to obtain matchable events lying on parallel and horizontal epipolar lines. The rectified events from the left and right imager are stored in a ring buffer structure to exploit the asynchronous behaviour of the camera. The algorithm reads from the ring buffer and starts processing the stereo data. At first, all events are matched and each matched event gets a weight corresponding to their time difference and is stored into a buffer afterwards. The time correlation and weighting of the matches found is followed by an aggregation, used for the improvement of confidence of the matched events. After the aggregation the search of the maxima starts and the corresponding disparity of the highest weight is written into a result buffer, the so-called disparity map. This approach does not use collecting and image building mechanisms as the approaches described in [13], because the data is directly used from the buffer and matched against the opposite side. Early prototypes showed that by using this algorithm the advantage of the SR technology could be fully exploited at very high processing rates.

## 5 Embedded System

The embedded system used for the described stereo vision approach to perform data acquisition, and pre-processing is based on a TMS320C6455 single-core fixed-point "Digital Signal Processor" (DSP) from Texas Instruments. Due to the high performance requirements of the stereo vision algorithms, the TMS320C6474 is dedicated to data processing. Both cores are based on a C64x+ DSP core from Texas Instruments. The most significant difference between both DSPs is that the TMS320C6474 consists of three C64x+ DSP cores rather than one. This has a noticeable effect on the Peak MMACS of each DSP. Another difference in terms of data acquisition is that the TMS320C6474 has no adequate parallel interface for connecting the parallel interface of the optical sensor. Fig. 7 depicts the schematics of the embedded system, which compromises two optical sensors that are connected to the adapter-boards to handle the data traffic.

The adapter-board is connected to the "External Memory Interface" (EMIF) of the TMS320C6455 DSP starter kit (DSK) for data acquisition. After the acquisition and pre-processing, the data is output over Serial RapidIO™ to the TMS320C6474 evaluation module (EVM), where it is further processed by the stereo algorithm.

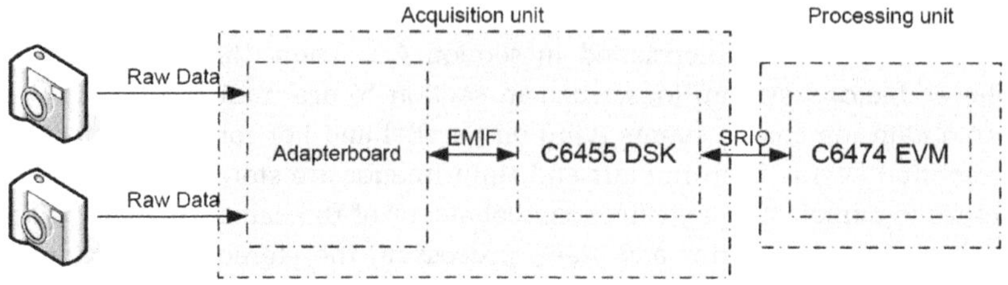


Fig. 7. Overview about the embedded system architecture

## 6 Conclusion

The paper shows the content of the project ADOSE and especially the task of the AIT in this project. For the chosen application of a side impact detection a stereo vision system with silicon retina cameras has been designed. Based on non-satisfying results from previous implementations a calibration and rectification step took place, but the accuracy of the results with these steps was still not satisfying and therefore new algorithm approaches have been searched and the first new approach has been outlined in this paper. Additionally the embedded hardware platform is shown where the final algorithm is executed.

## 7 Acknowledgment

The research leading to these results has received funding from the European Community's Seventh Framework Program (FP7/2007-2013) under grant agreement n° ICT-216049 (ADOSE).

## References

[1] Sivilotti, M., Wiring consideration in analog vlsi systems with application to field-programmable networks, PhD Thesis, California Institute of Technology, 1991.

[2] Fukushima, K., Yamaguchi, Y., Yasuda, M., Nagata, S., An Electronic Model of Retina, Proceedings of the IEEE (Volume 58 / Issue 12), pp. 1950-1951, 1970.

[3] Mead, C., Mahowald, M., A silicon model of early visual processing, Neural Networks Journal, Vol 1(1), pp. 91-97, 1988.

[4] Lichtsteiner, P., Posch, C., Delbrück, T., A 128x128 120dB 30mW Asynchronous Vision Sensor that Responds to Relative Intensity Change, IEEE International Solid-State Circuits Conference (ISSCC'06), San Francisco, USA, 2006.

[5] Ebers, O., Überblick über aktuelle Verfahren zur Tiefenschätzung aus 2D Video Sequenzen, Studienarbeit an der TU Berlin, Institut für Telekomminukationssyste me, FG Nahrichtenübertragung, 2004.

[6] Schreer, O., Stereoanalyse und Bildsynthese, Springer-Verlag, 2005.

[7] Zhang, Z., A Flexible New Technique for Camera Calibration, Technical Report, MSR-TR-98-71, Microsoft Research, 2002.

[8] Bouguet, J.Y., Camera Calibration Toolbox for Matlab, Computer Vision Research Group, Department of Electrical Engineering, California Insitute of Technology, http://www.vision.caltech.edu/bouguetj/calib_doc/index.html, Last Access: 07.12.2008.

[9] Hess, P., Low-Level Stereo Matching using Event-based Silicon Retinas, Semester Thesis, Institute of Neuroinformatics, ETH Zurich, 2006.

[10] Scharstein, D., Szeliski, R., A Taxonomy and Evaluation of Dense Two-Frame Stereo Correspondence Algorithms, International Journal of Computer Vision, Vol. 47(1-3), pp. 7-42, 2002.

[11] Shi, J., Tomasi, C., Good Features to Track, Proceedings of the IEEE Computer Vision and Pattern Recognition Conference (CVPR'94), Seattle, USA, 1994.

[13] Kogler, J., Sulzbachner, C., Kubinger, W., Bio-inspired stereo vision system with silicon retina imagers, Proceedings of the 7th International Conference on Computer Vision Systems (ICVS'09), Liege, Belgien, 2009.



**Jürgen Kogler, Christoph Sulzbachner, Erwin Schoitsch,**
**Wilfried Kubinger, Martin Litzenberger**
AIT Austrian Institute of Technology GmbH
Donau-City-Str. 1
1220 Vienna
Austria
juergen.kogler@ait.ac.at
christoph.sulzbachner@ait.ac.at
erwin.schoitsch@ait.ac.at
wilfried.kubinger@ait.ac.at
martin.litzenberger@ait.ac.at


**Keywords:** safety, embedded, stereo vision, silicon retina, bio-sensors

# Improving Pedestrian Safety in Urban Scenarios Through Autonomous Collision Avoidance

M. Roehder, S. Humphrey, B. Giesler, Audi Electronics Venture GmbH
K. Berns, Universität Kaiserslautern

## Abstract

More than 8000 pedestrian deaths per year in the European Union alone have prompted research into active pedestrian protection systems for motor vehicles. This paper introduces requirements of a collision avoidance system and a practical realization in the form of an experimental vehicle prototype capable of emergency braking and steering maneuvers. The main foci of the paper are the description of the environmental model, its implementation using approaches from the field of computer graphics and suitable testing methods allowing for the assessment of overall system quality. Finally we present the results obtained in both artificial test scenarios as well as actual real world environments.

## 1 Introduction

In the European Union more than 8000 pedestrians are killed in accidents involving motor vehicles every year. As a consequence, guidelines and legislation, such as the EU mandated Brake Assist System, were introduced to reduce the number of pedestrian fatalities and severity of injuries [1]. Due to technical and legal constraints, current active and passive systems however focus primarily on collision mitigation. While this is reasonable for the short term, it lacks potential as a long term solution. Therefore future active pedestrian protection systems focus on avoiding accidents. To further research in this area, the German Federal Ministry of Economics and Technology created the subproject "active safety" of the publicly funded research project "aktiv – Adaptive and Cooperative Technologies for Intelligent Traffic". Analyses of accident data from the EU show that that in over 90 % of accidents, pedestrians are struck from the side [2]. Based on data from the German In-Depth Accident Study (GIDAS) project, accidents were divided into seven major subclasses. Pedestrian crossing accidents with and without occlusion account for 84,3 % of all relevant incidents. The pedestrian protection system introduced in this paper was designed to avoid pedestrian injuries and fatalities in this dominant class of car / pedestrian crashes.

## 2 Active Pedestrian Protection System

This section deals with the design and a possible implementation of an active pedestrian protection system focusing on the accident classes described above. The three major phases necessary for a successful active intervention were defined as Perception, Comprehension and Action.

### 2.1 Perception

Perception, the act of gaining awareness of the elements of the environment through physical sensation, is the most fundamental precondition on which all succeeding phases are predicated. The sensor used for the vehicle prototype is a time of flight distance sensor comprised of two components: an optoelectronic camera providing 3D information and light sources illuminating the scene with modulated infrared light. The camera is mounted behind the windshield and the infrared light is mounted in the vehicle grill. Due to the reflection of the infrared signal, the sensor pixels receive a damped and phase shifted version of the original wave. Only phase shifts of less than $2\pi$ can be interpreted correctly. As a consequence, our current sensor has a non-ambiguous coverage range of 150 m. The outputs of the camera are distance and amplitude images with a frame rate of up to 100 Hz. Additional details concerning the underlying functional principles and operating modes of the sensor system can be found in [3, 4, 5].

Experiments confirming a sensor range of about 30 m were conducted, due to pedestrian detection range being dependent on the clothing reflectivity. There are two factors determining the coverage limit in the lateral dimension: due to the illumination characteristic, only a certain region reflects the modulated signal and additionally the camera's aperture angle is limited to 53°. Focusing on the addressed scenario however, the sensor is able to observe a significant part of the road and sidewalk ahead the car. While this is not sufficient to avoid every possible collision, the majority is handled. Therefore collision avoidance is defined from the action phase's point of view: all collisions which are detectable early enough by the sensor should be prevented by the pedestrian protection system.

### 2.2 Comprehension

Comprehension is the phase in which the pedestrian protection system must interpret the information provided by its sensors and arrive at a decision concerning commensurate action. The conception of a suitable environmental

To ease calculation, the continuous domain $D$ is divided into $m$ timeslots. Object positions during these are assumed to conform to the linear interpolation of the positions immediately prior to and following the regarded timeslot.

The complexity of detecting vehicle / pedestrian collisions depends on the complexity of the involved object geometry and the total number of objects present in a given scene. Since all objects are represented by rectangles at any given point in time and all movement is predicted for identical timeframes, the number and complexity of object segment geometries is identical. Therefore computational complexity scales linearly with the total number of objects in the scene.

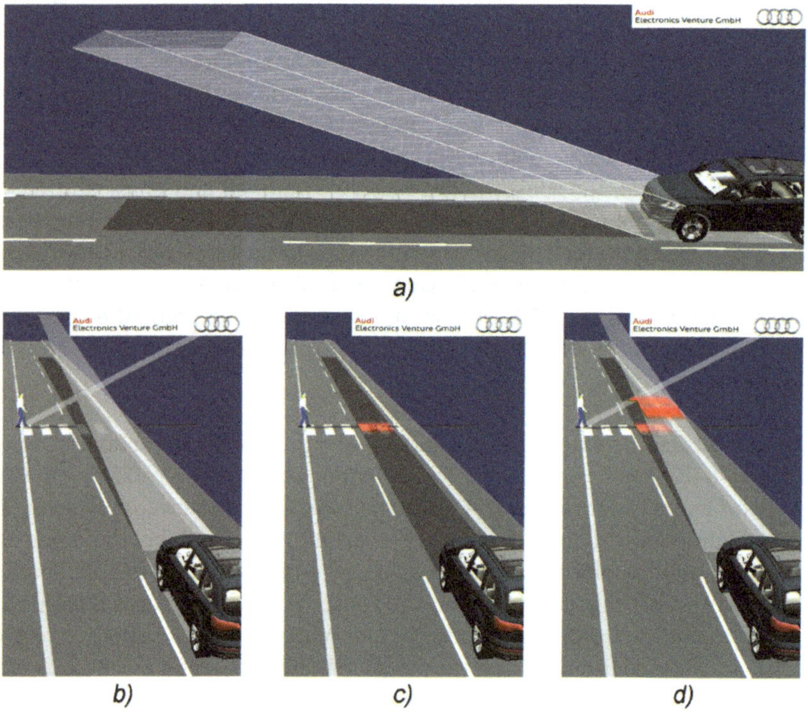


Fig. 1. Stepwise illustration of the collision prediction algorithm

Due to the need for real-time performance on contemporary hardware, the collision detection problem is solved in two steps: the first being a projection of all object tube segments onto the $x/y$ plane (figure 1b/c), followed by a more costly three dimensional intersection test if necessary (figure 1d). While both steps use the same basic principle to determine if the objects overlap, the first is not only executed in the $x/y$ plane, but also utilizes an incomplete set of axes to verify a lack of collision. As a result the preliminary test features higher performance, while yielding a superset of the colliding segment pairs. The second step serves to eliminate non colliding pairs.

Since our objects are rectangular (and therefore convex), two-dimensional cross sections of the trajectory yield convex shapes. These can be checked for overlap using the separating axis theorem, an established approach from the field of computer graphics [4]. The theorem states that two convex objects intersect if and only if no separating axis can be found where the projections of both objects onto this axis are disjoint. As mentioned above, only a subset of all possible axes is checked for disjoint segment projections, to reduce the associated computational costs. The corresponding approach for the three dimensional problem, the separating plane theorem, is then run in its entirety for all collision candidates as the second step: Let $P_i$ be the set of normal vectors for the limiting planes of object $i$ and $E_i$ be object $i$'s set of edge vectors. In accordance with the separating plane theorem:

Two objects $M_1$ and $M_2$ do not overlap if and only if at least one vector $n \in P_1 \cap P_2 \cap (E_1 \times E_2)$ exists for which the projections of both objects' vertices onto $n$ are disjoint.

Figure 2 shows the application of the separating axis theorem to detect an overlap. If there is no overlap for any of the $m$ timeslots, no collision is predicted. In this first iteration of the prototype, the underlying prediction model is relatively simple: the vehicle is assumed to feature a constant yaw rate, while pedestrians have constant velocities. The limitations caused by this simplification of reality are discussed in greater detail below.

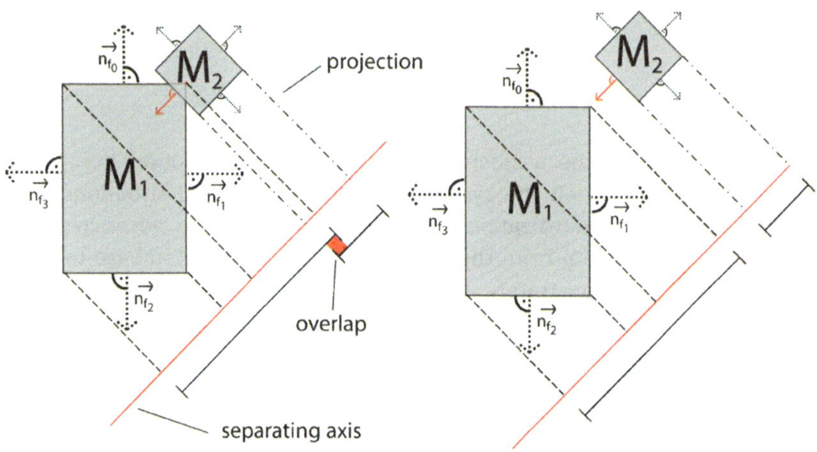


Fig. 2. Two examples of the separating axis theorem

### 2.3 Action

The action phase finally encompasses all measures employed to avoid an impending collision. The prototype presented in this paper is capable of autonomous emergency braking maneuvers and steering suggestions. Decisions are based on information concerning the maximum available braking deceleration, maximum practical lateral velocity and the relative velocity of car and obstructing object. Initially the time to collision (TTC), calculated in the perception phase, is compared to the time required to avoid a collision (TAC) by reducing the relative speed to zero. If the TAC is sufficiently close to or even less than the TTC, an emergency braking maneuver is initiated immediately. Otherwise should the time of intersection between car and object trajectory be within a predefined steering suggestion window, a torque of approx. 3 Nm is applied to the steering wheel. This alerts the driver and prompts him to perform an evasive maneuver should he deem one appropriate.

## 3 Testing and System Performance Evaluation

Since suitable methods of evaluating a system's performance are critical to guiding the development process, this section focuses on describing the used testing procedures. The two primary design goals of the performance evaluation process were measuring both basic system functionality and everyday practicality. Due to the nature of active pedestrian collision avoidance systems, appraisal of the former in real world scenarios proved problematic and necessitated the introduction of synthetic test scenarios. This, however, did not prove an issue in the evaluation of everyday practicality.

### 3.1 Assessment of Basic System Functionality

As mentioned above, the assessment of basic system functionality, i.e. the quantification of how well the system intervenes in critical situations, made the definition and standardization of artificial test scenarios necessary. Their design was based on data from the aforementioned GIDAS database to reflect relevant situations as accurately as possible. To this end, parameters such as pedestrian and vehicle speed were chosen to approximate typical urban settings.

The artificial tests discussed in this paper will be limited to those conducted at the pedestrian-bridge on Audi testing grounds. This testing facility allows for pedestrian dummies to stand or cross in front of an approaching vehicle.

Immediately prior to impact, the dummy is pulled upward with great acceleration and thus removed from the vehicle's path. This allows for the realistic, reproducible simulation of accident scenarios, while protecting the dummy and vehicle prototypes from damage caused by impact. Dummies of differing sizes and shapes with varying clothing can be used to simulate worst case scenarios. The dummy used during testing was developed according to the RAMSIS model and is a realistic, rigid model of a standing human being wearing street clothing. While vehicle velocity may not exceed 22 m/s and dummy velocity is limited to 4 m/s, these constraints pose no problem in simulating urban scenarios.

Fig. 3/4. Audi pedestrian bridge testing facility

The test rig described above allowed for the testing for type II errors, i.e. the evaluation of system intervention performance under the premise of guaranteed impact. The diagrams below show the relative and absolute reduction of kinetic energy at collision caused by the protection system's interventions. A total of 43 test runs were conducted using a Volkswagen Passat Variant weighing 1789 kg at vehicle speeds of approximately 30 km/h (~62 kJ) and 50 km/h (~173 kJ). In the tested scenarios, the system avoided collisions completely in 29 cases, while averaging a 96% reduction of kinetic energy. While the incomplete reduction of kinetic energy, due to delayed object recognition and consequential delayed intervention, is subject to further research, the results are quite encouraging.

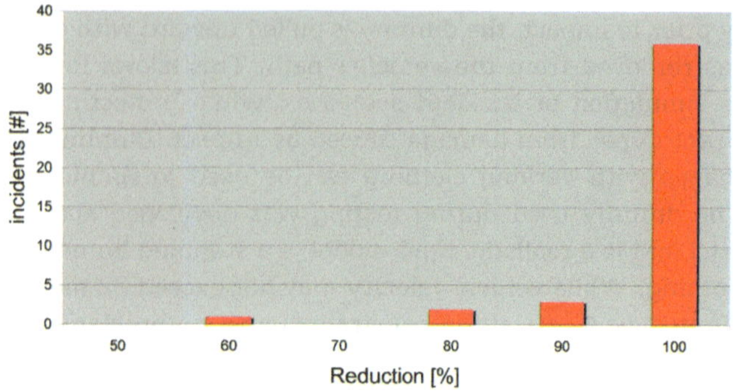


Fig. 5 Relative reduction of kinetic energy at collision

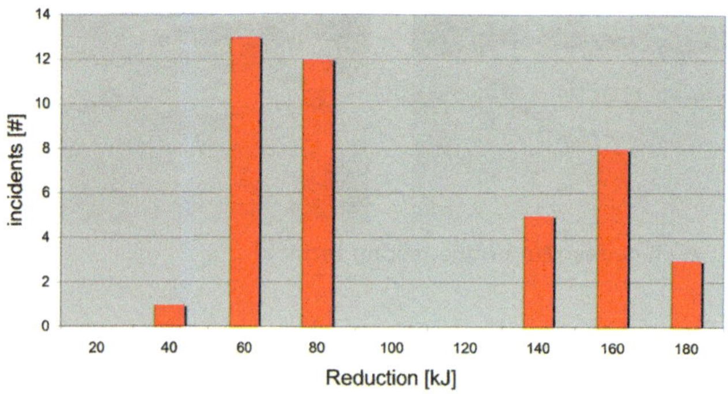


Fig. 6. Absolute reduction of kinetic energy at collision

## 3.2 Assessment of Real World Practicality

In the context of this paper, the extent to which the pedestrian protection system interferes with normal vehicle operation is defined as real world practicality. An ideal system will only intervene by initiating an emergency braking procedure, if an accident would otherwise be highly probable. To asses the real world practicality of the developed prototype, a series of tests in urban settings were run, with all active intervention components deactivated. The chosen urban settings were highly complex, featuring tight, winding roads as well as high pedestrian and stationary obstacle concentrations. Equipped with the a-posteriori knowledge, that no critical incident occurred in the absence of interventions during the test runs, we stipulate an ideal intervention rate of zero. Unfortunately reality deviates from this ideal.

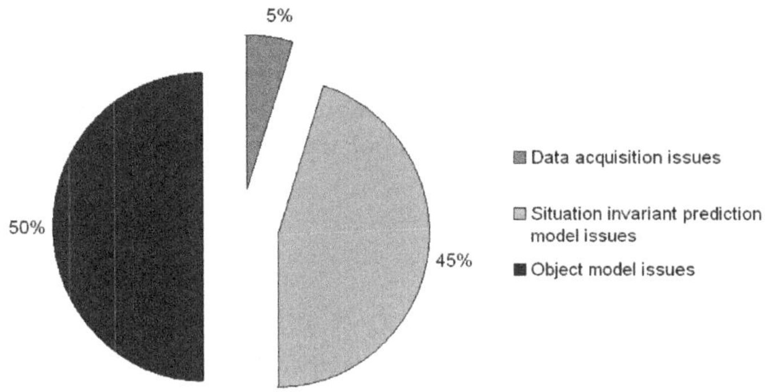


Fig. 7. Share of the three major error categories

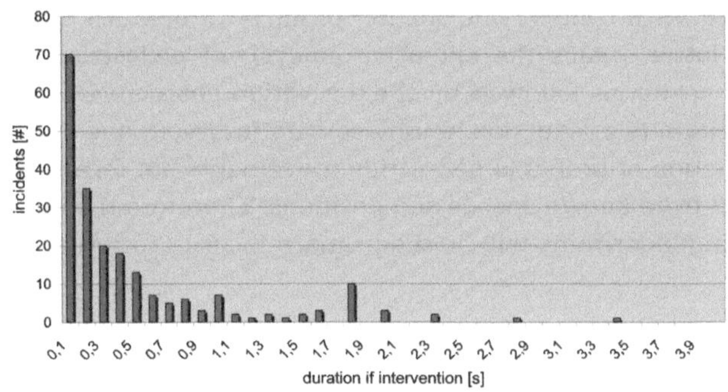


Fig. 8. Distribution of interventions by duration

Type I errors encountered in the context of this analysis can be clustered into three major categories:

**Data acquisition issues:** These relatively infrequent problems occur when the sensor looses an already detected object, or fails to detect an object altogether.

**Object model issues:** These errors are due to the object model being optimized for less complex highway scenarios. Therefore it does not feature object rotation or ego compensation and exhibits problems with large objects leaving the field of view.

**Situation invariant prediction model issues:** As mentioned in section 2.2, the environmental model uses a situation invariant prediction model. This simple approach, however, does not tap the potential held by common behavior patterns of pedestrians, other vehicles or a-priori knowledge of road design.

As illustrated in figure 8, the distribution of unnecessary interventions during the real world tests is strongly skewed toward very short interventions with over 45% of false positives lasting less than 200 ms. In total there were however approximately 16.3 type I errors per traveled urban kilometer.

## 4 Conclusion and Outlook

While the prototype introduced in this paper features robust basic system functionality, its high false positive error rate renders real world use impractical. Primarily this can be attributed to the two major classes of type I errors presented in section 3.2. Adaptation of object segmentation and tracking algorithms to urban requirements and the conception of a situational model capable of dealing with context information promise to produce drastic improvements. The latter entails the extensive analysis of pedestrian movement in well-defined scenarios followed by the derivation characteristic behavior patterns. In addition to a reduction in unnecessary interventions, this would allow for the prediction of probable pedestrian movements not foreseeable by mere extrapolation from known speeds and positions. Therefore the improvement of the underlying algorithms will be the primary focus of future research.

## References


[1] Commision of the European Communities, COMMISSION REGULATION (EC) No 631/2009, Official Journal of the European Union, 25.7.2009, L195/1, 2009.

[2] Kühn, M., Degener, S., Köppel, W., Fußgängerschutz im Straßenverkehr – eine interdisziplinäre Aufgabe, Z. f. Verkehrssicherheit, 53, 134, 2007.

[3] Heinol, H., Xu, Z., Schwarte, R., Olk, J., Klein, R., Electrooptical correlation arrangement for fast 3D cameras: Properties and facilities of this electro-optical mixer device, SPIE EOS -Sensors, Sensor Systems, and Sensor Data Processing, 245-253, 1997.

[4] Schwarte, R., Heinol, H., Xu, Z., Olk, J., Tai, W., Schnelle und einfache optische Formerfassung mit einem neuartigen Korrelations-Photodetektor-Array, GMA-Bericht 30 - Optische Formerfassung, 30, 199-209, 1997.

[5] Elias, B., Barke, A., Fardi, B., Douša, J., Wanielik, G., Obstacle Detection and Pedestrian Recognition Using A 3D PMD Camera, Proceedings of IEEE International Conference on Intelligent Vehicles, 225-230, 2006.

[6] Möller, T., Haines, E., Real-time rendering, Peters, Wellesley, 2008.

[7] Meinecke, M., Roehder, M., Nguyen, T., et al., Motion Model Estimation for Pedestrians in Street-Crossing Scenarios, Proceedings of the International Workshop on Intelligent Transportation, 7, 2010.

**Martin Roehder, Sean Humphrey, Björn Giesler**
Audi Electronics Venture GmbH
Sachsstr. 18
85050 Gaimersheim
Deutschland
Martin.Roehder@audi.de
Sean.Humphrey@audi.de
Bjoern.Giesler@audi.de

**Karsten Berns**
Universität Kaiserslautern
Gottlieb-Daimler-Str., Gebäude 48
Postfach 3049
67653 Kaiserslautern
Deutschland
berns@informatik.uni-kl.de

**Keywords:** automotive safety, pedestrian safety, situation awareness, collision avoidance, active and passive safety, vulnerable road users, testing safety systems

# Saferider On-Bike Information System

R. Montanari, A. Spadoni, University of Modena and Reggio Emilia
M. Pieve, Piaggio s.p.a.
M. Terenziani, MetaSystem s.p.a
S. Granelli, AvMap s.r.l.

## Abstract

In the last 15 years, the In-Vehicle Information System (IVIS) availability in the automotive domain widely increased especially to meet the needs of the community. What is missing in the IVIS literature are studies, indications and developments of IVIS tailored for PTW (Powered Two Wheelers). Within this context, SAFERIDER, an European project founded by DG Information and Society within 7th Framework Programme, aims to enhance comfort and safety level for motorcycle riders by introducing several functionalities, with particular attention on IVIS in the PTW domain that, as a matter of fact, have been renamed OBIS (On-Bike Information System). This paper introduces in the literature a short and effective review of the nowadays functionalities developed on OBIS side, namely, eCall, Telediagnostic Service, Navigation and Route Guidance, Weather Traffic and Black Spot Warning [1].

## 1 Functionality Description

During the last decade, IVIS development is one of the main research areas of the automotive industry, in order to increase safety and comfort of four-wheel vehicles [1, 2]. However, such technologies in motorcycles and even clean motorbikes (electric) are currently lacking behind and should undoubtedly be studied further. Such technologies should be designed and developed in a way that will not interfere with driving and/or annoy the rider.

SAFERIDER intends to learn from the problems of IVIS application in cars, developing key functionalities for PTW under a unified HMI concept [3]. The system as a whole is able to inform the rider about incoming risky weather conditions, unexpected traffic jams, dangerous locations in which statistically there is a high accident probability or the need of a particular service.

### 1.1 eCall

Over 1.2 million accidents require medical help in Europe every year, and many more need other types of assistance. After an accident, the occupants in the vehicle may be in shock, not know their location, be unable to communicate or to use a mobile phone. In all these cases, wherever they are in Europe, eCall could make the difference: it can drastically cut the emergency response times, save lives and reduce the severity of injuries. When fully implemented in Europe, the socio-economic benefits of eCall will be huge [4].

The eCall system requires the capacity of PTW to detect and remotely provide information, as the location of a crash or fall. As stated in the prCEN/TS 15722 document by the European Committee for Standardisation (2007) [5] the eCall, considered in the context of "Road Traffic and Transport Telematics" (otherwise known as "Intelligent Transport Systems" or "ITS") , is a user instigated or automatic system to provide notification to public safety answering points, by means of wireless communications, that a vehicle has crashed, and to provide coordinates and a defined minimum set of data. The document provides also the specifications for the data that have to be transferred by such an in-vehicle eCall system in the event of a crash or emergency. These data are crucial information in order to assist the provision of the most appropriate services to the crash site and to speed up the response. Data makes it possible for the PSAP (Public Safety Answering Point) operator to respond to the eCall even without the voice connection. In order to develop the eCall functionality, crashing sensors and a remote communication unit are needed, as well as a GPS for the localisation. An On-Board System manages the generation of different types of alert, manual and automatic (e.g. crash or fall), in line with eCall communication protocols and standards. The e-Call system must recognise type, number and location of the sensors on-board, which will trigger the chain of events (sensing the crash, identifying it as a crash, and communicating to the system on-board) that will alert the e-Call system, in order to effectively detect a crash or, in case of PTW, a fall of the vehicle and provide information needed to timely dispatch the emergency services. The information about the location of the crashed vehicle will be sent to the e-Call system by the GPS-based navigation system. In case an on-board communication system is present (e.g. in-helmet phone speaker), the e-Call system may be used for establishing communication between the rider and incoming riders: it is also important to consider whether and how the rider (if conscious) should be informed about the e-Call being activated.

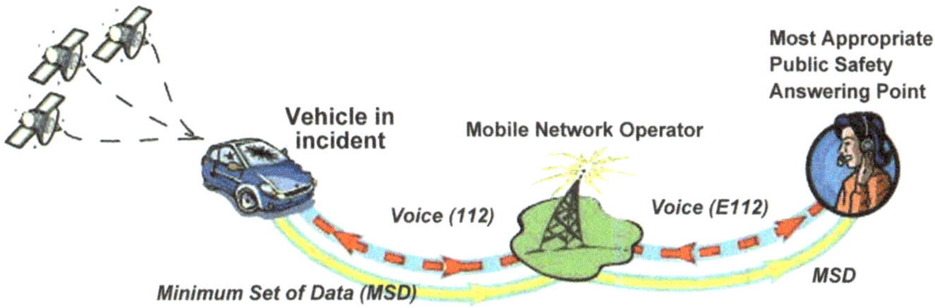


Fig. 1. eSafety eCall functional scheme [6]

Within SAFERIDER, the eCall functionality has been improved, taking into consideration various sources of warnings. Actually, the SAFERIDER eCall system could be activated by 3 sources placed in the helmet, in the PTW and via manual activation. In fact, a 3-axis accelerometers placed in the helmet is able to detect the accelerations of the head during the crash (figure 2). On the other hand, the PTW could detect the crash thanks to a 3-axis accelerometers placed in the Telematic and Security Services Unit (TSU). Finally a manual activation is possible via the Navigation Unit touch screen which provides the relevant information to the TSU via CAN bus (figure 2) starting with the procedure. Basically, the action workflows of the eCall functionality is composed by the following steps:

- Detection of the crash - the detection of the crash is performed in 2 different ways:
  - **the first one** via 3-axis accelerometers placed in the helmet which detect an accelerations over a certain threshold. This warning is provided to HMI Manager which writes it on the CAN bus to TSU.
  - **the second way** to detect the crashes is via a 3-axis accelerometers placed in the vehicle, and more precisely in the TSU.
- After the detection, TSU generates the message and sends it to Navigation Unit for rider potential abortion.
- If the rider stops the procedure via the Navigation Unit display, no emergency call is forwarded to the Server Infrastructure (Service Center). If the rider confirms, or no abortion is made, the procedure continues.
- The TSU sends via GPRS data regarding the typology of the vehicle.
- The Service Centre receives the messages and opens the emergency call to TSU.
- TSU receives the emergency call and forwards it to Navigation Unit via BT; if the BT connection is active, Navigation Unit forwards it to in-helmet speakers/microphone via BT. If it is not possible, the Service Center calls directly the rider mobile phone.

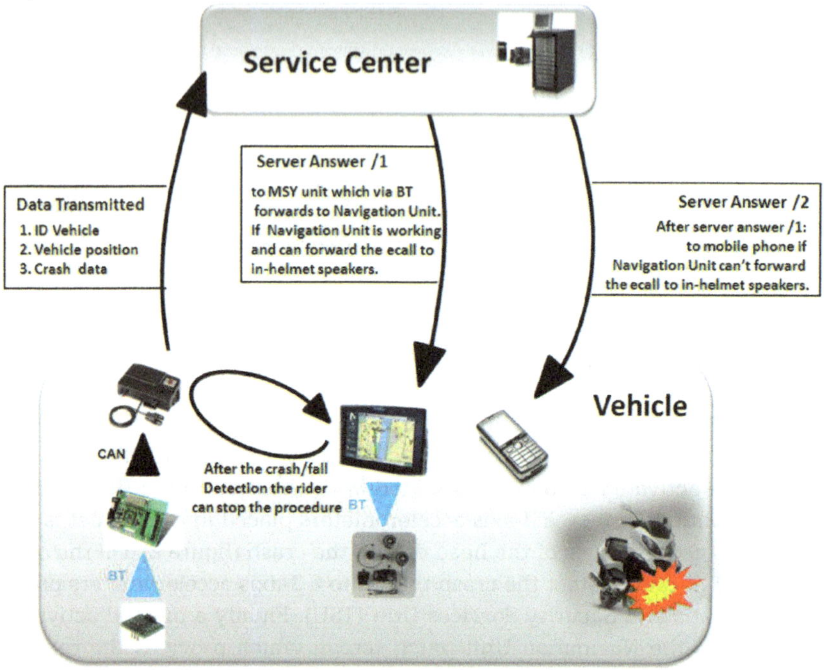


Fig. 2. eCall sketch that shows the aforementioned workflow

In any case, the rider has always the situation under control. The rider has the possibility to stop the automatic eCall routine if it is not necessary; for example if the crash is serious but the rider is not injured the eCall could be manually stopped by pushing a button on the Navigation Unit. The time that the system leaves to rider is around 30 seconds. After this time, the system starts the eCall routine. On other hands, if the crash is not serious enough to activate the automatic eCall routine but the rider is injured, the eCall routine can be manually activated by pushing a button.

## 1.2 Telediagnostic Service

The Telediagnostics service provides monitoring of, use and functioning conditions of the vehicle and offers early warnings about the next vehicle service or imminent failure of any vehicle subsystems. The system will be able to perform remote telediagnostic of the most critical operational parameters of the PTW. These parameters are related to: security concerns, performance, maintenance and upgrading [6]. The implementation of selected ITS technologies on clean PTWs will enhance interest on the use of clean PTWs and more importantly their comfort and safety on the road. The main actors involved within the telediagnostic logical framework are: TSU, Navigation Unit, VIF (Vehicle Interface providing relevant vehicle information), Server/Infrastructure (intended as Service Center). The server receives the message and elaborates the information stored in it, in order to perform the post processing analysis. After the post processing analysis the most relevant information will be accessible on line for the rider. The telediagnostic HMI consists of the navigator device and of the web interface access [6]. Especially the web access gives the basis to take advantage of remote services and information. The remote diagnostics service is focused on the collection and delivery of information concerning the vehicle status, available on the CAN bus as VIF messages. Some example are listed: Engine speed (rpm), that may be used to identify that too high rotation is used; Front and rear wheel speed (km/h). Besides detecting large values of speed, the difference between these two speeds may represent conditions where adhesion to the ground was possibly lost [6].

In the SAFERIDER application the same device that manages the eCall functionality is designed in order to be integrated in the vehicle information flow on CAN bus. Thanks to such an integration it collects all vehicle parameters, elaborates them and sends the results by GPRS connection to a remote server. On the server the data are arranged in order to extract critical information about the vehicle on board system status and to present to the rider indications on his or her behaviour or important statistics about vehicle mode of use. Piaggio Mp3 Hybrid has been selected as vehicle platform for the most extensive testing of telediagnostic capabilities because of its intrinsic complexity and number of parameters to be controlled. In particular, the remote diagnosis allows for the acquisition of essential information about advanced integrated system like lithium-ion battery packs (subject to potential issues of ageing and temperature management). SAFERIDER system enables the rider to be early warned through HMI of possible incoming failures by real-time monitoring of parameters evolution such as the temperature gradient under duty loads. In addition the rider behaviour is evaluated by analysing the instantaneous fuel consumption and comparing it with the ideal figures for different hybrid modalities. In figure 3, for example, the acquired instantaneous values are reported as a function of vehicle speed and displayed in green if equal or better than the claimed vehicle figures. With such an instrument the rider can easily check its trips and try to modify the riding attitude aiming at minimization of emissions and improvement of the vehicle eco-compatibility.

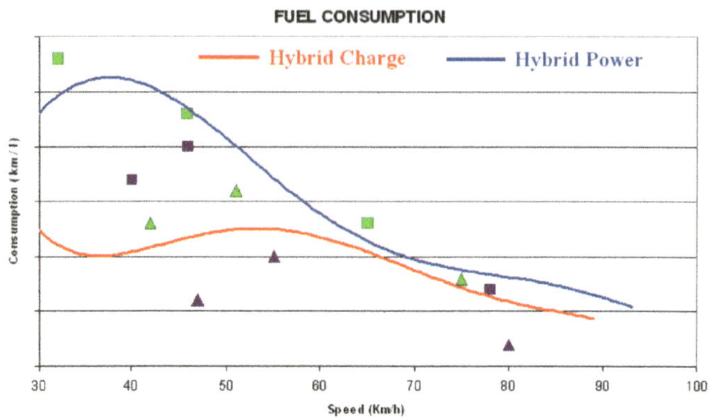


Fig. 3. Eco compatibility evaluation.

### 1.3 Navigation and Route Guidance

Navigation and Route Guidance is a function that allows a driver or a rider to set a destination, reaching it following the system turn by turn indications. As a matter of fact, in the automotive domain there is a huge availability of types and models of Navigator. In the PTW domain only a very few examples of navigator are available on the market instead. One of the aim of SAFERIDER was to design and develop a Route Guidance that fits the rider needs related to the head position and the high task demand. The rider can choose a destination and the reach it following simple vocal instructions and viewing this route on the map. In this way the route guidance allows the rider focusing on the road and looking at the dangerous situations in the street, in order to avoid accidents.

The navigation function offers several security features. First of all, the "Curve approaching bar", that is a bar placed in the map layout which enlightens step by step the rider while he / she is going to approach a curve. In this way he / she knows how far the curve is, making himself ready to approach it in a safe way.

The second function is the "Lane assistance", a panel placed in the dashboard Whenever a bifurcation along a selected route occurs an arrow indicating the direction is visualised, in order to help the user to follow the right direction.

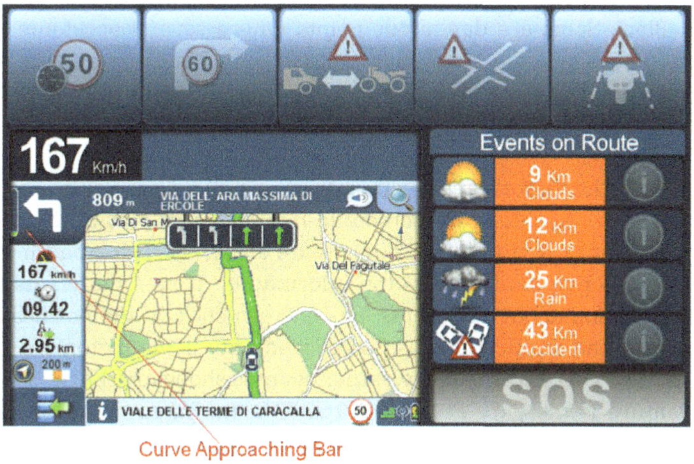


Fig. 4. The SAFERIDER Navigator and the Curve Approaching Bar

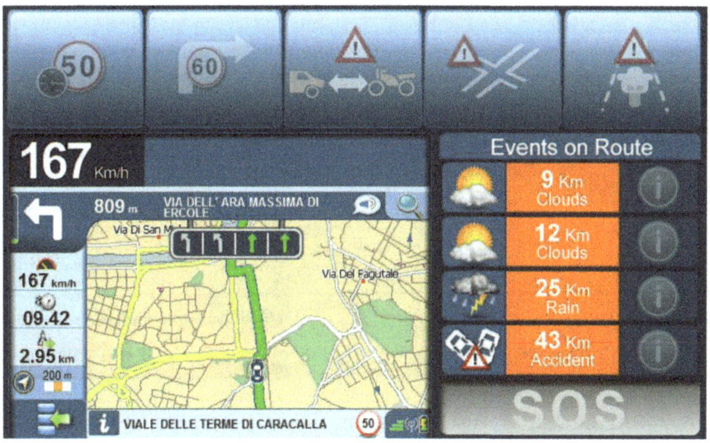


Fig. 5. The SAFERIDER lane assistance function

Other useful functionalities are in the route options choice. For example, it is possible to avoid in the route calculation some kind of streets, selecting the proper option in the "Route Preferences" menu. An example are the unpaved roads, that are really dangerous to be passed through riding a motorbike. Finally, it is possible to change the map display colors, enabling the night vision when it is needed. This way the clash is increased, allowing an optimal vision during the night. Besides the standard navigation the rider is provided with additional information, such as Traffic and Weather information and Black Spot warnings. Black Spots are dangerous points where several accidents occured in the past. These points are stored in a database in the Secure Digital device: when a Black Spot appears an icon is displayed on the map together with a few word description positioned in a bar in the screen of the map.

Traffic and weather information comes from a remote service centre and is displayed in the right side of the screen. In fact, in the display of the navigator device both the map and the OBIS advices are shown at the same time, so the user can see the map and receive the information. If he / she wants to have further information about the events, he / she can click on the notification: for safety purpose, the notifications can be clicked only if the motorbike speed is zero. An icon for every traffic and weather event is shown on the map. Another security feature is the „My Position Button", in the main menu: it allows to know the own position according to GPS data and it allows to search for several useful places, that are, in the most part emergency hospitals.

Fig. 6. SAFERIDER MyPosition button

### 1.4 Weather, Traffic and Black Spot Service

Within SAFERIDER, weather, traffic and black spot warnings allows the user to be constantly informed about the weather and traffic condition and on the next black spot. The service consists in displaying on the screen information about traffic, weather and black spots along the route. The system architecture is mainly composed by TSU and NAV module and by Infrastructure Server (Service Center). The TSU exchanges information with the Infrastructure server, typically a service provider, and forwards it along the CAN network.

The Infrastructure Server will be able to upload and refresh weather and traffic information. The transactions associated to this service are driven by an on-board Navigation Unit that, according to defined principles, repeatedly submits requests for weather information to a TSU via appropriate CAN messages. Requests include the geographical location of the vehicle (latitude and longitude – they might be different from the actual current location as they may be the result of a predicted journey). Via GSM/GPRS network the TSU submits the request to a central service infrastructure. Weather information is available at the central service infrastructure. The central service infrastructure retrieves the weather information relevant to the geographical point nearest to the location requested by the TSU. Such information is sent to the TSU via GSM/GPRS, and then from the TSU to the navigation unit via CAN bus. The general principles for traffic information service are quite similar to those of a weather information service. Transaction is driven by the Navigation Unit that repeatedly submits requests for traffic information to the TSU. The TSU retrieves, via GSM/GPRS network, traffic information stored in a file available at the central service infrastructure. However, a couple of significant differences exist with respect to the weather information scenario. The traffic information returned are not relevant to one point only, but to a whole area surrounding the location specified in the request. The shape and size of such an area is fixed (circle, 10 km radius). The quantity of traffic information returned as a response to a single request may be variable, depending on the number of active traffic events falling within the area in subject.

## 2 Architecture

Figure 7 shows the functional architecture developed within the project [7]. The hardware modules involved on OBIS side are the Navigation Unit, the HMI Manager, TSU, the Vehicle InterFace (VIF) and the Infrastructure. Basically, the wired connection follows a CAN philosophy: on wireless side, BT connection and GPRS connections are expected to connect the helmet with the Navigation Unit (and with the TSU) and the Service Infrastructure (Service Center) with the TSU. A CAN based approach is a new entry within the PTW domain. In fact, several automotive applications follow a CAN based architecture as a de-facto standard, but in the 2-wheelers domain a is an innovation [7].

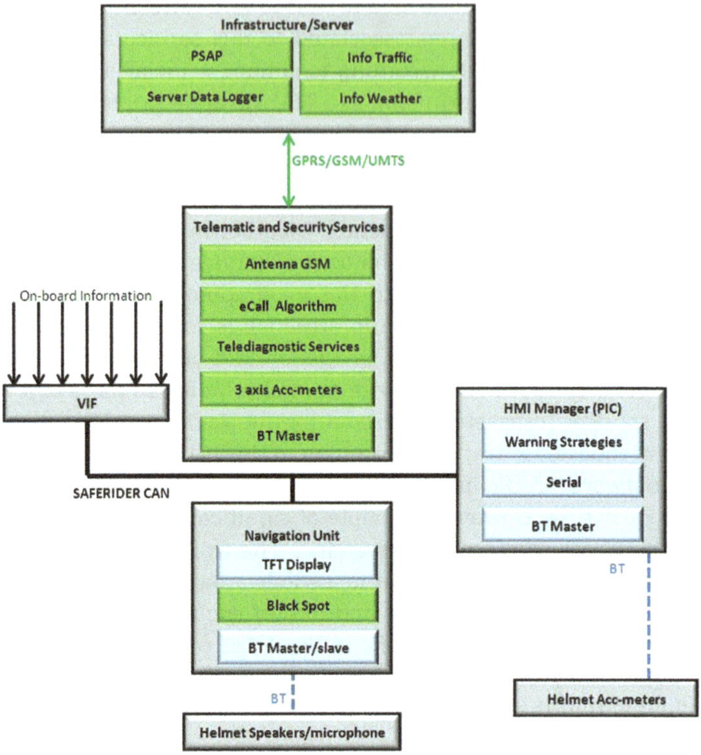


Fig. 7. Architecture scheme [7]

## 3 Conclusion

The development and implementation of relevant OBIS functionality might contribute to the significant enhancement of riders' safety and comfort. The SAFERIDER project, launched in January 2008, is approaching the third year of development: waiting for the its conclusion, the draft results seem to be very interesting for these important rider needs. The added value of this project is certainly the difficulty of the particular and undiscovered motorcycle domain. The work results lay the basis of a future implementation of on-board functionalities which don't exist in the global market yet [8].

## References

[1] Santos, J., Merat, N., Mouta, S., Brookhuis, K., de Waard, D., The interaction between driving and in-vehicle information systems: Comparison of results from laboratory, simulator and real-world studies, Transportation Research Part F, pp. 135–146, 2005.

[2] Vashitz, G., Shinar, D., Blum, Y., In-vehicle information systems to improve traffic safety in road tunnels, Transportation Research Part F: Traffic Psychology and Behaviour, Volume 11, Issue 1, pp. 61-74, 2008.

[3] SAFERIDER Project http://www.saferider-eu.org.

[4] eSafety Group: http://ec.europa.eu/information_society/activities/esafety/index_en.htm.

[5] Road transport and traffic telematics prCEN/TS 15722:2007- eSafety – ECall minimum set of data (MSD).

[6] SAFERIDER "Description of Work", copyright SAFERIDER Consortium, October 2007.

[7] Montanari, A., Spadoni, A., Bekiaris, E., Nikolaou, S., Pieve, M., Borin, A., Saferider: Can Based Architecture On 2-Wheelers Domain in Proceedings of 15th International Conference Road Safety on Four Continents, Abu Dhabi, The United Arab Emirates, 2010.

[8] Bekiaris, E.D., Spadoni, A., Nikolaou, S.I., SAFERIDER Project: new safety and comfort in Powered Two Wheelers, in Proceedings of HIS 09, 2nd International Conference on Human System Interaction, May 21-23, 2009, Catania, Italy, 2009.

**Roberto Montanari, Andrea Spadoni**
University of Modena and Reggio Emilia
Via Amendola 2
Modena
Italy
roberto.montanari@unimore.it
andrea.spadoni@unimore.it

**Marco Pieve**
Piaggio s.p.a.
Viale Rinaldo Piaggio 26
Pontedera 56025
Italy
marco.pieve@piaggio.com

**Maurizio Terenziani**
MetaSystem s.p.a.
Via Galimberti 9
Reggio Emilia
Italy
maurizio.terenziani@metasystem.it

**Sara Granelli**
AvMap s.r.l.
Viale Zaccagna 6
Carrara-Avenza (MS)
Italy
sgranelli@avmap.it

**Keywords:** SAFERIDER, motorcycle, rider safety and comfort, traffic management, navigation, information system

# The Multi-Functional Front Camera Challenge and Opportunity

A. Tewes, M. Nerling, H. Henckel, Hella Aglaia Mobile Vision GmbH


## Abstract

Since front cameras were firstly deployed in vehicles there has been a movement towards the integration of an increasing number of functions utilising the same camera. While this is obvious with regard to both costs and limited installation space, the integration of several functions into one camera represents both a challenge and an opportunity. The opportunity apart from reduction with respect to costs and space is the exploitation of synergies among different functions which is a crucial prerequisite for realising a more realistic environment model which in turn forms the basis for future functions. These advantages, however, are accompanied by several challenges concerning the functional, testing and integration concept. This article is meant to explain this stress field with respect to these three concepts and points out the neuralgic points.


## 1 Introduction

Camera based driver assistant functions have had a remarkable spread during the last couple of years. Even though reserved to the luxury class at first, camera-based functions have already made their way to the medium-class in the meantime and are likely expected to be available also in compact cars within the next couple of years. There have been several concepts of applying camera-based driver assistant functions in the car over the last years, each focussing on a very specific task like blind spot detection or rear-view assistant [1]. In consequence of the specific tasks, the mounting position, the optical setting and the hardware have been chosen accordingly. Among the first camera-based assistant systems, however, there has also been Lane-Departure-Warning (LDW) assistant functions, which are meant to alert the driver in case of leaving his lane unintentionally, that require video signals taken by a front camera. Having a camera looking at the scenery in front of the car in a way very much the same as the human driver does, immediately awakens demand for additional functions handling tasks that have been triggered by the driver himself so far. The range of such functions is diverse, among them Traffic-Sign-

Recognition (TSR), Adaptive-Lighting-Functions (ALF) ranging from a High-Beam-Assistant (HBA) which is expected to switch high beam on or off without glaring other road users up to more sophisticated applications in combination with headlights capable of supporting dynamic curve lighting. Apart from these comfort functions there is a growing tendency towards safety-functions like Vehicle-Detection (VD) and Pedestrian-Detection (PD) [2]. While the former is meant to detect other vehicles like passenger cars, trucks and two-wheelers, detecting human beings, which is more complicated due to the fact that human beings are articulated objects rather than rigid ones, is up to the latter function. Additional sensors like RADAR or LIDAR, however, usually support these safety-functions, in order to guarantee the required reliability, which is essential for systems meant to automatically intervene the driving process. This demand for additional functions without having a new camera for each of them marked the beginning of the development of the multi-functional front camera (MFFC). This trend is intensified also because of increasing computational power and memory of current electronic control units (ECU).

While there was realised a two-function camera in the Opel Insignia which offers assistance for TSR and LDW, there are already three functions available in the latest models of BMW and Mercedes including TSR, LDW and HBA. The next generation of front cameras is supposed to include up to four functions and there is no end in sight yet.

This article is organised as follows. The consequences of realising a multi-functional front camera for the integration task are discussed in chapter 2. This includes dealing with requirements given by the single functions, other sensors like navigation systems as well as additional functions realised by the original equipment manufacturer (OEM) himself, however, meant to be deployed on the same ECU. Moreover, answers to the question of how to remain flexible with respect to different bundles of applications for meeting the variety of customer wishes is also addressed in this chapter. Chapter 3 is meant to discuss the consequences from the functional point of view. This includes analysing the impact on the optical setting and hardware as well as the consequences for unifying pre- and post-processing. It is up to chapter 4 to examine the consequences for testing. Apart from discussing the general requirements associated with testing vision-based applications, the challenges of testing MFFC-functions shall be discussed. Finally, a vision for the future shall be sketched in chapter 5. Based on an increasing number of single functions and sensors, a concept for creating an environment-model is motivated which in turn represents the basis for both more robustness for current functions as well as more sophisticated functions in general.

## 2 System Integration

The architectural concept must be of such kind that the synergy between the functions reaches a maximum while remaining as flexible as possible with respect to the combinability of individual functions. It is this combinability, which eventually ensures that individual functions can be shaped both customer-specific as well as cost-efficient in order to become integrated into a multifunctional front camera. At first, subsets of fundamental functionalities, which are required by several assistant functions, are outsourced in independent components which are integrated into the Service-Function-Layer as shown in figure 1. These service functions though not visible for the driver are essential for the actual assistant functions. Among these service functions there is the optical diagnosis which is responsible for analysing the optical path and the camera calibration which measures the camera's current position and orientation with respect to the vehicle's coordinate system. The optical diagnosis is meant to identify situations of limited sight caused by weather conditions or occlusions in the optical path. It is up to each individual assistant function to decide based on this information whether a reliable assistance can still be guaranteed. Several processes that usually occur on different time scales influence the camera's orientation with respect to the vehicle. While bad road conditions can lead to vibrations which imply significant changes already from one image-frame to the other, the vehicle's loading for instance can have an influence on the long-term base. The camera calibration is meant to consider these effects and is usually used by all assistant functions. The next important step is to separate the customer-specific assistant function into one part which is responsible for image processing (see layer Image Processing in figure 1) and another part dealing with the application itself (see layer Assistant Function in figure 1). The image-processing tasks can be seen as sensors for detecting objects having certain properties (among those traffic signs, vehicles or light sources). The sensor which is meant to detect light sources for instance will be mandatory for each kind of camera-based lighting function. The application tasks, however, are customer-specific and might also be supplied by the OEM himself. A maximum of flexibility regarding combinability is achieved by establishing standardised interfaces between all relevant components. In this way a toolbox of accurately fitting pieces of software can be realised. Customer-specific projects use this toolbox in order to reduce their efforts for achieving the project's objectives significantly. The look and feel of an assistant function is essential for the driver's sense of security. Therefore, some OEMs prefer to develop the actual human machine interface (HMI) on their own, however, want to integrate their software into the front camera ECU as well. The software meant to be integrated into the front camera is therefore delivered by the camera supplier, the OEM and possibly also by other suppliers being retained by the OEM.

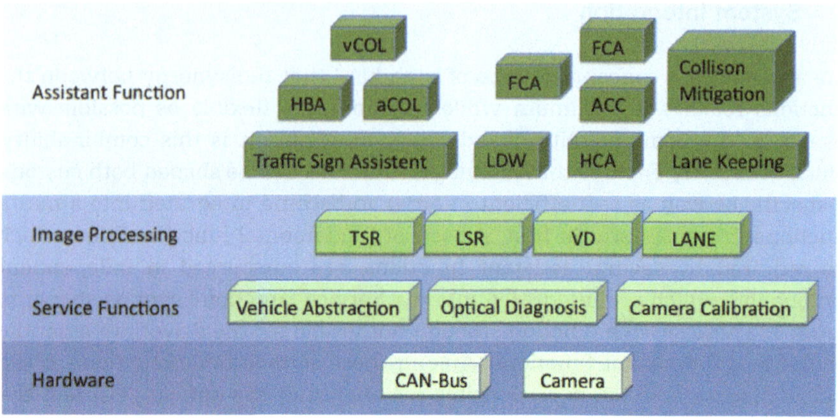


Fig. 1. Layered software architecture within current up to date multi-functional front cameras

Apart from the HMI this can also include specific components dealing with the input from additional sensors. Software development for multi-functional front cameras is therefore profoundly parallelised and distributed among different participants. While it is usually up to the camera supplier to finally integrate all pieces of software, the system's overall complexity, however, has to be taken into account by every participant.

Well-established communication channels among the participants on different layers (development, management, quality assurance, etc.) are the key to successful projects. In order to minimise the dependencies between distributed teams of developers it is essential to have clear agreements on stable interfaces from the outset. For handling subsequent amendments it is crucial to have a well-established change management which is also optimized with respect to short response times. This puts the camera supplier in a position to integrate external components systematically and to realise short and automated integration processes. On the basis of these processes it is also possible to establish short and automated testing methods which ensuring short feedback loops across all participants provided the camera supplier has the necessary tools and competence. Short cycles of integration (just-in-time integration) are preferable to the classical big-bang-integration [3] in case of interface modifications or functional extensions since error analysis in distributed software development is a technical and organisational challenge and might be a source of potential conflict. All in all, the synchronisation of the participating teams should not be limited to a few releases or samples but being interpreted as a continuous process of adjustment and integration. Simplifying the integration significantly and minimising sources of potential conflict can be achieved by restricting the development of chains of effects to one participant alone. A chain of effects in this connection consists of all pieces of software involved in realising a specific assistant use-case. Thus, the basic competence of the participant can be exploited in order to coordinate the single components optimally and to improve the overall quality for the end-user. If one considers for instance a traffic-sign application which consists of two components, one meant to recognise traffic-signs and another one which merges this information with that delivered by a navigation system, improving the availability of the overall system concerning each single component means having each component to level the weaknesses of the other. This can only be achieved by an honest and extensive analysis of their specific weaknesses. Such an analysis across participants of different companies, however, is almost impossible to become realised.

## 3 Functional Aspects

The biggest advantage of a multi-functional front camera at the same time is also its biggest drawback. Having only one camera being utilised by several assistant functions at once reduces the costs as well as the necessary installation space, however, does also limit all functions using the same optical setting. This relates in particular to technical parameters among which the angle of aperture, the kind of image sensor (colour sensor versus a black-and-white sensor), the image sensor's resolution and response curve are the most prominent ones. Since each assistant function has unique requirements with respect to the optical setting a multi-functional front camera is always a compromise. The crucial factor is to find the optimum under these constraints. Fig. 2 gives an overview of the longitudinal ranges today's assistant functions are expected to cover. This ranges from about 800 metres for detecting oncoming vehicles in order to switch off high beam down to about 10 metres for detecting traffic signs. To cover also the close-up range (range down to the vehicle's front) what becomes necessary for applications like pedestrian detection means having a greater angle of aperture which has to be corrected for by increasing the sensor's resolution for still supporting long range functions. The requirements for the image sensor's sensitivity are high due to the fact that the lighting conditions can vary from direct sunlight up to total darkness where the only light source is the vehicle's own headlamps. Finding an exposure time for dark lighting conditions is another trade-off that has to be considered when setting up the image sensor's response curve. This is because assistant functions which are expected to detect objects in greater distance are less sensitive to motion blur than those which are meant to detect objects at a shorter distance.

Realising several assistant functions on the same piece of hardware does also have an impact on the pre- and post-processing algorithms. Due to limited resources it is necessary and obvious to identify tasks, which are common among all assistant functions. This does usually comprise routines associated with pre-processing like edge-detection and blob-detection but can also be extended to typical post-processing tasks like pattern recognition and object tracking. These synergies, however, have to be identified at an early development stage in order to become finally realised.

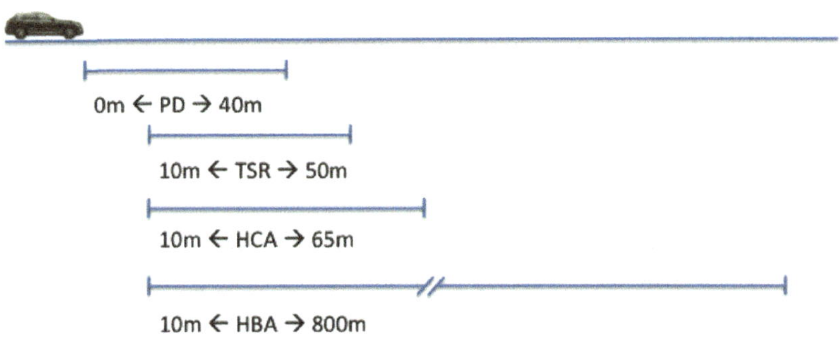


Fig. 2. Typical longitudinal ranges of today's assistant functions

## 4 Testing and Validation

It characterises camera-based applications that the significance of testing results depends strongly on the control sample being used. It is a priori impossible to anticipate every situation such a system might encounter when being used in daily routine. The test report can therefore only contain statistical measures based on criteria that were predefined. These criteria do usually cover several use cases, which are application specific as well as proper success rates. In order to avoid conflicts concerning the later interpretation of testing results it is therefore highly recommended to specify these use cases in close coordination with the OEM. In particular the importance of specific use cases as well as the distribution of covered environment and traffic situations represent important criteria. Each additional assistant function increases the number of combinations of use cases considerably while adding additional levels of complexity to existing ones due to functions' interaction. Since these use cases have to be recorded on the road, the effort of data acquisition increases to the same extent.

Since it is not possible to exactly reproduce relevant situations with respect to camera based assistant functions (think for instance of lighting conditions, other road users etc.), every situation has to be recorded together with all relevant vehicle signals. Based on these recordings as well as an adequate testing environment it is possible to accurately reproduce tests afterwards which is a fundamental demand. Fig. 3 shows the architecture for integrating the multi-functional front camera into a test environment. But this does also mean that properly equipped vehicles for data acquisition have to be available already at an early stage. This, however, might be difficult in case of a new platform where the camera is usually only one new component among several others.

In order to evaluate several vehicles using the same test data it is necessary to implement an appropriate converter tool that translates the vehicle messages from one vehicle model to the corresponding messages of another vehicle model. With each additional assistant function both the overall efforts as well as the likelihood of having relevant data not recorded so far do increase.

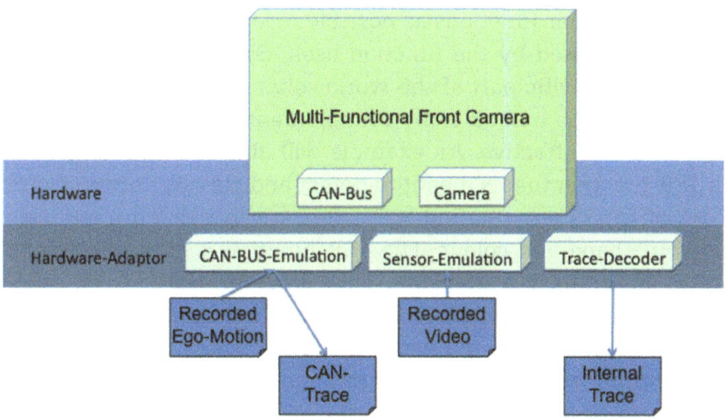


Fig. 3. Interface adaptors needed for testing

The functionality of embedded systems can only be proven by signals transmitted on the vehicle's CAN-bus. It is therefore necessary to record and archive these signals. In case of having external components too it is furthermore required to record also the internal communication traffic in order to analyse the overall workflow.

A further problem is the scarce resources as well as the software's real-time requirements. The more functions are executed in parallel the more complex are the mutual dependencies. Therefore it is rarely possible to identify them exactly so that a regression test is usually accompanied by a complete test as well. This has to be considered while planning costs and efforts.

## 5 Towards More Sophisticated Applications – A Vision

So far there have been more or less isolated functions on multi-functional front cameras. Apart from outsourcing and sharing fundamental functionalities like service functions (see chapter 2) each function is expected to operate without or only with minimal interaction with other functions. While this is a benefit with respect to a high flexibility in creating bundles of functions in order to meet different customer demands it also represents an artificial bottleneck. This bottleneck can be analysed in at least two different ways.

The first perspective is that of improving a single function's robustness. Being as independent of other functions as possible means that each and every problem has to be addressed by the function itself. Since each function does only "perceive" a very specific part of the world solving specific problems can only be realised up to some extend or by reimplementing parts of other functions. Both of which is not effective. An example will illustrate this subject. Assume there is a TSR function that is meant to detect and classify certain traffic signs. On the back of trucks one can find very often stickers that are quite similar to traffic signs in shape and colour. The TSR function is expected to not recognise them as an ordinary traffic sign. This, however, can only be achieved by analysing the context. One might think of using tracking over time in order to estimate their 3D-position in the world. But what's about a parked truck? One might further think about the traffic sign's size as well. But there are so-called repeater signs in Great Britain whose size can also be very small and which are meant to be recognised as well. On the other hand if there is a function for detecting and classifying vehicles executed in parallel, using the information of the coincidence of detecting a traffic sign in an area that is also associated with a truck does immediately resolve the ambiguity. This kind of cooperation between single functions can significantly reduce the efforts of finding special solutions otherwise. Furthermore, apart from solving specific problems by exploiting the information from other functions, this can also lead to significant savings of computing time and memory. One might think about a reliable lane model for example that can be used in order to constrain the region of interest when looking for vehicles in the VD-function. The price to pay, however, is to increase the interdependency of the single functions and therefore to loose some part of flexibility which has been discussed in chapter 2 as well as raising the complexity for testing as discussed in chapter 4.

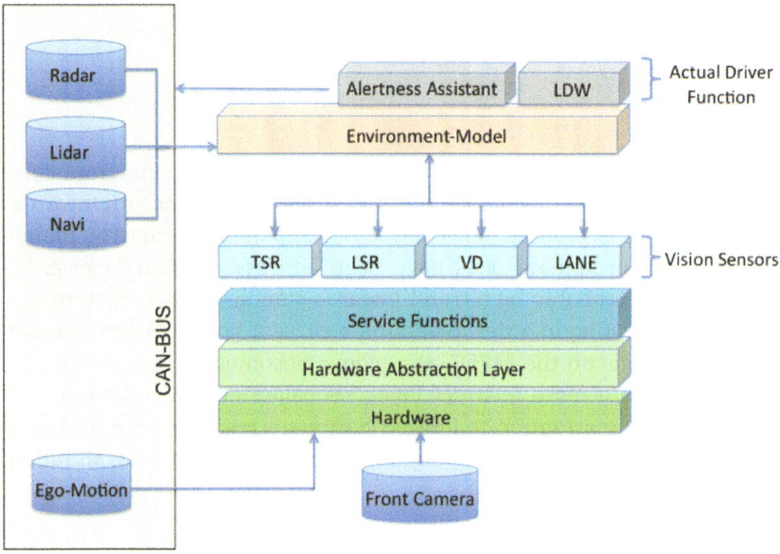


Fig. 4. The architecture of an environment model

The second perspective is even more interesting and future-oriented. The idea is to abstract from single functions that shall now be treated as being vision sensors. Each of these vision sensors depicts a certain aspect of the scenery in front of the vehicle. By combining the output of all these vision sensors perhaps enhanced by the output of ordinary sensors like RADAR, LIDAR, navigation systems etc. an environment model can be deduced as it is depicted in figure 4. Actual assistant functions like LDW for example could now be put on top of this environment model instead of being related to the LANE function exclusively. So far we have only sketched the architecture in order to realise what has already been pointed out in the last paragraph. Having an environment model, however, is more than just having an architecture for realising cooperation between single functions. On top of this model it also becomes possible to build assistant functions that are meant to evaluate the scene in front of the camera to derive recommendations for the driver's behaviour. One example might be an alertness assistant that can anticipate approaching hazardous situations like for instance children playing on the pavement of a play street (TSR recognises the play street's traffic sign, PD recognises little pedestrians moving behind parked vehicles which have been recognised by VD). Such an environment model is depicted in Fig. 4. It might also be the entity for integrating the input from other cameras (rear view camera, panoramic camera on top of the car etc.) or sensors too. It is therefore flexible with respect to both extensibility and the potential for future developments.

## 6 Conclusion

Even though driver assistant functions based on multifunctional front cameras have only recently been available, there is already a momentum towards further functions. This process, however, is accompanied by several challenges where the most prominent ones have been addressed in this article. In the future there will probably be a trend towards standardisation with respect to integration and testing in order to eliminate current hot spots. The greater the variety of functions on the MFFC, the bigger becomes the advantage of integrating them into a more abstract layer we called environment-model. This would not only pay attention to limited hardware resources but also to assistant functions that are capable of handling more complex traffic situations.

## References

[1] Preiß, R., Gruner, C., Schilling, T., Winter, H., Mobile Vision – Developing and Testing of Visual Sensors for Driver Assistance Systems, In: Valldorf, J., Gessner, W., [Eds.], Advanced Microsystems for Automotive Applications 2004, Springer, 2004.

[2] Schiele, B., Wojek, C., Kamerabasierte Fußgängerdetektion, Handbuch Fahrerassistenzsysteme, Vieweg+Teubner, 2009.

[3] http://en.wikipedia.org/wiki/Integration_testing

**Andreas Tewes, Matthias Nerling, Harry Henckel**
Hella Aglaia Mobile Vision GmbH
Treskowstr. 14
13089 Berlin
Germany
andreas.tewes@hella.com
Matthias.nerling@hella.com
harry.henckel@hella.com

**Keywords:** multifunctional front camera, driver assistant functions, computer vision

# An Occupancy Grid Based Architecture for ADAS

O. Aycard, T. D. Vu, Q. Baig, Université of Grenoble 1


## Abstract

Perceiving or understanding the environment surrounding a vehicle is a very important step in advanced driving assistance systems (ADAS). The task involves both Simultaneous Localization and Mapping (SLAM) and Detection and Tracking of Moving Objects (DATMO). In this context, we have developed a generic architecture based on occupancy grid to solve SLAM and DATMO in dynamic outdoor environments. In this paper, we give an overview of this architecture and results obtained in different European projects: PReVENT-ProFusion2, INTERSAFE-2 and Interactive.


## 1 Introduction

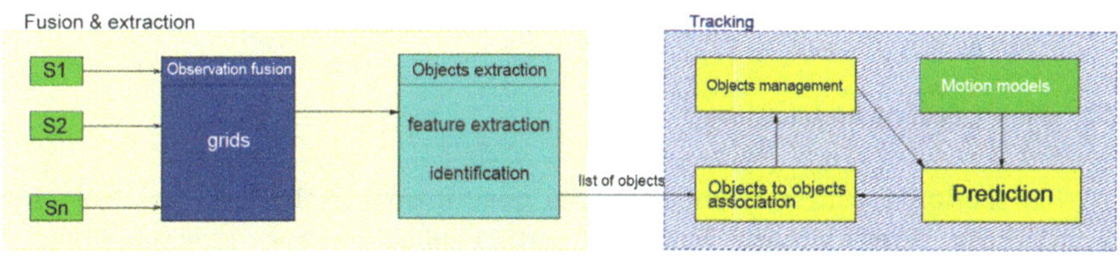


Fig. 1. General architecture of our perception system

Perceiving or understanding the environment surrounding a vehicle is a very important step in advanced driving assistance systems (ADAS) or autonomous vehicles. The task involves both Simultaneous Localization and Mapping (SLAM) and Detection and Tracking of Moving Objects (DATMO). In this context, we have designed and developed a generic architecture based on occupancy grid to solve SLAM and DATMO in dynamic outdoor environments (Fig. 1). In this paper, we give an overview of this architecture and results obtained in different projects: PReVENT-ProFusion2 with Daimler [1], CMU public datasets [2] and also preliminary results in INTERSAFE-2 [3].

## 2 PReVENT-ProFusion2 Project

### 2.1 Demonstrator

The Daimler demonstrator car is equipped with a camera, two short range radar sensors and a laser scanner (Fig. 2). The radar sensor is with a maximum range of 30 m and a field of view of 80°. The data of radar are processed and deliver a list of moving objects. The maximum range of laser sensor is 80 m with a field of view of 160° and a horizontal resolution of 1°. The laser data are not processed. In addition, vehicle odometry information such as velocity and yaw rate are provided by the vehicle sensors. Images from camera are for visualization purpose.

Fig. 2. Daimler demonstrator

Fig. 3. CMU demonstrator

### 2.2 First Level of Architecture

In this section we first summarize the description of the first level of our architecture: Environment Mapping and Localization and Moving Objects Detection. In the last subsection, we describe the fusion between objects detected by laser and radar data.

#### 2.2.1 Environment Mapping and Localization

We propose an incremental mapping approach based on a fast laser scan matching algorithm in order to build a consistent local vehicle map. The map is updated incrementally when new data measurements of laser arrive along with good estimates of vehicle locations obtained from the scan matching algorithm. The advantages of our incremental approach are that the computation can be carried out very quickly and the whole process is able to run online.

Using occupancy grid representation, the vehicle environment is divided into a two-dimensional lattice $M$ of rectangular cells and each cell is associated with a measure taking a real value in the interval [0,1] indicating the probability that the cell is occupied by an obstacle. A high value of occupancy grid indicates the cell is occupied and a low value means the cell is void. Suppose that occupancy states of individual grid cells are independent, the objective of a mapping algorithm is to estimate the posterior probability of occupancy $P(m| x_{1:t}, z_{1:t})$for each cell of grid $m$, given observations $z_{1:t} = (z_1, ..., z_t)$from time $1$ to time $t$ at corresponding known poses $x_{1:t} = (x_1, ..., x_t)$ from time $1$ to $t$.

In order to build a consistent map of the environment, a good vehicle localization is required. Because of the inherent error, using only odometry often results in an unsatisfying map. To solve this problem, we used a particle filter. We predict different possible positions of the vehicle (i.e. one position of the vehicle corresponds to one particle) using the motion model and compute the probability of each position (i.e. the probability of each particle) using the laser data and a sensor model.

### 2.2.2 Moving Objects Detection

After a consistent local grid map of the vehicle is constructed, moving objects can be detected when new laser measurements arrive by comparing with the previously constructed grid map. The principal idea is based on the inconsistencies between observed void space and occupied space in the local map.

- If an object is detected on a location previously seen as void space, then it is a moving object.
- If an object is observed on a location previously occupied then it probably is static.
- If an object appears in a previously not observed location, then it can be static or dynamic and we set the unknown status for the object in this case.

Figure 4 illustrates the described steps in detecting moving objects. The leftmost image depicts the situation where the vehicle is moving along a street seeing a car moving ahead and a motorbike moving in the opposite direction. The middle image shows the local static map and the vehicle location with the current laser scan drawn in red. Measurements which fall into void region in the static map are detected as dynamic and are displayed in the rightmost image. After the clustering step, two moving objects are identified (in green boxes) and correctly corresponds to the car and the motorbike.

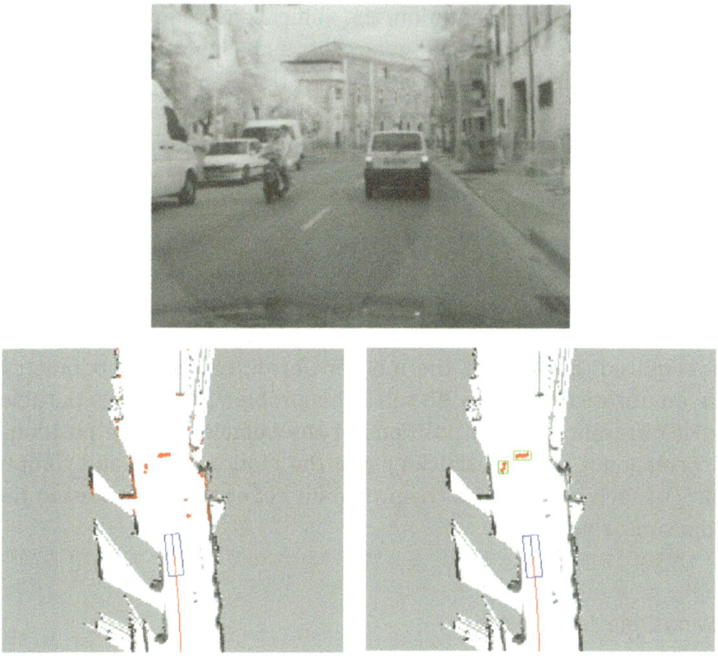

Fig. 4. Moving object detection example

### 2.2.3 Fusion With Radars

After moving objects are identified from laser data, we confirm the object detection results by fusing with radar data and provide the detected objects with their velocities. For each moving object detected from laser data as described in the previous section, a rectangular bounding box is calculated and the radar measurements which lie within the box region are then assigned to the corresponding object. The velocity of the detected moving object is estimated as the average of these corresponding radar measurements.

## 2.3 Second Level of Architecture

In this section we briefly summarize the four different parts of the second level of our architecture (Fig. 1) to solve the different parts of multi-objects tracking:

- The first one is the gating. In this part, taking as input predictions from previous computed objects, we compute the set of new detected observations which can be associated to each object.
- In a second part, using the result of the gating, we perform observa-

tions to objects association and generate association hypothesis. Output is composed of the computed set of association hypothesis. To solve this problem, we use the MHT algorithm.

- In the third part called objects management, objects are confirmed, deleted or created according to the association results and a pruned set of association hypothesis is output.
- In the last part corresponding to the filtering step, estimates are computed for "surviving" objects and predictions are performed to be used the next step of the algorithm. In this part, we use an adaptive method based on Interacting Multiple Models (IMM).

Fig. 5. Experimental results for different environments

## 2.4 Experimental Results

The detection and tracking results are shown in figure 5. The images in the first row represent online maps and objects moving in the vicinity of the vehicle being detected and tracked. The current vehicle location is represented by a blue box along with its trajectories after correction from the odometry. The red points are current laser measurements that are identified as belonging to dynamic objects. Green boxes indicate detected and tracked moving objects with corresponding tracks displayed in different colors. Information on velocities is displayed next to detected objects if available. The second row represent images for visual references to corresponding situations.

Our architecture has been validated in complex crash and non-crash scenarios and compared with the Daimler architecture [4]. To conduct the experiments, we built up a comprehensive database that consists of short sequences of measurements recorded during predefined driving maneuvers. To measure the quality, we counted the false alarms that occurred in non-crash scenarios and the missed alarms in case a collision was not detected by the application. As a general result it can be stated that a reliable collision detection is achieved with both perception modules. Whereas the module of Daimler enables a lower false alarm rate, the crash detection rate of our module is very high (98.1%) in urban areas.

## 3 CMU: Navlab Dataset

### 3.1 Demonstrator

In this part we detail the public dataset obtained on the CMU demonstrator. This dataset was collected using a SICK laser scanner mounted on a moving vehicle (Fig. 2). The vehicle was driven in real-life traffic. The maximum laser range of the scanner is 80 m with the horizontal resolution of 0,5°. We only use laser data and odometry vehicle motion information such as translational and rotational velocity (speed and yaw rate) are computed and provided by internal sensors. Images from camera are only for visualization purpose.

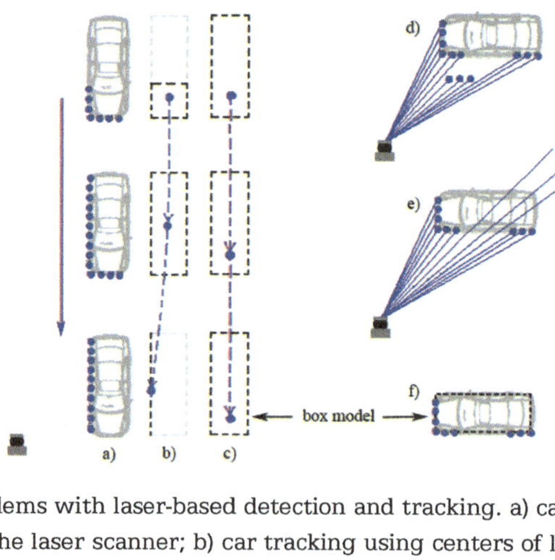


Fig. 6. Problems with laser-based detection and tracking. a) car approaching the laser scanner; b) car tracking using centers of laser impact bounding-boxes; c) correct tracking using car model; d), e) objects divided into several segments; f) car model

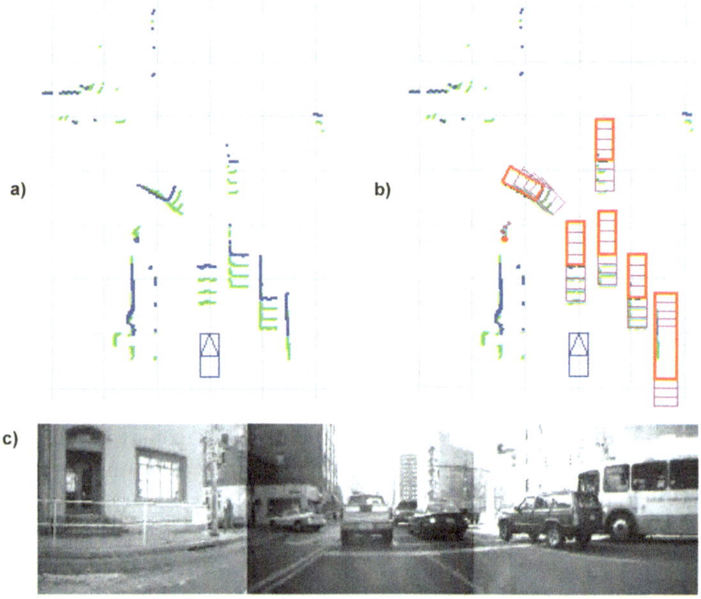

Fig. 7. Example of an interpretation of moving objects from laser data. a) four consecutive scans: in blue is the current scan and in green are previous scans; b) one possible solution including seven tracks of four cars and one bus represented by red boxes and one pedestrian represented by red dots which are imposed on the range data; c) situation reference

### 3.2 Improvements of the Architecture

Conventionally, as introduced in [1], we separate detection and tracking as two independent procedures: the detector and the tracker. At each time step, the tracker takes a list of observations about moving objects returned by the detector together with observations in the past to solve the data association over the observation space to find correct object trajectories (or tracks), and applying filtering techniques to estimate dynamic states of moving objects. In this conventional approach, moving object detection at one time instant usually results in ambiguities that makes the data association become more difficult with missing detections or false alarms. Here we introduce another approach which combines detection and tracking together. We present a probabilistic method for simultaneous detection and tracking of moving objects taking history of measurements that allows the object detection process to make use of temporal information and facilitates a robust tracking of the moving objects. Moreover, noting that moving objects are detected as void form in the detection process presented previously. An advantage of this method is that it can be used to detect any kind of objects without prior knowledge about that objects. However, this suffers from well-known problems with laser-based tracking as explained in figure 6. Here we take a model-based approach and introduce predefined models to represent typical moving object classes. We distinguish four classes of moving objects: bus, car, bike, pedestrian (motorcycles and bicycles belong to the bike class). We use a box model of fixed size to represent bus, car, bike and a point model to represent pedestrians.

Our algorithm to solve DATMO is summarized as follows. We formulate the detection and tracking problem as finding the most likely trajectories of moving objects given data by measurements over a sliding window of time (Fig. 7). A trajectory (track) is regarded as a sequence of object shapes (models) produced over time by an object which staisfies the constraint of both an underlying object motion and the consistency with measurements observed from frame to frame. In this way, our approach can be seen as a batch method searching for the global optimum solution in the spatio-temporal space. Due to the high computational complexity of such a scheme, we employ a Markov chain Monte Carlo (MCMC) technique that enables traversing efficiently in the solution space.

### 3.3 Experimental Results

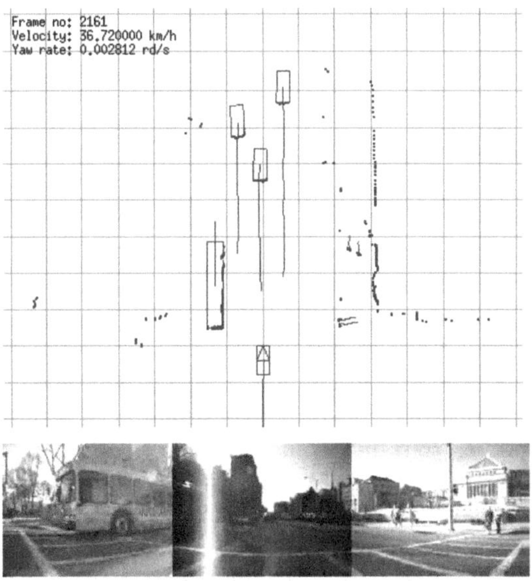


a)

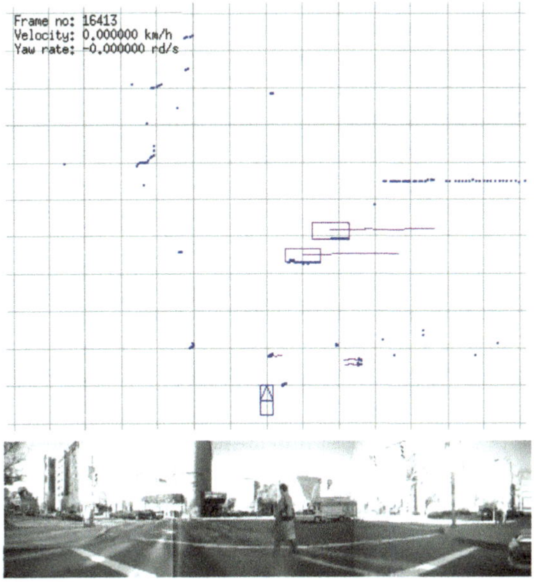


b)

Fig. 8. a) Different class of objects b) Occlusion example

Figure 8a shows an example of our detection and tracking algorithm in action. In the ego-vehicle's view, the detected moving objects and their trajectories are shown in pink color with current laser scan is in blue color. Moving objects in the situation include a bus moving in the opposite direction on the left, three cars moving ahead, two pedestrians walking on the left pavement and the other two pedestrians passing the intersection. Figure 8b shows an example of our detection and tracking algorithm when an occlusion occurs. Even if only a part of the second car is detected by the laser, we are able to track this occluded car.

With initial evaluations, the MCMC detection and tracking outperforms the detection and tracking using MHT in our previous work [1] in terms of a higher detection rate and less false alarms.

## 4 INTERSAFE-2

### 4.1 Demonstrator

Figure 9 illustrates the chosen sensor set and coverage area. The sensor set-up includes sensors which are already available in serial cars, namely front

ACC radar and rear-looking radar for lane change support. These sensors are accompanied by a stereo camera system to the front with high field of view of about 60°. A scanning laser with a field of view of about 160° and dedicated radar sensors directed to +90° and -90° respectively are foreseen for measuring the objects coming from the side. These sensors are able to measure position, velocity and some geometrical parameters of the relevant objects at intersections.

### 4.2 Current Architecture and Future Improvements

Our primary goal was to perform laser processing for local SLAM and detection of moving objects. We have applied methods developed in [1]. Ongoing work is about the fusion of data from laser and stereovision. Currently, we are planning fusion at detection level, object segments detected in vision and laser scans will be fused together to get a more consolidated view.

### 4.3 Preliminary Results

In figure 10 the first column shows a left turn scenario where the demonstrator car is turning left and a cyclist is going to right in front of the vehicle. This cyclist is detected and tracked. In the second column we have a scenario where a moving vehicle is coming from opposite direction and the demonstrator is crossing a vehicle on its right. In both cases precise trajectories of the demonstrator are achieved and local maps around the vehicle are constructed consistently.

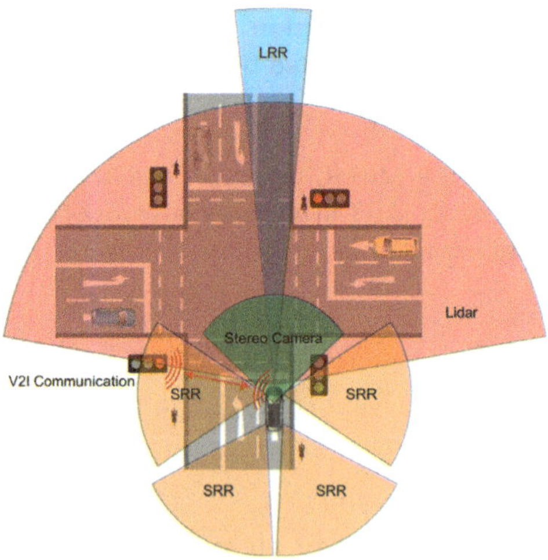


Fig. 9. INTERSAFE-2 demonstrator

Fig. 10. Local SLAM and moving objects detection

## 5 Conclusion and Perspectives

In this paper we presented the architecture developed in our group for perception systems for ADAS. This architecture has been used in the European project PReVENT-ProFusion2 in cooperation with Daimler [1] and also with Volvo Truck [5] (not presented here). It has been also used with a CMU demonstrator [2], and is currently used in the European project INTERSAFE-2 [3]. More results and videos can be found at http://emotion.inrialpes.fr/~tdvu/videos. We are currently involved in the European project Interactive-Perception. In this project, we will extend our architecture for the Daimler demonstrator, and cooperate with TRW on low level fusion between radar and vision.

## Acknowledgements

This work is supported by the European projects INTERSAFE-2 and Interactive.

## References

[1] Vu, T.D., Burlet, J., Aycard, O., Grid-based localization and local mapping with moving objects detection and tracking, International Journal on Information Fusion, 2009.

[2] Vu, T.D., Aycard, O., Lased-based detection and tracking moving object using data-driven markov chain monte carlo, IEEE ICRA, 2009.

[3] Baig, Q., Vu, T.D., Aycard, O., Online Localization and Mapping with Moving Object Tracking in Dynamic Outdoor Environments, IEEE ICCP, 2009.

[4] Pietzsch, S., Vu, T.D., Burlet, J., Aycard, O., Hackbarth, T., Appenrodt, N., Dickmann, J., Radig, B., Results of a precrash application based on laser scanner and short range radars, IEEE Transactions on Intelligent Transport Systems, 2009. To be published.

[5] Garcia, R., Aycard, O., Vu, T.D., Ahrholdt, M., High level sensor data fusion for automotive applications using occupancy grids, IEEE ICARCV, 2008.

**Olivier Aycard, Trung-Dung Vu, Qadeer Baig**
Université of Grenoble 1
Avenue Marie Reynoard
38100 Grenoble
France
Olivier.Aycard@imag.fr
Trung-Dung.Vu@imag.fr
Qadeer.Baig@imag.fr

**Keywords:** perception, ADAS, occupancy grid, sensor data fusion

# A Concept Vehicle for Rapid Prototyping of Advanced Driver Assistance Systems

R. Schubert, E. Richter, N. Mattern, P. Lindner, G. Wanielik,
Technische Universität Chemnitz

## Abstract

The ongoing development of Advanced Driver Assistance Systems (ADAS) requires new prototyping concepts. In this paper the concept vehicle Carai is presented as a generic vehicle with a variety of sensors, computing units, and different HMI components in order to allow the fast implementation and evaluation of different ADAS applications. In addition to the description of the technical components, a software framework is presented which enables fast software prototyping by providing basic data acquisition and processing modules including sophisticated data fusion algorithms. Finally, the usage of the Carai is demonstrated on the example of different ADAS applications.

## 1 Introduction

The ongoing development of Advanced Driver Assistance Systems (ADAS) is promising a further increase of safety, efficiency, and comfort for road traffic. However, while the ADAS functionalities are steadily improving, also the complexity of the developed systems is growing. Thus, the prototyping process requires new concepts for evaluating different sensor sets, algorithms, and Human-Machine-Interface (HMI) concepts.

The current design process often includes dedicated solutions for hardware, software, and test environments. For instance, new functions are often written as a specific piece of software for a certain Electronic Control Unit (ECU). For the test, usually a specific demonstrator vehicle is set-up which includes the required sensors, actuators, and other electronic components. This dedicated approach may require an increased effort for e.g. including additional sensors which were not specified at the beginning of the design process.

In this paper, a concept vehicle for ADAS prototyping is presented which is called Carai (see Fig. 1). The idea is to set up a generic vehicle with a variety of sensors, computing units, and different HMI components in order to allow the fast implementation and evaluation of different functions [1]. The overall objective is to decrease the time from the first idea to the prototypical implementation of a new Advanced Driver Assistance System.

The paper is structured as follows: Section 2 reviews the wide ranging set of complementary sensors which are integrated into the vehicle. In order to rapidly set up software applications, a data processing & fusion framework has been developed which is presented in section 3. Finally, section 4 discusses an innovative HMI concept which is used in to evaluate the benefits of new functions for the driver. The paper concludes with the presentation of selected ADAS applications which have been implemented and tested in different research projects using the Carai.

Fig. 1. The concept vehicles Carai 1 (right) and Carai 2 (left)

## 2 Sensor Equipment and Components

The Carai is equipped with a wide variety of different types of sensors. There are vision sensor, range sensors, localization sensors, and communication devices. Depending on the current application, the Carai may be equipped with:

A) Vision systems:

- VGA Stereo System (Hella Aglaia INKA)
- Megapixel Stereo System (Kamera Werke Dresden)
- Near-Infrared Cameras (Bosch)
- Far Infrared Cameras (FLIR Photon)

B) Range sensors:

- Range Cameras (PMD A2, PMD 3k-s)
- Imaging Radar (Prototypical, Continental ARS300 – see Fig. 2)
- Laserscanner (IBEO Lux, SICK, Denso)
- NanoLOC ranging devices (IEEE802.15.4)

C) Communication devices:

- Car-2-Car-Communication compliant devices (WLAN, 802.11p, IEEE 1609.3/4, WAVE)
- GSM Modules

D) CAN gateway for obtaining ego motion data, indicator status, etc.

E) Localization equipment (which can also be used as reference system for evaluating less accurate technologies):

- Inertial navigation system (NovAtel SPAN Differential GPS with RTK and HG1700 Inertial Measurement Unit)
- Leica 1200 DGPS with RTK

The sensors are connected to a 19" dual quad-core server PC, which provides appropriate computational resources for the rapid prototyping algorithm development. The Carai can be equipped with a maximum of two of those Server PCs. All sensors work hand in hand for the particular applications. The flexibility of the software framework and the modular sensor mounting system allows a quick integration of new sensors.

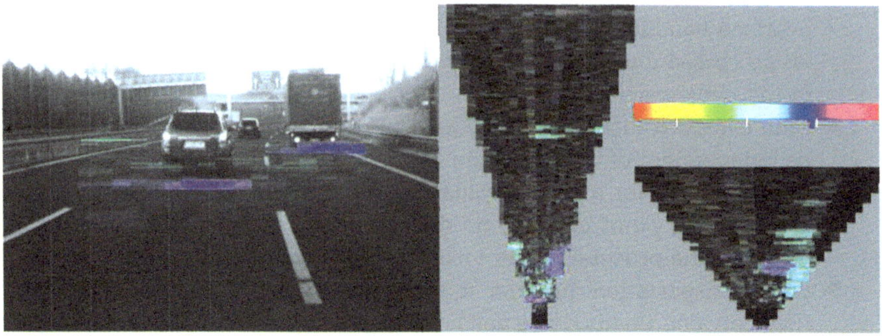

Fig. 2. Example output of the Continental Imaging Radar ARS200 (right) and its transformation into the image (left)

## 3 Rapid Prototyping Software Framework

The development of data fusion algorithms and their evaluation is usually driven by real sensors requiring lots of software infrastructure and integration. This integration is often very resource consuming and a repeating process, which can be supported and massively accelerated using the proposed framework.

### 3.1 Data Acquisition and Processing

The processing of different data types originating from several sources requires a software framework which contains the following main functionalities:

- Data acquisition: The data from different data sources or sensors need to be acquired using a usually hardware specific driver. From that, a common sensor representation is needed in order to abstract from the concrete hardware. Furthermore, a common globally time stamped data format enables a unified handling between the software modules.
- Data recording: Different data types like images, GNSS data, and odometry need to be stored in a common stream file format, so they can be replayed in the right order during offline processing (We denote online if "real" data sources or sensors are connected and offline if the data are taken from an associated file stream).
- Data synchronization: As the data from different sources are not time synchronized in general, they need to be synchronized in order to replay them in the right order during offline processing. In the case of online processing, data need to be processed asynchronously. Therefore, an overload handling is required, which usually discards data which cannot be processed due to high processing load and allows for the processing of the newest available data.
- Data source transparent processing: The processing of the data has to be independent of the used data sources. Therewith, sensors can be easily exchanged for a rapid evaluation or comparison. Additionally, from a processing point of view, no difference arises while switching from on- to offline processing and vice versa.
- Data transportation: In order to allow distributed applications, the data flow between the modules needs to be generalized enabling both simple in-memory and more sophisticated transportation modes like network connections (e.g. TCP, UDP).
- Toolbox: To further ease the development of Advanced Driver Assistance Systems, a large set of tools like user interface supported stream editing, labeling, and calibration is needed.

## 3.2 Algorithmic Components

Most of nowadays Advanced Driver Assistance Systems rely on a common set of algorithms, which can be reused if implemented in a generic way. Among others, the proposed framework consists of the following components:

- Bayesian Filtering: The widely used Bayesian filtering framework is implemented in several ways, e.g.
  - Gaussian filters: Kalman Filter (Linear/ Extended/ Unscented)
  - Gaussian Sum filters (including the Interacting Multiple Model (IMM) filter)
  - Histogram filters: Occupancy and 4D Grid [2]
  - Particle filters: Sequential importance sampling, (Reversible Jump) Markov Chain Monte Carlo
- Existence estimation: Especially for multiple object tracking applications (e.g. ACC), it becomes necessary to estimate the existence of an object. Therefore, several algorithms are implemented within the framework, including:
  - Integrated/ Joint Probabilistic Data Association (IPDA/ JPDA)
  - Sequential Probability Ratio Testing (SPRT)
  - Probability Hypothesis Density (PHD) filter
- Situation Assessment: A complex behavior modeling of the driver and the objects in the surrounding is often necessary to improve the performance of several applications. Therefore, several approaches can be chosen:
  - Knowledge-based expert systems
  - (Dynamic) Bayesian Networks for situation assessment and decision making systems [3]
  - Fuzzy Logic [4]
  - Belief Theory (Dempster-Shafer) [5]

# 4 Human Machine Interfaces

In current driver information and driver assistant systems, the implementation of the Human Machine Interface (HMI) providing visual advices and warnings usually works by means of head-down displays. They are prevailingly integrated into the instrument board (speed indicator etc.) or into the vehicle's center console. Head-up displays (HUD) are increasingly applied for further display. They project the information onto the windshield and generate a virtual image directly in front of the vehicle by means of an elaborate optical characteristic. Those two concepts of displaying information differ with respect to the driver's level of distraction. Head-up display systems operate less distracting than head-down displays, as they are spatially closer to the traffic situation.

The new multifunctional Human Machine Interface in the Carai vehicle is supposed to bridge the gap between head-up displays and future virtual projection screens. It consists of a LED matrix, which is applied beneath the windshield (see Fig. 3). The Multifunctional HMI is able to display information and warnings which support the driver on various levels of driving. The extension over the driver's entire frontal field of view is what thereby distinguishes it from conventional head-up or head-down displays. As an assisting function on the level of navigation, the driver can be given indications concerning route guidance. On the level of maneuvering, the display offers the possibility to depict information concerning road signs (e.g. no passing) or the occupancy of neighboring lanes (e.g. lane change assisting systems). In terms of turning at a crossroad, the display can indicate pedestrians or bicyclists within the blind spot.

Fig. 3. Concept of the Multifunctional Human Machine Interface (left) and a first prototype in the Carai (right)

## 5 Selected Applications

The presented concept vehicle has been used in a variety of research projects and industry collaborations. In the following, selected examples of successfully implemented functions are presented.

### 5.1 Radar- and Vision-Based Vehicle Tracking

In the subproject Saspence of the European Integrated Project PreVent, a data fusion application was implemented in order to detect and track vehicles using radar and vision sensors. The evaluation showed that by the data fusion approach, an improved estimation of the lateral position and the width of tracked vehicles could be achieved. Furthermore, the detection rate of stationary objects could be increased [6].

### 5.2 Situation Assessment for Automatic Lane Change Maneuvers

In an industry collaboration project, an approach towards automatic maneuver decisions on highways was developed. The system is able to track vehicles in front of and behind the ego vehicle using cameras and radar sensors. Furthermore, the course of the lanes is estimated from image measurements (see Fig. 4). Finally, all those data are combined in a probabilistic situation assessment module in order to derive a lane change recommendation for the driver [7].

### 5.3 Pedestrian Recognition by Combining Image Processing Techniques and Communications

In the European research project WATCH-OVER, a system for the detection of vulnerable road users (in particular pedestrians) has been implemented using a data fusion approach based on image processing and communication-based ranging devices [8].

### 5.4 Image-Landmark-Based Vehicle Localization Using Highly Precise Digital Maps

In the European research project SAFESPOT, an innovative approach for vehicle localization has been developed. The system is based on the detection of line landmarks (e.g. lane markings, curbstones) in the camera image and the matching of those features with the content of a digital map [9].

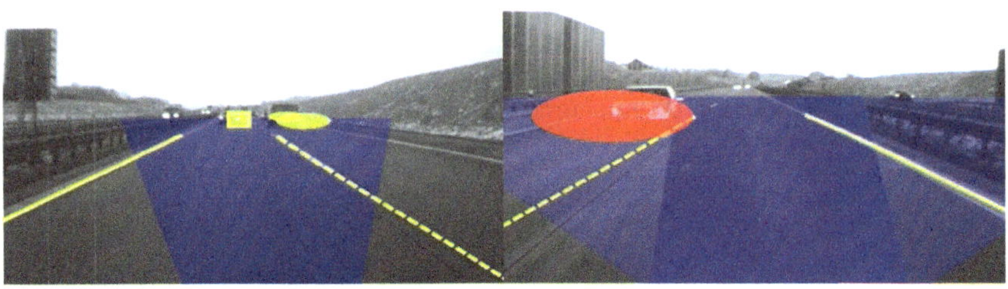

Fig. 4. Front and rear view from the ego vehicle including results from vehicle tracking and lane recognition

## 6 Conclusions

In this paper a generic concept vehicle has been presented as a step towards a faster prototyping of new ADAS functions and applications. In order to evaluate also cooperative systems (based on V2X communications), a second vehicle (Carai 2) is currently set up. The prototypical implementation of cooperative functions will be an important aspect of the future work of the authors.

## Acknowledgements

The authors would like to thank Michael von der Trenck and Marco Conrad for their commitment in integrating all required components into the Carai.

## References

[1] Chemnitz University of Technology, "I-FAS," last checked: 15. February 2010, http://www.tu-chemnitz.de/projekt/I-FAS/content/view/42/28/lang,en/.

[2] Richter, E., Lindner, P., Wanielik, G., Takagi, K., Isogai, A., Advanced occupancy grid techniques for lidar based object detection and tracking, 12th International IEEE Conference on Intelligent Transportation Systems ITSC '09, pp. 1-5, 2009.

[3] Schubert, R., Wanielik, G., Unifying Bayesian networks and IMM filtering for improved multiple model estimation, 12th International Conference on Information Fusion, pp. 810-817, 2009.

[4] Scheunert, U., Lindner, P., Cramer, H., Multi Level Fusion with Fuzzy Operators using Confidence, 9th International Conference on Information Fusion, pp. 1-6, 2006.

[5] Skutek, M., Linzmeier, D., Appenrodt, N., Wanielik, G., A precrash system based on sensor data fusion of laser scanner and short range radars, 8th International Conference on Information Fusion, 2005.

[6] Tango, F., Richter, E., Scheunert, U., Wanielik, G., Advanced multiple objects tracking by fusing radar and image sensor data - Application on a case study, 11th International Conference on Information Fusion, 2008.

[7] Schubert, R., Schulze, K., Wanielik, G., Fahrzeugumfelderkennung und probabilistische Modellierung von Manöverentscheidungen, In: Stiller, C., Maurer, M. (ed.) 6. Workshop Fahrerassistenzsysteme, Freundeskreis Mess- und Regelungstechnik Karlsruhe e. V., pp. 147-156, 2009.

[8] Fardi, B., Neubert, U., Giesecke, N., Lietz, H., Wanielik, G., A Fusion Concept of Video and Communication Data for VRU Recognition, 11th International Conference on Information Fusion, 2008.

[9] Mattern, N., Schubert, R., Wanielik, G., Image landmark based positioning in road safety applications using high accurate maps, IEEE/ION Position, Location and Navigation Symposium, pp. 1008-1013, 2008.

**Robin Schubert, Eric Richter, Norman Mattern, Philipp Lindner, Gerd Wanielik**
Technische Universität Chemnitz
Reichenhainer Str. 70
09126 Chemnitz
Germany
robin.schubert@etit.tu-chemnitz.de
eric.richter@etit.tu-chemnitz.de
norman.mattern@etit.tu-chemnitz.de
philipp.lindner@etit.tu-chemnitz.de
gerd.wanielik@etit.tu-chemnitz.de

**Keywords:** advanced driver assistance systems, rapid prototyping, surrounding field sensors, data fusion, tracking, situation assessment, image processing, localization

# Intersection Safety

# INTERSAFE-2: Progress on Cooperative Intersection Safety

B. Roessler, D. Westhoff, Sick AG


## Abstract

This paper presents the intermediate results of the intersection reconstruction work performed in the European funded project INTERSAFE-2. It summarizes the current achievements and findings in this topic that were gained in the first half of the project.


## 1 Introduction

The INTERSAFE-2 project aims to develop and demonstrate a Cooperative Intersection Safety System (CISS) that is able to significantly reduce injury and fatal accidents at intersections.

Vehicles equipped with communication means and onboard sensor systems cooperate with the road side infrastructure in order to achieve a comprehensive system that contributes to the EU-25 and "zero accident" vision as well as to a significant improvement of efficiency in traffic flow and thus reduce fuel consumption in urban areas. By networking state-of-the-art technologies for sensors, infrastructure systems, communications, digital map contents and new accurate positioning techniques (intersection reconstruction), INTERSAFE-2 aims to bring Intersection Safety Systems much closer to market introduction.

The novel Cooperative Intersection Safety System combines warning and intervention functions demonstrated on three vehicles: two passenger cars and one heavy goods vehicle. Proof of concept with extended testing and user trials will be shown based on these demonstrators. Furthermore, a simulator is used for additional research and develeopment

These functions use novel cooperative scenario interpretation and risk assessment algorithms. The cooperative sensor data fusion is based on:

- state-of-the-art and advanced on-board sensors for object recognition and intersection reconstruction and
- a standard navigation map.

Additional information supplied over a communication link from:

- other road users via V2V if the other vehicle is so equipped and
- infrastructure sensors and traffic lights via V2I if the infrastructure is so equipped.

## 2 Technical Objectives

Three demonstrator were set up which introduce new functionalities ranging from warning systems to active vehicle intervention. The vehicles are equipped with:

- on-board sensors that are able to perceive the environment and
- communication means that complement the overall environment perception from the infrastructural side.

The vehicle support functions comprise:

1. Right turning assistance for heavy goods vehicles and passenger cars to prevent accidents especially with VRUs at intersections.
2. Left turning assistance to prevent left-turn related fatal accident types.
3. Crossing assistance to prevent accidents that may occur when trying to cross a road with priority at an intersection.
4. Traffic light assistance to prevent red light running accidents and to improve traffic flow.

Secure V2V and V2I communication and advanced sensor data fusion techniques result in reliable perception of environmental data and scene interpretation. Using data not only from vehicle sensors but also from additional data sources (like traffic lights or infrastructure sensors) an advanced and more reliable interpretation of the intersection scene can be realised. Furthermore, real-time traffic management applications become possible. This will contribute on the one hand to traffic safety and on the other hand, to less turbulent traffic flow and therefore, to decreased energy consumption and reduced emissions. Another result will be the enhancement of standardisation of V2x communication by the cooperative approach developed in this project.

An overall intersection environment perception even under challenging conditions like occluded field of view is analysed and the improvement will be demonstrated. One intersection will be equipped with infrastructure sensors like Laserscanners and video cameras. This will enhance the perception area for the Intersection Safety System at wide intersections.

Based on state-of-the-art technologies, accurate intersection reconstruction is performed by the demonstrator vehicles which serves as basis for further situation analysis. The focus on state-of-the-art in sensor technologies and standard map data enables a more rapid market introduction of such Intersection Safety Systems. Objects within the intersection are reliably classified and tracked both from the demonstrator vehicles and from the equipped intersection. Based on relative intersection localisation and on digital maps, a background elimination process is performed. Thus, the focus is on road users present in the foreground at the intersection, which are robustly tracked and classified.

Based on the above-mentioned techniques and on improved sensor data fusion, a comprehensive scenario interpretation becomes possible for the demonstrator vehicles. The cooperative environmental model is established in the demonstrator vehicles based on information from the on-board sensors and from the infrastructure transferred via V2x communication. Thus, a cooperative scenario interpretation and risk assessment can be performed in the demonstrator vehicles. Once the situation is clearly analysed, the warning or intervention strategies can be accomplished.

## 3 Intersection Identification and Reconstruction

The focus of this paper is the intersection reconstruction which can be found in the perception layer of the vehicle demonstrator architecture shown in Fig. 1. The purpose of this intersection reconstruction is to build a model of the intersection, the demonstrator vehicle is approaching to, in order to perform scenario interpretation and risk assessment in the application layer. The reconstruction is done solely based on vehicle sensors. Therefore, a Laserscanner and video camera cluster is used as a standard intersection reconstruction module in each of the three INTERSAFE-2 demonstrators. Optionally, a standard digital map can be utilized in order to identify a broad region where an intersection is expected to appear (see section 3.2). The intersection reconstruction procedure and especially the part of the Laserscanner system are described in the following.

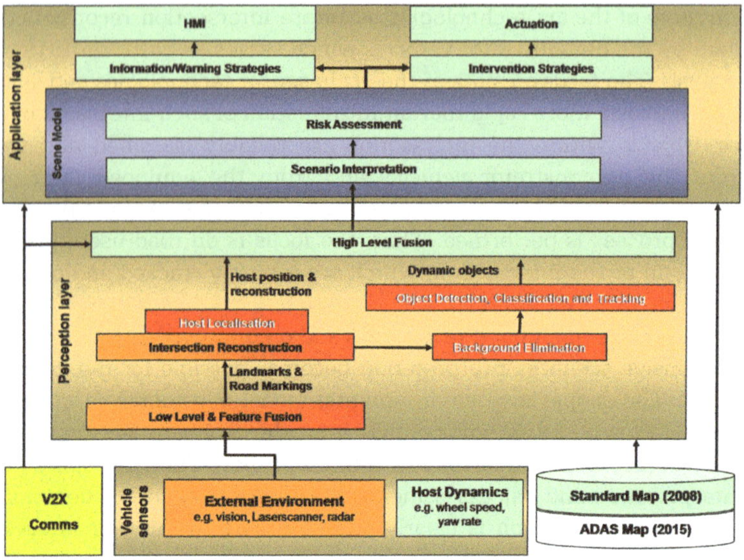


Fig. 1. INTERSAFE-2 vehicle demonstrator architecture [1].

### 3.1 Feature Detection

The video camera system is able to detect the following features of an intersection:

- White line markings – fixed and dashed
- Arrows
- Stop line
- Vertical posts
- Kerbs
- Road edges
- Signposts

Some of these features are illustrated in Fig. 3.

For the Laserscanner system several Ibeo LUX Laserscanners are used. Each sensor scans simultaneously in four scan planes. This four layer technology makes it possible to receive reflections from objects within the intersection (e.g. cars or pedestrians) as well as from the road surface at the same time. Therefore, the Ibeo LUX Laserscanner is able to even detect line markings or other flat features (e.g. kerbstones or swards). An example of the results of the filtered reflection points from the intersection surface is shown in Fig. 2.

In this figure a history of filtered ground point reflections is shown within a defined area.

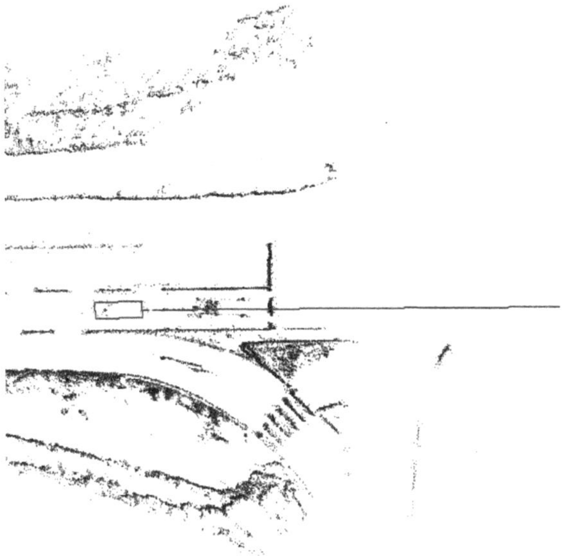

Fig. 2. Laserscanner reflections from the intersection surface.

The described technology and algorithms make it possible for the Laserscanners to detect similar and additional features compared to the video system. Features which can be extracted from the ground map are:

- Some white line markings – fixed and dashed
- Stop line
- (Opposite) kerbstones

In addition the Laserscanners are well suited to detect all kind of vertical objects from which the following intersection features can be extracted:

- Vertical posts (e.g. from traffic lights)
- Signposts

Some of the described features are exemplarily shown in Fig. 3.

Fig. 3. Video detected intersection features (left) and Laserscanner detected intersection features (right).

### 3.2 Intersection Identification

When searching for the above mentioned features in the Laserscanner raw data, it is helpful if one knows in advance that in some area an intersection is expected to appear. Otherwise, the search for the intersection features is quite challenging since for example even in non intersection areas a lot of Laserscanner raw data points can look like vertical posts (e.g. trees). This can be solved by using a standard digital map along with a normal GPS receiver which indicates that an intersection will appear in the surrounding of the vehicle. However, the accuracy of GPS does not allow the full procedure of the intersection reconstruction to be performed with a standard digital map alone.

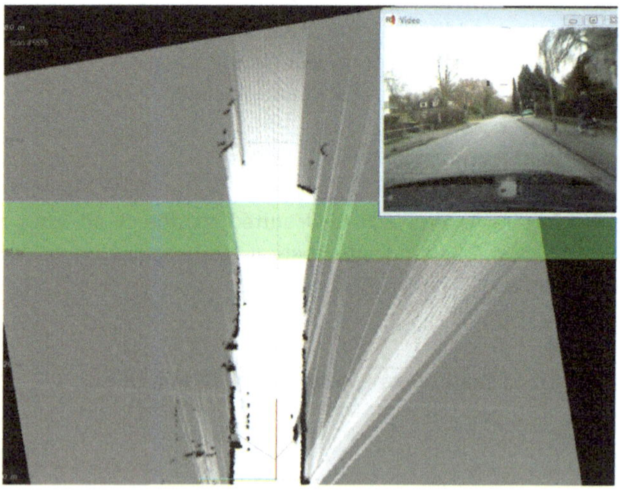

Fig. 4. Free space detection in an occupancy grid map.

One important indicator for the presence of an intersection which can be detected with the used Laserscanners is the free area of the crossing roads. Using an occupancy grid map technique, the free area can be found at places where the occupied cells at the roadside show a gap of a defined size of a typical crossroad (see Fig. 4).

This is a very fast approach in order to identify intersection candidates. Of course, these candidates have to be verified with the additional intersection features described above.

### 3.3 Feature Fusion and Reconstruction

Both separately extracted features of the video system and the Laserscanners have to be fused afterwards into a common feature representation as is exemplarily shown in Fig. 5, left.

Fig. 5. Fused intersection feature (left) and intersection reconstruction (right).

Features that were detected in both sensor systems can be used to align the individual parts of the intersection. In order to simplify the fusion process the Laserscanners are synchronized to the video camera so that with every video frame the Laserscanners point to the front of the demonstrator vehicle.

### 3.4 Reconstruction

Finally, the fused intersection features are used to gain a reconstruction (model) of the intersection where the demonstrator vehicle is approaching to. Such a model contains all relevant parts for the application layer like e.g. number and position of lanes, free drivable area or infrastructure components. This is schematically shown in Fig. 5, right.

This reconstruction process is a very challenging one since the intersection model has to be build out of some scattered intersection features. The missing parts which were not seen by the sensor cluster have to be reconstructed based on some prior knowledge of the layout of an intersection.

## 4 Object Detection, Classification and Tracking

In addition to the intersection feature detection, the Laserscanner is able to detect and classify other road users on the intersection. As can be seen in the architecture of Fig. 1, the intersection reconstruction builds the basis for the object detection, classification and tracking task of the Laserscanner perception system. The purpose is to perform at first a background elimination algorithm which extracts all the relevant data (for the application layer) from the intersection (i.e. relevant road users).

This is done in order to set the focus on those objects for a better classification and tracking performance. The result of this background elimination is schematically shown in Fig. 6.

Fig. 6. Background elimination.

If the complete Laserscanner raw data would be used for object detection and tracking a lot of performance would have to be spent on objects which are not relevant for an intersection safety system. Such Laserscanner raw data (from one scan) is shown in Fig. 7, left.

After performing the described background elimination process the Laserscanner raw data can be reduced to only the relevant objects (see Fig. 7, right) so that the detection, classification and tracking can be performed more reliably.

Fig. 7. Laserscanner raw data at an intersection (up) and data after background elimination algorithm (bottom).

## 5 Conclusion

INTERSAFE-2 is the follow-up project of the EU funded project PReVENT-INTERSAFE. INTERSAFE-2 started in June 2008 and will end in May 2011. During this period a comprehensive Cooperative Intersection Safety System is developed which is able to significantly reduce injury and fatal accidents at intersections. By using state-of-the-art technologies a much closer market introduction of Intersection Safety is aspired.

Novel algorithms are developed in order to build a comprehensive model of the considered intersections. These algorithms build the basis for all the INTERSAFE-2 demonstrators and can also be used in any other kind of advanced Intersection Safety Systems.

## References

[1] PReVENT-INTERSAFE consortium, Deliverable D4.1: Specification and architecture documentation, 2009.

**Bernd Roessler, Daniel Westhoff**
Sick AG
Merkurring 20
22143 Hamburg
Germany
bernd.roessler@sick.de
daniel.westhoff@sick.de

**Keywords:** intersection safety, cooperative driver assistance systems

# Intersection Safety for Heavy Goods Vehicles Safety Application Development

M. Ahrholdt, G. Grubb, E. Agardt, Volvo Technology Corporation

## Abstract

At urban intersections, safety of vulnerable road users (VRUs) such as pedestrians and cyclists in connection with heavy goods traffic is a relevant question. In the project INTERSAFE-2, Volvo Technology is currently developing a traffic safety application for heavy goods vehicles in urban intersections that is particularly addressing safety of VRU during right turning manoeuvres. In this paper, an overview of the subsequent steps in the development is given, including an overview of sensor set-up, the communication and map units and the initial development of the traffic safety application.

## 1 Introduction

Even for heavy goods vehicles driver assistance systems are an important factor for increasing traffic safety. First functions already commercially available, as Adaptive Cruise Control (ACC) or Lane Departure Warning (LDW), provide good support to the truck driver mainly in highway and even rural environments.

In urban traffic, however, with complex traffic situations and a high traffic density, additional driver support systems are needed. Within the European research initiative INTERSAFE-2 [1], Volvo is investigating how to support truck drivers in urban intersection situations. Urban traffic of heavy goods vehicles is a vital factor in today's cities to ensure distribution of supplies or collection of waste. A lot of efforts are currently undertaken to make this traffic as safe and eco-friendly as possible.

The INTERSAFE-2 project is a European research initiative, started in June 2008 and will be ongoing until May 2011. It consists of partners from 6 European countries, amongst them 3 vehicle manufacturers, 3 suppliers, 1 traffic signal manufacturer and 4 research bodies.

Particularly pedestrians and cyclists, so-called vulnerable road users (VRU), have a high risk of injury when involved in an accident with a heavy goods vehicle. In a previous contribution [2], this challenge has been investigated thoroughly, and the scenario of trucks turning in intersections has been identified as relevant. For this reason, the VRU safety has been selected as the main target scenario for the intersection assistance system being developed in INTERSAFE-2.

This paper will give an overview of the addressed intersection scenario, describe the safety application to being developed in INTERSAFE-2 in an overview, and discuss the derived system architecture.

## 1.1 Addressed Scenario

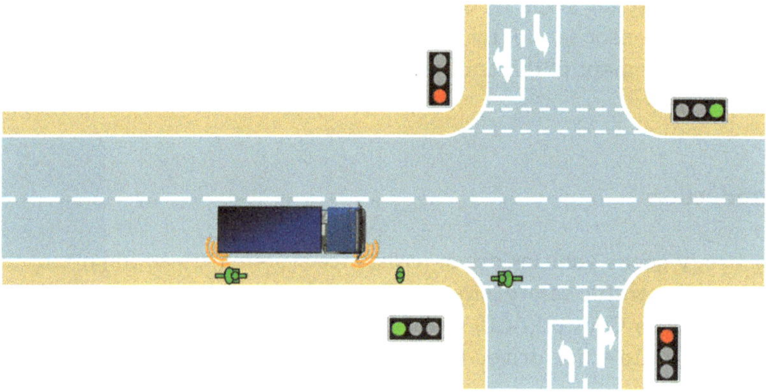

Fig. 1. Right turning scenario

Different kinds of intersection assistance functions are being developed in the INTERSAFE-2 project. For the heavy goods vehicles, Volvo Technology concentrates on one specific scenario: Right turning in intersections and the safety of pedestrians and cyclists. For the driver of a truck, objects and road users are not well visible on the right side of the vehicle which has caused a number of serious accidents. In this type of accident, even minor impacts at slow speeds can easily lead to significant injuries of vulnerable road users.

Typical attributes of the addressed situations can be characterized as:

- The truck is turning right in an urban environment.
- The vehicle speed is low.
- The work load of the truck driver is at a quite high level, both during and before the turning manoeuvre.

- The VRU is not visible in the mirrors, alternatively not observed by the driver.
- The VRU is not aware that the truck driver does not observe him.
- The VRU may even have the right of way.

This type of accidents has previously been addressed, first in the 1990ies by improving passive safety through mandatory deflector frames, reducing the risk of the VRU falling in front of the vehicles rear axis. Then in 2003 and 2005, regulations for mirrors and visibility have been toughened, also illustrated in [2].

## 2 Safety Application

For the described turning scenario, a traffic safety application is being developed by Volvo Technology together with the partners of the INTERSAFE-2 project consortium. The following sections give an overview of the sensor system, the overall system architecture derived, the roadside installation at the test site Gothenburg and of situation assessment and HMI strategies.

At the time of writing of this contribution, definitions of the requirements and system architecture have been concluded, whereas development of the safety application is still ongoing, such that no final results are presented here yet.

### 2.1 In-Vehicle Perception System and System Architecture

The perception system to be developed in INTERSAFE-2 comprises a set of different components:

- A front and side vicinity monitoring system for object detection, consisting of:
  - an IBEO laser scanner system
  - an ultrasonic sensor system
- An intersection reconstruction system to determine the relative position of the host vehicle with respect to the intersection and extracting intersection features, using:
  - a TRW vision system, recognizing intersection land marks and road markings
  - the beforementioned laser scanner system, detecting intersection landmarks
- A wireless communication link between traffic light and vehicle, provided by NEC, to inform the vehicle system about signal phases and

push-button requests of pedestrians or cyclists, using the proposed IEEE 802.11p standard for vehicle-to-vehicle/infrastructure (V2X) communication.

- An ADAS map, so-called e-horizon, comprising information such as intersection location, number and angle of connecting roads, presence of traffic lights and right-of-way regulations.
- A sensor data fusion module aggregating the information originating from the different sources and providing an environment model to the application.

The desired optimal coverage area for the perception system is shown in Fig. 2. However, good sensor coverage in the close vicinity of the host vehicle is a challenge. Two different sensor types are used for object detection, a laser scanner system and ultrasonic sensors. The laser scanner gives good lateral resolution, some classification capability and a contribution to the intersection reconstruction task. The ultrasonic sensors in contrast offer a wide lateral coverage area with cost-efficient technology. Fig. 3 shows the achieved field of view of the object sensors.

The derived system architecture is depicted in Fig. 4. The system is built up in a scalable way, such that functionality also is achieved in intersections not equipped with a communication device or with missing map information. The equipped vehicle is shown in Fig. 5, respectively.

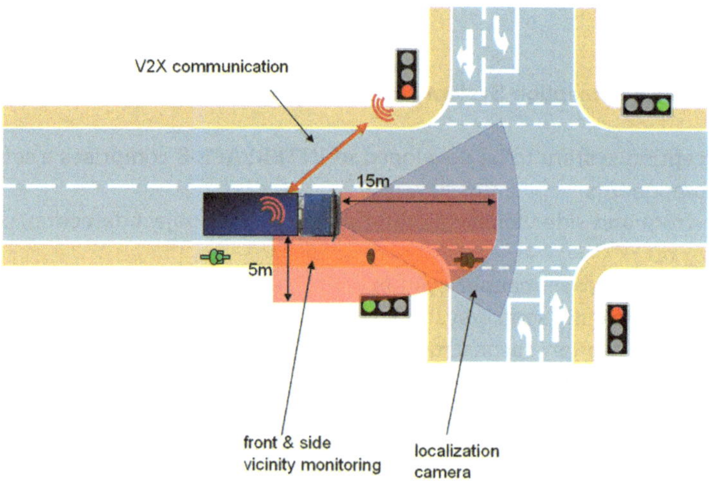


Fig. 2. Perception system desired optimal coverage area

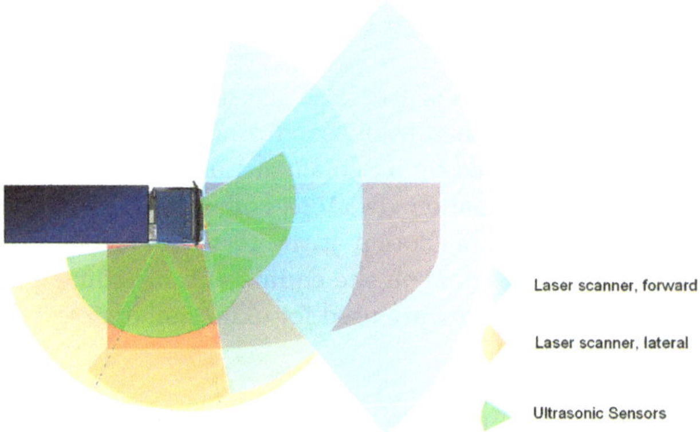


Fig. 3. Object sensor field of view

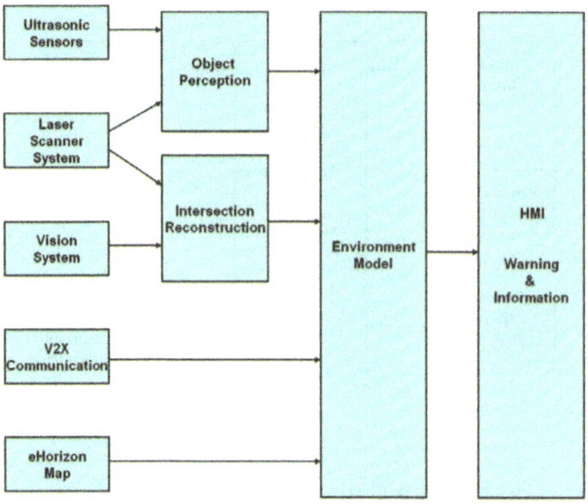


Fig. 4. System architecture overview

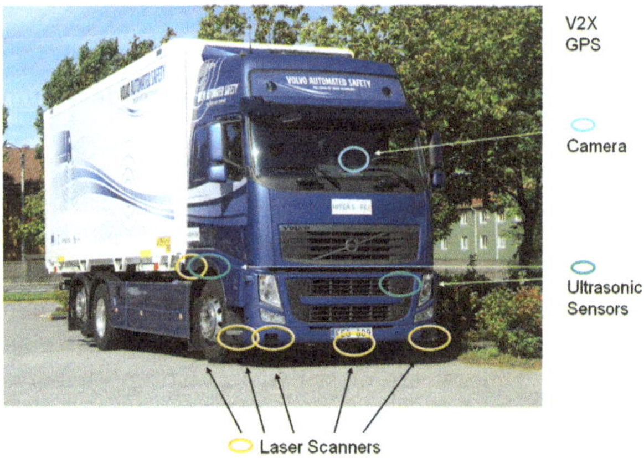


Fig. 5. Demonstrator truck

## 2.2 Infrastructure-Based System and Test Site Gothenburg

In addition to the vehicle-based sensor, also a test site at a public intersection in Gothenburg, Sweden has been established. The existing traffic signal controller has been complemented by a Signalbau Huber traffic signal controller capable of transferring information via a V2X communication link. Fig. 6 gives an architectural overview of the set-up at the test intersection. Note that the parts denoted in grey are original on-site equipment, including a pair of radar sensors to detect pedestrian, that are used for extension of green time for slow moving traffic pedestrians.

The information to be transferred from the test intersection includes traffic signalling phase, pushbutton requests, and pedestrian presence observed by the roadside radar detector units.

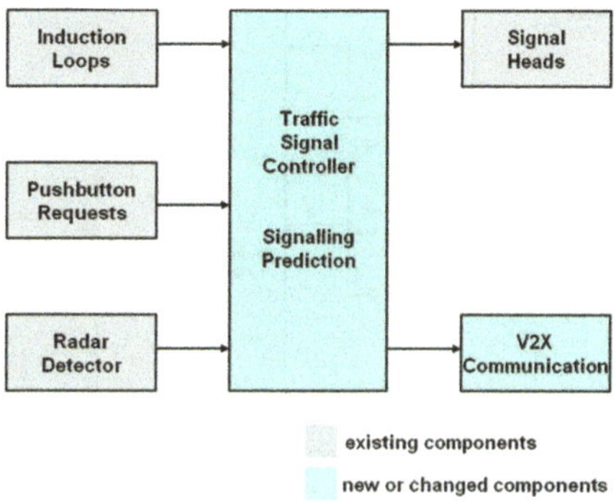


Fig. 6. Intersection test site Gothenburg: architecture overview

## 2.3 Situation Assessment

The situation assessment is currently still under development in the INTERSAFE-2 project at the time of writing.

For the situation assessment, the host vehicle path is predicted and evaluated against the sensor detections of VRU motion. During the right turning manoeuvre, the environment on the front and right side is constantly monitored. Stationary and moving objects are distinguished, but both are taken into account for the risk assessment, although particular regard is given to moving objects, such as VRU. Objects are classified to distinguish between VRU or other objects, like infrastructure elements, by the sensor system. From the path of the host vehicle and the motion of the VRU (or other object) the risk

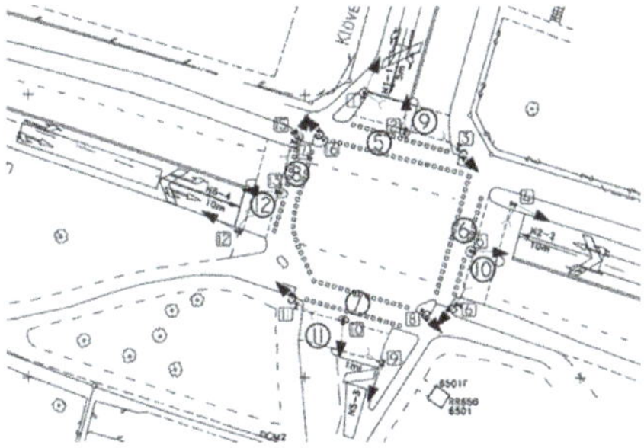

Fig. 7. Test site Gothenburg

of collision is calculated and information and warning output to the driver are triggered accordingly. In order to better understand the current driving situation, additional sources of information are used. These include the ADAS map for an a-priori knowedge about the presence of the intersection and an overview about the road layout. The V2X communication module gives information about traffic signal states, pushbutton requests, and roadside observations. The intersection reconstruction module gives for example stop line position or lane direction arrow indication.

### 2.4 Driver Interface

For the application's HMI (Human-Machine Interface), a two-step driver interface is being developed, based on the continuous assessment of presence of VRU and the risk of a collision:

- Information (about the presence of a VRU in the in the relevant area)
- Warning (about an upcoming collision risks due to the host vehicle's and the VRU's path of motion).

For the HMI, an overview of strategies and guidelines has been given in [3]. Several methods are currently under investigation and implementation. However, the primary project focus in the Intersafe-2 project is on developing the sensing and assessment system, rather than the HMI.

## 3 Conclusion

This contribution describes the intersection safety system currently under development in the INTERSAFE-2 project. Intersection safety is a very important factor for heavy goods vehicles, particularly in urban environments together with vulnerable road users (VRU).

In this paper, the current development is described with an overview about system set-ups and architectures both for the in-vehicle system and for the roadside system at the project's test site Gothenburg. The concept derived allows for a modular functionality both being able to use additional information from roadside sensors and to work solely based on the on-board sensor system.

## Acknowledgement

The work described in this contribution has been co-funded by the European Commission, in the project INTERSAFE-2 within the Seventh Framework Programme.

## References

[1] INTERSAFE-2 project web site http://www.intersafe-2.eu/

[2] Ahrholdt, M., Grubb, G., Agardt, E., Intersection Safety for Heavy Goods Vehicles, Meyer, G. et al. [Eds.], Advanced Microsystems for Automotive Applications 2009, Springer, Berlin 2009.

[3] INTERSAFE-2 Deliverable D3.1, User needs and requirements for INTERSAFE-2 assistant system, available via http://www.intersafe-2.eu/

**Malte Ahrholdt, Grant Grubb, Erik Agardt**
Volvo Technology Corporation
40508 Gothenburg
Sweden
malte.ahrholdt@volvo.com
grant.grubb@volvo.com
erik.agardt@volvo.com

**Keywords:** intersection safety, heavy goods vehicle, INTERSAFE-2

# Progress of Intersection Safety System Development: Volkswagen Within the INTERSAFE-2 Project

J. Knaup, S. Herrmann, M.-M. Meinecke, M. A. Obojski, Volkswagen AG

## Abstract

A high number of accidents in the European countries happens at intersections. The Volkswagen AG is building up a demonstrator for developing intersection safety related driver assistance functions within the project INTERSAFE-2. These functions will be a contribution to make driving in urban intersection scenarios safer. The sensor setup of the demonstrator, the system architecture with its environmental models and the single functions are described in this paper.

## 1 Introduction

In the current decade environment perception became more and more important for automotive applications such as driver assistance systems, autonomous driving, pre-crash etc. After addressing various applications of longitudinal vehicle control (like ACC) support at intersections came into the focus of research activities nowadays, which is addressed in this paper.

Volkswagen AG is developing a demonstrator with on-board sensors as well as communication with infrastructure units to show driver assistance functions in selected scenarios at intersection. This work is developed within the project INTERSAFE-2, which is co-financed by the European Commission. First the accident analysis is presented, which was used to identify significant risks at intersections. In the third chapter the sensor setup and the system architecture of the Volkswagen demonstrator are described. Based on the accident analysis scenarios were classified and assistance functions were created, which are delineated in chapter four.

## 2 Accident Analysis

The project INTERSAFE-2 began with a detailed accident analysis. The conducted accident analysis included results from other projects like INTERSAFE and SAFESPOT. This data was updated and adapted to the frame conditions of INTERSAFE-2, namely accidents with passenger cars and trucks.

The available accident databases on European level are quite different concerning their quality of data. The degree of details and volume of database entries are different as well. Besides official statistics, which are supported by police accident reports and published by the national and European Authorities, databases like GIDAS (German in-depth accident study) [1] were considered. GIDAS offers a wide spectrum of accident describing attributes since 1973. Other useful databases in this context are ONISR from France, OTS from UK, and STRADA from Sweden.

It is a fact that national or local databases are valid for the area in which accidents are monitored and cannot be extrapolated to the complete area of EU-27. Nevertheless it was possible to find out interesting characteristics of accidents, which happened at intersection-related scenarios.

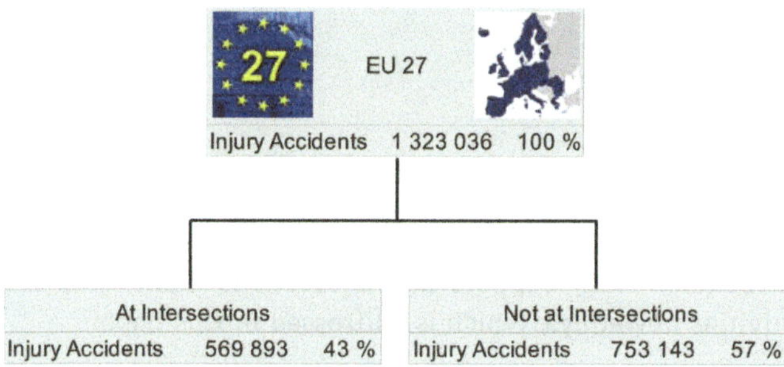


Fig. 1. Statistics about counted accidents at and not at intersections (EU-27)

In Fig. 1 can be seen that a very high proportion of accidents occurs in intersection-related scenarios – found to be 43%. Consequently, this is an important argument to focus safety research activities on these situations.

After screening all accident scenarios which have a significant probability of occurrence clustering is to be conducted to provide a clearer view on what topology safety systems require. The analysis of collisions of passenger cars and trucks were conducted separately, because the characteristics is completely different. Source for the analysis for passenger cars was GIDAS. Accident hotspots are scenarios like straight crossing, turning left and turning right in which collisions very often occur. Sometimes this goes along with a right of way violation or red traffic light violation. Accidents with pedestrians involved occur mainly in turning to left and right scenarios. A detailed description of the accident analysis, user needs and requirements is given in [2].

## 3 Volkswagen Demonstrator

This chapter describes the scenarios addressed by Volkswagen Demonstrator and the resulting system design. A short overview of the used sensors is given. Finally the system architecture and the fusion design are briefly outlined.

### 1.1 Addressed Scenarios

In Fig. 2 the addressed scenarios of the Volkswagen Demonstrator are shown. From these scenarios five functions are derived: Left Turning Assistance, Crossing Traffic Assistance, Right Turning Assistance, Right-of-Way Assistance and Stop Line / Traffic Light Assistance. These functions are described in section 4.

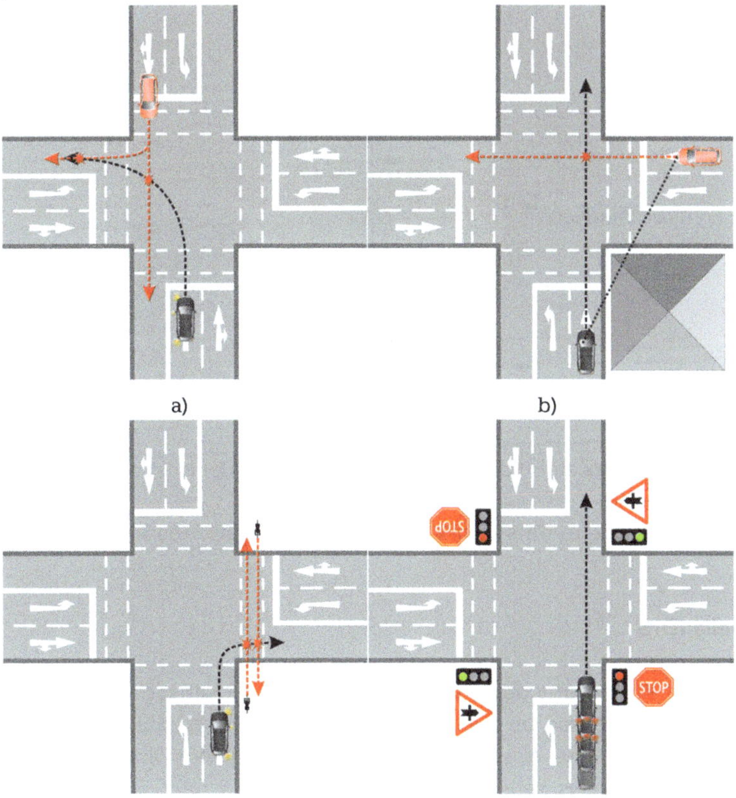


a) b) c) d)

Fig. 2. Addressed scenarios: a) left turning, b) crossing, c) right turning, d) right of way/stop line/traffic light

### 3.2 Sensor Setup

Based on the addressed scenarios a sensor setup was developed. The combination of the five scenarios results in a sensor setup which nearly covers 360° around the car. The selected sensors are sketched with their mounting positions and the field of view in Fig. 3.

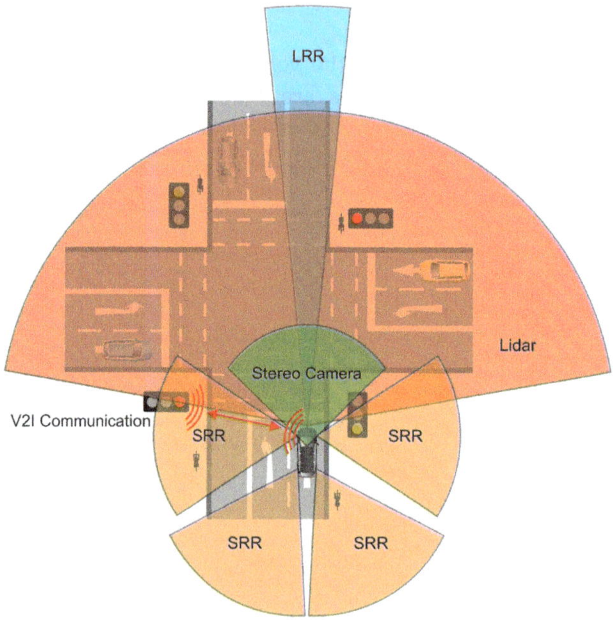


Fig. 3. Sensors and their coverage of Volkswagen demonstrator

In the following the main sensors of the setup are described.

#### Lidar Hella IDIS

The selected lidar (light detection and ranging) is a sensor from Hella KGaA Hueck & Co. It is based on a sensor, which proves its reliable performance in series application. The sensor uses the laser technology and the measuring principle of the series sensor. For reaching a wide field of view a scanner principle is used. The lidar is used for measuring the static and dynamic environment of the intersection. Objects are tracked in lateral and longitudinal direction. This data is used in all addressed scenarios of Fig. 2.

### Stereo Camera

In order to cope with the complex nature of the intersection a stereo camera system is used. It perceives the environment in two modes. Firstly, a structured approach for scenarios, where the road geometry is estimated from lane delimiters providing the parameters of the lane and the position, size, speed and class of the static and dynamic objects. Secondly, an unstructured approach, where the road geometry is estimated from elevation maps providing an occupancy grid having the cells classified as free space, obstacle areas, curbs and islands etc. The fusion of the two methods will provide completeness, increased accuracy and robustness of the environment representation. A complete overview of the stereo camera system is given in [3]. The stereo camera system is used for detecting road users in all addressed scenarios in Fig. 2 and for relative localisation of the Volkswagen demonstrator.

### Short Range Radar

The selected short-range radar (radio detection and ranging) is a sensor from Smart Microwave Sensors GmbH (SMS) in Brunswick, Germany [7]. The sensor is dedicated for short and mid range applications and is shown in Fig. 4. It operates in the 24 GHz ISM band and has been deployed on vehicles for blind-spot assistance. The short-range radars are used for detecting crossing traffic in crossing scenarios without occlusion (Fig. 2b) and bicyclists coming from behind in right turning scenario (Fig. 2c).

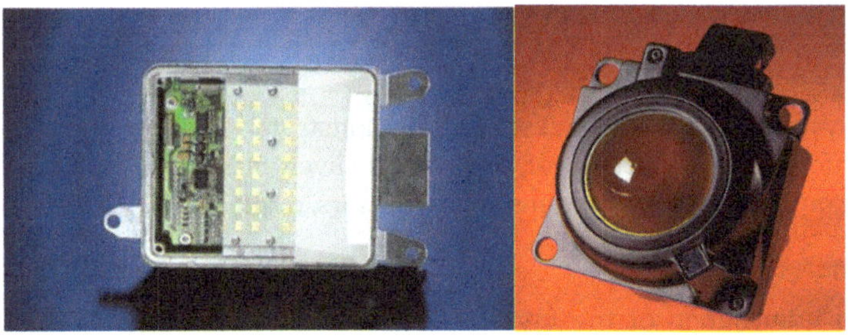

Fig. 4. SMS 24 GHz mid range radar (left) and TRW 77 GHz long range radar AC20 (right)

### Long Range Radar

The long-range radar selected for the Volkswagen demonstrator is a sensor from TRW. It operates in the 77 GHz band, which is designed for long range detection like oncoming traffic in the left turning scenario. It is in series production for ACC and is shown in Fig. 4.

### GPS and ADAS Map

An absolute localisation of the host vehicle while approaching and within the intersection area can be achieved using GPS and onboard maps. GPS (Global Positioning System) navigation is the de facto standard navigation technology currently in use and is widely spread amongst a variety of applications including vehicle navigation. While its accuracy is generally sufficient for route navigation purposes, it is not sufficient for accurate localisation. To overcome this limitation the standard GPS system is often supplemented by the use of a locally transmitted correction signal to increase accuracy – this system is known as D-GPS (Differential GPS) and is widely used in modern ADAS (Advanced Driver Assistance System) applications. Typical accuracies using GPS are 15 to 25 m. For DGPS the expected accuracies in standard automotive traffic are 0.5 to 15 m dependent on the number of visible satellites.

The ADAS map technology selected for INTERSAFE-2 is the Navteq ADAS-RP navigation map [4] that has been used by a number of similar ADAS projects. Such ADAS maps can be used along with a global localisation system to provide an "electronic horizon". In addition to the features available within standard maps, ADAS-RP maps also include extended attributes for each road segment like curvature, slope or right-of-way. Especially the right-of-way information is used for dealing with the right-of-way and stop line scenarios. Topology information supports the host localisation based on map-matching technologies.

## 3.3 System Architecture

Several system parts of the Volkswagen demonstrator can be grouped in three layers: Information Source Layer, Perception Layer and Application Layer. The Information Source Layer contains single sensors, digital maps and communication link.

Based on information of the Information Source Layer the Perception Layer fuses different information to get one consistent representation of the environment. Therefore different fusion modules on low and high level are under development. They are responsible for host localisation and object detection, classification and tracking.

The Application Layer is based on the environmental model of the Perception Layer. In this layer the Scenario Interpretation and the Risk Assessment for each function are applied. Information and Intervention Strategies use the output of the Risk Assessment to control the actuators and the HMI.

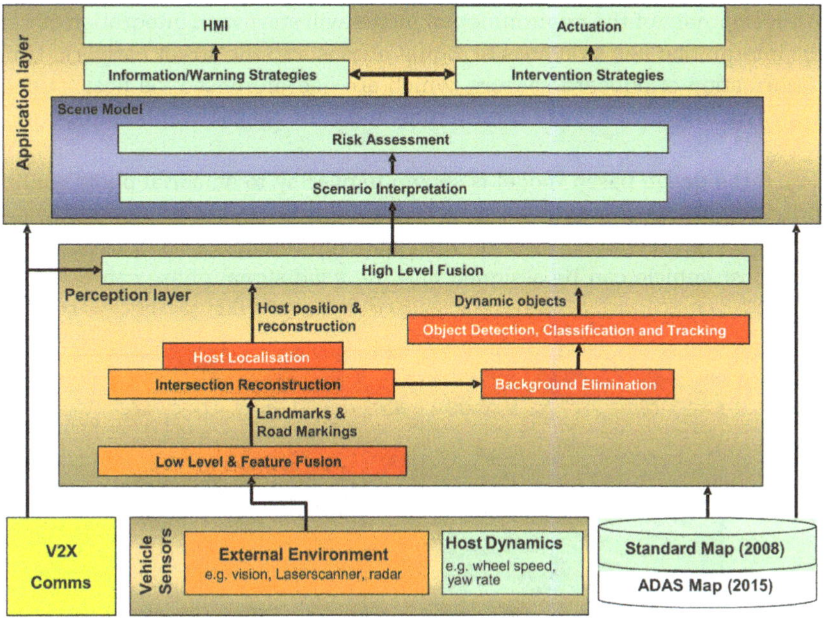


Fig. 5. System architecture overview

### 3.4 Environmental Models

Complex urban intersection scenarios addressed by INTERSAFE-2 require an extension of the state-of-the-art environmental models. In addition to well known models like object based [5] or grid based [6] models a graph based model are being investigated. Object based and grid based models process the information of the onboard sensor system. For urban scenarios it is very helpful to use a-priori knowledge of digital maps and information send via communication.

For integration of a-priori knowledge of digital maps and online information from onboard sensors and communication this graph based model is under development. This model uses a directed graph for modelling the environment on lane-level. For initialisation the directed graph is build up based on the digital map of the ADAS-RP. The detailed intersection topology is integrated if corresponding information is transmitted via a communication channel. Additional data of onboard sensors like positions of stop lines or dynamic objects is integrated into the graph. All this information has to be georeferenced for storing it in the graph. From this it follows that the level of complexity for integration of the information is very different.

The development of the environmental model will start with integration of geostatic information like stop lines or signal phases into the street network. Later on information of onboard sensors, which are georeferenced via host localization and relative distances of information and host vehicle will be integrated.

In Fig. 6 the graph based model is shown in overlay to an aerial photo. Online information of actual signal phase of traffic light is integrated and associated with the street network. Based on map matching methods the relevant edge for the host vehicle can be assigned and the valid signal phase can be identified.

Fig. 6. Graph based model with integrated online information about traffic light phase (green and red cylinders)

## 4 Functions

This chapter depicts the five functions left turning, crossing traffic, right turning, right of way / stop line and traffic light assistance that are addressed within the INTERSAFE-2 project by Volkswagen AG. The selection of these functions is based on scenarios with a high risk of major accidents and the analysis of human factors that may cause collisions at intersections. After a description of these scenarios the possible methods integrated in the Volkswagen demonstrator to avoid collisions are introduced.

### 4.1 Addressed Scenarios

The **left turning assistance** function addresses the scenario pictured in figure Fig. 2a. The driver has a green traffic light and enters the intersection to turn left. He has to yield oncoming traffic, i.e. road users turning right or going straight on from the opposite direction. The goal of this assistance function is to avoid collisions with oncoming traffic.

Fig. 2b shows the situation for the **crossing traffic assistance** function. The vehicle approaches an intersection without traffic lights and has to yield crossing road users coming from the left and right side. This function has the objective to avoid collisions with crossing traffic.

In Fig. 2c is the next scenario illustrated, which is addressed by the **right turning assistance** function. The driver has a green traffic light and enters the intersection to turn right. He has to yield road users going straight on at his right side. The goal of this function is to prevent collisions with pedestrians and cyclists going straight on from both directions.

Fig. 2d illustrates the last scenario, which is addressed by two functions. If the vehicle approaches an intersection without traffic lights and the traffic signs command the driver to yield crossing traffic the **right of way / stop line assistance** function acts in the situation. The goal of this function is the prevention of conflicts when road users do not respect right of way rules. The **traffic light assistance** addresses the situation when the driver is running into a red light violation. This function has the objective to avoid red light violations.

## 4.2 Collision Prevention Methods

The Volkswagen demonstrator is equipped with a LED strip that is mounted at the bottom of the windscreen. It is possible to enable parts of this optical **light information system,** for example the right area only. So this system can be used to guide the gaze of the driver from the left side to the scene at the right side of his car or vice versa. It is also possible to flash the middle of the windshield to draw attention to the area in front of car. The urgency of the situation can be expressed by the **acoustical information system** to help the driver understand that a collision might happen.

The braking system of the car can be used in three ways to inform the driver. First a short **brake jerk** can be engaged so that the driver receives a recommendation to brake. If the driver tries to stop the car, but the collision will still not be avoided the **brake support** system can help him by adjusting the brake pressure. If the driver is not braking and a collision is inevitable the system can engage an **emergency stop**.

The steering system of the Volkswagen demonstrator can initiate **steering wheel vibrations** to inform the driver or give **steering recommendations**.

The steering recommendation system directs the car to the right or left side to avoid a collision. Due to safety aspects the driver can always override the steering torque.

Depending on the time to collision and the reaction of the driver the assistance system uses different information techniques. Firstly it will engage the optical light system to notify the driver of a possible collision. If there is no driver reaction to avoid the crash the system uses acoustical information followed by haptic assistance functions. Finally it can engage an emergency stop and do a steering recommendation if the time to collision is very short or the collision is inevitable. At the moment the Volkswagen group research optimizes the selection of these prevention methods in the described scenarios.

## 5 Summary

Volkswagen is developing five functions in a demonstrator for pointing out the possibilities to support for the driver in intersection related scenarios. Dealing with the complex scenario of an urban intersection requires a highly sophisticated perception system and new approaches of data fusion and environmental modeling. First ideas of a graph based environment model in addition to object and grid based model were outlined. The selected assitance functions and the collision prevention methods were described. In the next few months all these concepts will be developed further. At the end of INTERSAFE-2 in May 2011 Volkswagen can contribute several features for making traffic in urban intersection scenarios safer.

## Acknowledgement

The work described in this contribution has been co-funded by the European Commission, in the project INTERSAFE-2 within the Seventh Framework Programme.

## References

[1] German In-Depth Accident Study - Accident Database 07/1999-12/2007, Dresden and Hannover, 2007.

[2] INTERSAFE-2 Deliverable D3.1 - User needs and requirements for INTERSAFE-2 assistant system, available via http://www.intersafe-2.eu/, 2009.

[3] Nedevschi, S., Danescu, R., Marita, T., Oniga, F., Pocol, C., Bota, S., Meinecke, M., Obojski, M., Stereovision-Based Sensor for Intersection Assistance, In: Meyer, G. et al., Advanced Microsystems for Automotive Applications 2009, Springer, Berlin 2009, pp. 129 - 163, 2009.

[4] EU-FP6 PREVENT integrated project, sub-project MAPS&ADAS.

[5] Becker, J.C., Fusion of Data from the Object-Detecting Sensors of an Autonomous Vehicle, IEEE International Conference on Intelligent Transportation Systems, 1999.

[6] Effertz, J., Autonome Fahrzeugführung in urbaner Umgebung durch Kombination objekt- und kartenbasierter Umfeldmodelle, Braunschweig, 2009.

[7] Mende, R., MMIC for a 24 GHz Automotive Radar Sensor, International Radar Symposium Hamburg, 2009.

**Jörn Knaup, Simon Herrmann, Marc-Michael Meinecke, Marian Andrzej Obojski**
Volkswagen AG
Postfach 1777
38436 Wolfsburg
Germany
joern.knaup@volkswagen.de
simon.herrmann@volkswagen.de
marc-michael.meinecke@volkswagen.de
marian-andrzej.obojski@volkswagen.de

**Keywords:** INTERSAFE-2, environment model, sensor, function, driver assistance, intersection, safety

# On-Board 6D Visual Sensor for Intersection Driving Assistance

S. Nedevschi, T. Marita, R. Danescu, F. Oniga, S. Bota, I. Haller, C. Pantilie, M. Drulea, C. Golban, Universitatea Tehnica din Cluj-Napoca

## Abstract

The problem of on-board intersection perception and modeling is complex and requires a wide field of view, dense and accurate data acquisition and processing, robust object detection and classification and fast response time. This goal can be achieved by using a large and redundant set of heterogeneous sensors and by fusing their information. Among the on-board sensors the visual sensors have the following main advantages: they are passive, and they provide the highest volume of information. The use of a pair of visual sensors in stereo configuration opens not only the possibility to infer the 3D coordinates for any image point but also the possibility to compute the 3D motion vector for any pixel. The exploitation of the motion information in driving assistance systems requires the estimation of the ego motion. This paper presents the architecture, implementation and use of a powerful on-board 6D visual sensor for intersection driving assistance.

## 1 Introduction

Intersections are the most complex traffic situations. With different crossing and entering movements by both drivers and pedestrians, intersections can be both confusing and dangerous. Due to the highly demanding nature of the scenario, and the consequences thereof intersection safety became a state priority and the core of several local or international, cooperative research projects.

Dealing with the intersection specific situations (detailed in the next section) imposes substantial changes in the physical setup and in the processing algorithms of a stereo sensor. The use of a wider field of view lenses (shorter focal length) comes at the cost of larger distortions and reduced depth accuracy. For this reason an increased accuracy in calibration and dense stereo reconstruction is required. Supplementary, motion information from optical flow must be used for a more accurate obstacle detection process. Static and dynamic obstacles can be distinguished if the 3D rigid motion (rotation and translation) of the stereo camera are known.

A 6D visual sensor was described by Franke et al in [1]. The proposed solution tracks points with depth known from stereo vision over consecutive frames and fuses the spatial (3D position) and temporal information (3D motion) using extended Kalman Filters. Sparse optical flow is computed using an optimized version of the KLT tracker. Improved accuracy is limited to a set of points of the image, points for which both stereo reconstruction and optical flow estimation is possible.

This paper presents a new architecture and implementation of a dense 6D visual sensor together with its use for driving environment description. The proposed 6D visual sensor consists of an improved accuracy stereo reconstruction engine, an accurate dense optical flow engine, and a visual odometry module. For this purpose an improved version of the Semi-Global Matching (SGM) method [2] was developed. For the dense optical flow computation the TV-L1 algorithm was selected, as one of the top-ranked modern methods for the flow estimation [3]. The visual odometry is implemented exploiting the available mono, stereo and optical flow information [4]. Taking advantage of the large computation power of current generation GPUs, a real time solution focused on quality is provided.

## 2 Requirements for the Stereo Sensor

The analysis of the user needs for Intersection Safety Assistance Systems identified the following driving assistance functions [5]: left turn assistance (LTA), intersection crossing assistance (ICA), right turn assistance (RTA), right of way and stop line assistance (SLA). The main roles of the stereovision sensor in an intersection driving assistance system are related to the sensing and perception in the front of the ego vehicle in a region up to 35 m in depth and a 70° horizontal field of view. The driving assistance functions are aimed to both static and dynamic elements of the intersection.

Static road and intersection environment perception functions are:

- Current and side lanes detection based on lane delimiters (lane markings, curbs, fences, traffic poles and trees);
- Target lane and side lanes detection based on same lane delimiters;
- Lane turning based on lane delimiters turning (or lane parameters changing);
- Stop line detection, pedestrian and bicycle crossing detection;

- Painted signs (stop, give way, turn right, turn left, and go ahead) detection;
- Lane turnings, stop lines, pedestrian crossings and painted signs will generate intersection hypotheses and will allow intersection corners detection;
- Static obstacle detection including parked vehicles, road construction sites delimiters.

Dynamic road and intersection environment perception functions are:

- Preceding, oncoming and crossing vehicles detection and tracking;
- Preceding, oncoming and crossing vulnerable road users detection and tracking.

Tracking of the dynamic objects will provide relative position, speed and acceleration information. Using them, the movement history of the ego and tracked vehicle is inferred.

The sensing and perception functions implementation depend on the type of elements of interest. The elements having a significant 2D nature like lane markings, painted road signs, stop lines, road crossings, and poles are first detected in the intensity images and localized using the stereo 3D information. The elements having a significant 3D nature like vehicles and vulnerable road users (pedestrians, bicyclists) are better detected in the 3D space. Here, the use of motion and intensity information is necessary for the refinement of the detections.

Due to variability of the driving scenarios complexity, three perception paradigms are necessary:

- The structured environment paradigm in which the road geometry is estimated from the lane/road delimiters and the obstacles are modeled as cuboids, having position, size, orientation and speed. This paradigm corresponds to un-crowded environments with visible lane delimiters.
- The unstructured environment paradigm in which the road geometry and obstacles are detected from the digital elevation map. This paradigm corresponds to overcrowded environments like intersections in which the lane delimiters are not visible. Starting from the representation of the dense 3D data as an elevation map, an occupancy grid which defines the drivable area, 3D curbs, traffic isles and obstacles is generated.
- The fusion of the results provided by the above mentioned paradigms will refine the results both in qualitative and quantitative parameters.

## 3 Stereo Sensor Architecture for Intersection Assistance

Based on the requirements analysis a two level architecture of a 6D stereo sensor is proposed (Fig. 1 and Fig. 2).

### 3.1 The Low Level 6D Stereo Sensor Architecture and Implementation

The low level architecture controls the image acquisition process and provides, after the sensorial data processing, the primary information needed by the high level processing modules: 6D point information (3D position and 3D motion), ego motion estimation and intensity images at a rate of 20 frames per second.

The image sensors can be integrated in a dedicated stereo-head or built from stand-alone cameras mounted rigidly on a stereo-rig. The cameras can be interfaced trough a frame-grabber or a common PC interface (Firewire, USB). The image acquisition is controlled by a software interface customized for the used cameras (Fig. 2). The quality of the acquired images is tuned through the Gain and Exposure parameters of the cameras for automatic adaptation to the lighting conditions [6]. The stereo images are acquired at the full resolution of the image sensors.

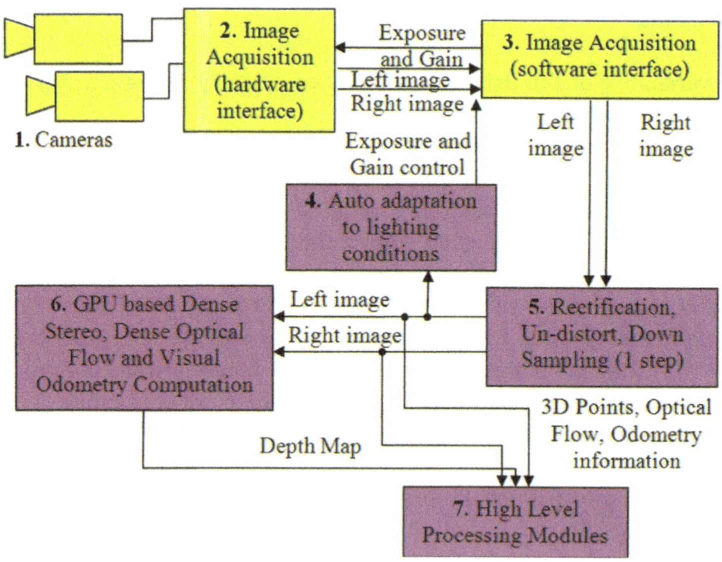


Fig. 1. Stereo sensor low level architecture

They are further rectified (to meet the canonical configuration), corrected (to eliminate the lens distortions) and down-sampled (to meet the processing capacity of some hardware dense-stereo reconstruction engines – e.g. 512 pixels in width). The down-sampling step is optional in the case of our GPU based stereo engine as it is able to accommodate a wide range of (variable) image resolutions and up to 128 disparities.

The rectified (and possibly) down-sampled images are fed to the stereo reconstruction module which is implemented in hardware. From the depth map provided by the stereo-engine the 3D coordinates of the reconstructed points are computed knowing the camera parameters and are further provided to the processing modules (Fig. 1). For a stereo setup using 6.5 mm lenses, 2/3" imagers and a baseline of 350 mm a stereo detection area with the following parameters is obtained: reliable depth range: 0.7 ... 35 m relative to the ego car front, horizontal field of view: 72 deg (i.e. delimiters of the current lane are visible at 1.0 m).

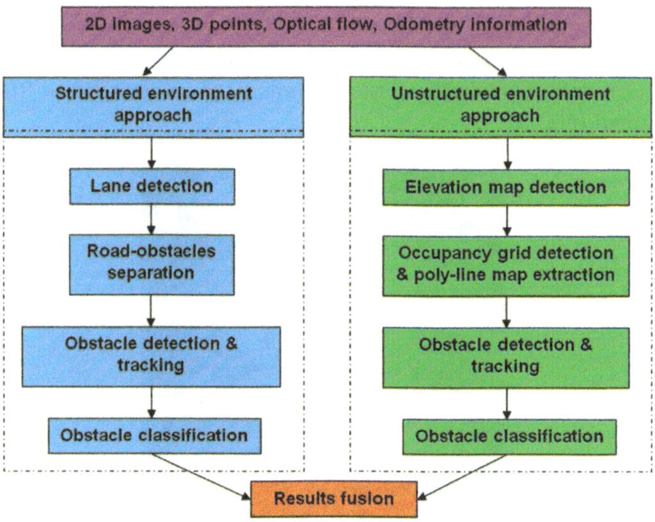


Fig. 2. Stereo sensor high level architecture

Real-time dense stereo reconstruction is a difficult problem due to the high computational complexity. In order to meet the accuracy requirements and execution time related constraints, we use a modified variant of the Semi-Global Matching (SGM) method implemented on a current generation GPU. SGM is able to handle difficult conditions better than the local methods used in other solutions because it can infer information in featureless or repetitive regions. This feature allows the system to be less dependent on the input data quality, a critical constraint for automotive systems.

The research was focused to improve the performances of the algorithm related to: running time, integer disparity accuracy and sub-pixel disparity accuracy. Optimizations were performed on both the algorithm and the implementation to reduce the running time. The modified SGM algorithm was verified to have the same accuracy as the original algorithm while reducing the computational cost of the Semi-Global step by a factor of two. Some solutions used by local methods to reduce the number of errors were also adapted. These include the matching metric and the error detection mechanisms. The sub-pixel accuracy is required because the system operates in the metric space. To improve the sub-pixel quality beyond existing solutions, a new interpolation technique was devised. It was created by modeling the output data and results in a significant reduction in depth error.

The final stereo processing requires less than 20 ms to generate the 3D points for the entire image. The running time can be further reduced by using a user specified Rectangle/Region of Interest (ROI). The density of the points is consistent across the image and even weakly textured areas are reconstructed. While reconstruction errors are below 2%, the density of reconstructed 3D points is almost doubled (Fig. 3).

a)
b)
c)

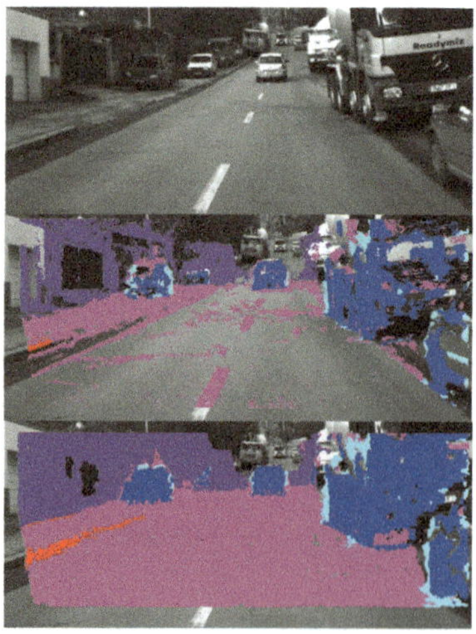

Fig. 3. Density of the reconstructed 3D points: a) intensity image, b) local method results, c) proposed SGM method results.

Dense optical flow estimation is highly complex and computationally intensive. Sequential implementations, for increased accuracy dense optical flow algorithms like TV-L1, cannot reach real-time performance. The high degree of parallelism of the problem favored a GPU-based implementation which boosted the performance to meet real-time requirements. Further more, optical flow and stereo reconstruction can be performed simultaneously, in one step, requiring less than 20 ms per frame.

Optical flow between two images represents the combination of scene motion and camera (ego-vehicle) motion. Ego motion aims to determine the 3D rigid motion (rotation and translation) of the camera from an image sequence in order to isolate it from actual motion in the scene. The vehicle motion can be estimated using information provided by the car sensors: yaw rate, speed and global time. There are some limitations however. The most severe one is that only one of the three rotation angles is accounted for. The use of an inertial sensor can overcome this shortcoming but has a lower accuracy compared to image processing methods. In order to reach maximum accuracy the six degrees of freedom of the ego-motion have to be determined precisely using visual odometry. When enough static points are present, the rotation matrix is computed via RANSAC exploiting the available optical flow information. The translation vector is determined from stereo-information. Similar to stereo reconstruction and optical flow, the ego motion estimation problem was parallelized and solved on using the GPU, in a second step, after performing the stereo reconstruction and optical flow computation.

### 3.2 The High Level 6D Stereo Sensor Architecture and Implementation

Using the rich output of the low-level architecture the two environment descriptions (structured and unstructured) are generated [7]. A complete structured description of the environment can be constructed using lane perception, painted road object perception, obstacle detection, obstacle tracking and obstacle class detection. The perception of the current and neighboring lanes is performed using a model-based probabilistic tracking technique [8] that uses the 3D information only as a raw filter selector (selection of road edges), the rest of the computations being performed in the image space. The 3D lane parameters are automatically updated through the filtering process, by mapping the 3D lane hypotheses in the image space and comparing them to the measurement features. When the lane geometry is unfitted to model-based tracking, the painted road objects can be directly detected and classified, and their position and class can be interpreted in order to help the system's understanding about the nature of the road environment and our position in relation to this environment.

Fig. 4. Detection in the image space and localization in the 3D space: painted road objects.

The detection of painted road objects (Fig. 4) starts from the road edges in the image space. First, the road edges are used to divide each horizontal line of the image into segments. For each segment, the average intensity is computed. The segments having the intensity lower than one of their neighbors are discarded, and so are the segments that are too wide or too narrow. The line segments that pass the tests can be used for preliminary image segmentation. This segmentation, however, can miss some of the object parts, or can add some noisy features. For this reason, an additional segmentation step, based on the intensity value of the pixels, is used. The initial segmentation is used to train the acceptable levels of intensity for road markings, and then the pixels are classified whether they comply with the marking levels or not. Following segmentation, a labeling algorithm is applied and the objects are identified. The objects are subjected to a classification algorithm, which assigns them a class such as: lane marking, right arrow, left arrow, forward arrow, stop-line, crossing etc. The 3D parameters of the object are recovered after labeling, by back-projecting the 2D image points in the 3D space. This can be done using the available 3D information or the road surface parameters.

Unlike painted road objects, vehicles and vulnerable road users have a significant 3D shape and are better detected in 3D space. The use of 2D features is necessary for the refinement of the detections. The obstacle detection algorithms analyze, on the compressed top view of the scene, the local density and vicinity of the 3D points, and determine the occupied areas which are then fragmented into obstacles with cuboidal shape: without concavities and only with 90° convexities [9]. Intensity, optical flow and ego motion information are used for discrimination at object boundaries where pure 3D based approaches fail (e.g. multiple pedestrians crossing the road: two pedestrians could be sometimes detected as a single obstacle; see Fig. 5). The fragmentation of the occupied areas into the individual cuboidal obstacles is the next step.

The orientation of the obstacles is determined in order to get a very good fitting of the cuboidal model to the obstacles in the scene ahead and, consequently, to minimize the free space which is encompassed by the cuboids. Using the motion information the orientation and speed of each obstacle is accurately determined and is less susceptible to noise.

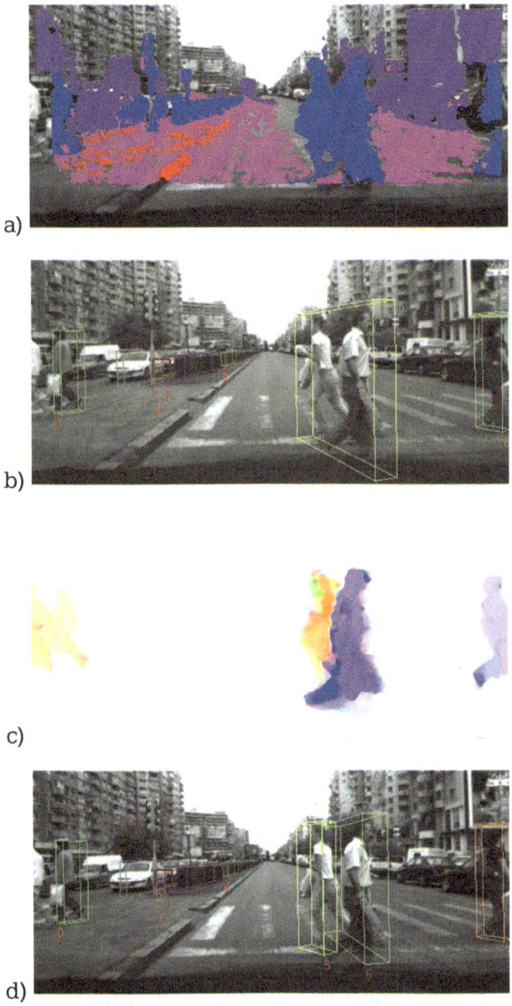


Fig. 5. 3D points grouping into obstacles: a) 3D points superimposed over the grayscale image, b) obstacles detected solely based on 3D information, c) color-coded representation of optical flow: hue encodes flow direction, intensity flow magnitude, d) obstacles detected based on 3D position, 3D motion and intensity information.

The intersections in urban areas often have no lane markings or these are occluded by crowded traffic. The lanes cannot be detected in such particular scenes, thus the obstacle/road separation becomes difficult. An alternative/complementary method must be used to detect elevated areas (obstacles), regions where the ego vehicle cannot be driven into. Complementary, the obstacle-free road areas can be considered as drivable. In these scenarios the unstructured approach provides better results.

Starting from the representation of the dense 3D data as a digital elevation map, an occupancy grid is computed. For each cell of the grid a label is associated: drivable (road), traffic isle, obstacle (static/dynamic, or others), etc [10]. Since the primary information used comes from reconstructed 3D points, the benefits of the new SGM based stereo are immediately harnessed. The high number of reconstructed 3D points increases the density of the computed elevation map, especially on the road surface where featureless regions lacked 3D reconstruction (Fig. 6). Another benefit is the increased accuracy of the DEM, coming from the use of more accurate input data. The use of optical flow together with ego motion information facilitates the discrimination between static and dynamic obstacles and gives the tracker valuable additional cues (speed, orientation). For further increasing the representation accuracy the 3D space or the occupancy grid can be integrated over time using the ego car's motion parameters.

a) b)

Fig. 6. Elevation map improvement: a) using local stereo algorithm, b) using improved SGM algorithm.

## 4 Conclusions / Discussions

The proposed 6D vision sensor provides a wide range of valuable information to the intersection driving assistance system. The first major benefit of using this approach is that all points are more reliably reconstructed. Thanks to several improvements made to the integer disparity accuracy and sub-pixel disparity accuracy of the original SGM method reconstruction errors dropped below 2%. Secondly the density of points is increased by a factor of two because this method is not limited to a finite region and can infer information in featureless or repetitive regions where local methods fail. The third benefit is coming from the dense motion vectors which can be used together with the ego motion parameters to discriminate between static and dynamic objects, different speed objects, object tracking and temporal fusion. Even though it lacks the accuracy of the radar and of the laser sensors, it compensates by the richness of data, and by the endless possibilities of high level processing. The fusion of the 6D visual sensor results with active sensors (e.g. laser), and with map information, will ensure a complete and accurate description of the intersection environment.

## Acknowledgement

This work was conducted within the research projects INTERSAFE-2 and PERSENS. INTERSAFE-2 is part of the 7th Framework Programme, funded by the European Commission. The partners of INTERSAFE-2 thank the European Commission for supporting the work of this project. PERSENS is a research grant funded by Romanian Ministry of Education and Research in the PN II IDEI program, ID 1522/2009.

## References

[1] Franke, U., Rabe, C., Badino, H., Gehrig, S, 6D-Vision: Fusion of Stereo and Motion for Robust Environment Perception, Pattern Recognition, DAGM Symposium 2005, pp. 216-223, 2005.

[2] Hirschmüller, H., Accurate and Efficient Stereo Processing by Semi-Global Matching and Mutual Information, IEEE Computer Society Conference on Computer Vision and Pattern Recognition CVPR '05, vol. 2, pp. 807-814, 2005.

[3] Zach, C., Pock, T., Bischof, H., A Duality Based Approach for Realtime TV-L1 Optical Flow, Pattern Recognition, pp. 214-223, 2007.

[4] Nedevschi, S., Golban, C., Mitran, C., Improving Accuracy for Ego Vehicle Motion Estimation using Epipolar Geometry, in Proceedings of 2009 IEEE ITSC, pp. 596-602, 2009.

[5] INTERSAFE-2 Consortium, User needs and requirements for INTERSAFE-2 assistant system, Deliverable D3.1, available via http://www.intersafe-2.eu/public/ cited 8th February 2010.

[6] Negru, M., Nedevschi, S., Camera Response Estimation. Radiometric Calibration, IEEE International Conference on Intelligent Computer Communication and Processing (ICCP), pp. 103-109, 2009.

[7] Nedevschi, S., Danescu, R., Marita, T., Oniga, F., Pocol, C., Bota, S., Meinecke, Obojski, M.A., Stereovision-Based Sensor for Intersection Assistance, in Meyer, G. et al., Advanced Microsystems for Automotive Applications (AMAA) 2009: Smart Systems for Safety, Sustainability and Comfort, Springer, Berlin, pp.129-164, 2009.

[8] Danescu, R., Nedevschi, S., Probabilistic Lane Tracking in Difficult Road Scenarios Using Stereovision, IEEE Transactions on Intelligent Transportation Systems, vol. 10, pp. 272-282, 2009.

[9] Pocol, C., Nedevschi S., Meinecke, M.M., Obstacle Detection Based on Dense Stereo for Urban ACC Systems, Proceedings of WIT 2008, pp. 13-18, 2008.

[10] Oniga, F., Nedevschi, S., Meinecke, M.M., Binh, T.T., Road Surface and Obstacle Detection Based on Elevation Maps from Dense Stereo, IEEE ITSC, pp. 859-865, 2007.

**Sergiu Nedevschi, Tiberiu Marita, Radu Danescu, Florin Oniga,**
**Silviu Bota, Istvan Haller, Cosmin Pantilie, Marius Drulea, Catalin Golban**
Universitatea Tehnica din Cluj-Napoca
15 C. Daicoviciu
400020 Cluj-Napoca
Romania
sergiu.nedevschi@cs.utcluj.ro
tiberiu.marita@cs.utcluj.ro
radu.danescu@cs.utcluj.ro
florin.oniga@cs.utcluj.ro
silviu.bota@cs.utcluj.ro
istvan.haller@cs.utcluj.ro
cosmin.pantilie@cs.utcluj.ro
marius.drulea@cs.utcluj.ro
catalin.golban@cs.utcluj.ro

**Keywords:** stereovision, optical flow, 6D vision, intersection, lane detection, obstacle detection, tracking, classification, elevation maps, occupancy map

# Cooperative Intersection Infrastructure Monitoring System

M. Kutila, P. Pyykönen, J. Yliaho, VTT Technical Research Centre of Finland
B. Rössler, Sick AG

## Abstract

This article proposes an advanced roadside monitoring sub-system for future cooperative intersection infrastructure. The sub-system consists of the camera for road condition monitoring, traffic light controller, Laserscanners, local dynamic maps database and V2I communication module. These components are interfaced to the high level fusion platform which is intended extract information and enable risk assessment if the situation is turning safety critical. The system is part of the European INTERSAFE-2 project which focuses on introducing cooperative traffic safety in intersection environment. Approximately every third of traffic incidences are intersection related. The project is expected emerge technology to influence beneficiary in 80% of the intersection related accidents with injuries and fatalities.

## 1 Introduction

Deploying cooperative traffic safety systems to vehicles has encountered kind of chicken-egg problem. Since amount of communication functionality in the vehicles will stay relatively small for next decades the motivation to pay extra due to communication is very low. Entering to positive deployment spiral is considered to be started by implementing such systems to the existing infrastructure first. Therefore, the first cooperative cars will use the infrastructure side detectors as additional equipment to support their own environment perception devices. This hopefully, brings critical mass to markets in order to have benefits of vehicle networks. The cooperative systems are seen one of key elements to improve traffic safety. Intersections are good example of local hot spots where the accident risk is high due to complexity of the environment.

The deployment of the INTERSAFE-2 system could provide a positive safety impact of 40% less injury accidents and up to 20% less fatalities in Europe. Under the INTERSAFE-2 project, the roadside equipment is developed in order to offer processed data for vehicles entering to the communication equipped intersection. The infrastructure sub-system merges data of the traffic light controller, road condition information, road user locations and speed, and the static map information to the package which assist vehicle to adapt speed and manoeuvring to approach intersection safely (see Fig. 1). The infrastructure side Laserscanners are used to detect pedestrians, vehicles and bicycles whereas camera system sniffs if the friction level of road surface is changing. Traffic light controller provides information of traffic signals and smartly controls traffic flow by minimising waiting times and vehicle accelerations. Static data (locations of curbs and poles) is available in the intersection model which is stored to the local dynamics maps database.

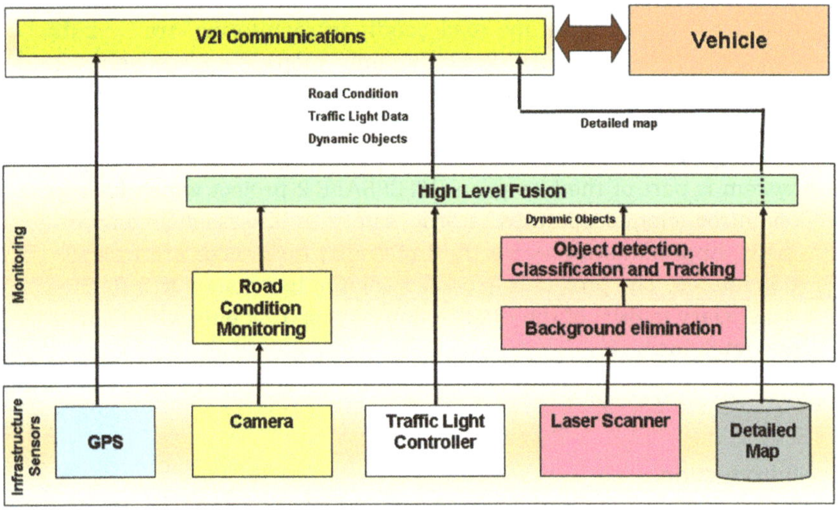


Fig. 1. Architecture of the infrastructure sub-system

## 2 State of the Art

FRICTION, SAFESPOT, TRACKSS and CVIS are examples of the prior projects funded by the European Commission which have engaged with development of cooperative traffic safety systems. One application area to support traffic safety is warning adverse road conditions which is related to over 3800 fatal accidents annually in EU-18 member states [1]. The FRICTION project developed an on-board technology in order to predict the available friction level [2]. On the other hand, the SAFESPOT project implemented technology which can be used also by roadside systems [3] to detect reduced friction levels. In both projects, VTT has developed the camera based low cost system since especially in the on-board case cost factor is the major constrain. Despite some similarities, specific solutions were needed for the two environments, which involve totally different operating conditions. New aspects to be considered for the intersection application are the complexity and dynamics of the scenarios, the geometry, and interaction between road systems and vehicle systems.

Casselgren et al. [4] have performed exhaustive investigations concerning the changes of light intensity with varying incident angle and the spectrum of light when asphalt is covered with water, ice or snow. The wavelength band in their experience varied from 1100 to 1700 nm. They observed that the snow reflection ratio drops consistently at wavelengths above 1400 nm, and as a conclusion they proposed two different bands in the NIR region (below and above 1400 nm) to classify road conditions. The main intention was to build a photodetector for automotive applications, called RoadEye. The advantage of this approach is that the equipment is robust and relatively cheap, but the drawback is the measuring area limited to one small spot on the road. This is not practical for the roadside system since whole intersection area needs to be monitored. The Finnish company Vaisala Oyj has a product which measures different road conditions like water, ice, snow and can even estimate a friction level [5]. Their DSC111 camera system uses a spectrometric measuring principle and thus, probably operates in the same wavelength bands as the Volvo's RoadEye system [4] which has been earlier used as a reference system.

## 3 Intersection Infrastructure Components

### 3.1 Road Condition Monitoring Data

Road condition monitoring system, called IcOR, bases on stereo camera arrangement. The reason to use stereo camera system (see Fig. 2) is not due to depth information but this is optimal solution for accurate spatial and temporal synchronization between captured images. This is key requirement for comparing polarisation differences between two images. The CMOS cameras are based on Micron MT9V032 imager element. For measuring illumination difference between cameras, horizontal and vertical polarization filters are mounted in front of the optics. In order to improve performance during dark time the IR-filter has been removed from the cameras. On the other hand the polarization filters allow only narrow spectral band to pass to the detectors, which eliminates most of the harmful ambient light. One camera with horizontal field of view of 65 degrees acquire monochrome image with a maximum resolution of 640 x 480 pixels. For outdoor use, cameras are enclosed in the IP67 standard meeting water proof and heated housing.

Fig. 2. A RayMax 100 IR light (left) and stereo camera (middle) inside housing

Without additional light source, the system detects different road condition types with a satisfactory reliability level. For demanding illumination conditions the system requires specific illumination arrangement as illustrated in Fig. 2. A IR light source, RayMax 100, is installed at the opposite side of the intersection. The illuminator produces eye-safe invisible near infrared illumination (850 nm) and it requires 50W / 230V power source. For optimising illumination level the camera and light source are installed in same height and in same angle. Optimum height for road monitoring camera is from 2.5 meters to 3.5 meters depending on diameter of the intersection. For large intersections camera must be installed higher.

From system interface, it is possible to select e.g. lanes to be monitored from intersection area. In Fig. 3 three different areas have been selected for classification. Lane selection enables to monitor individual lanes and detect different road condition more precisely. The right side of Fig. 3 shows four different classification areas for graininess and polarisation values. Graininess and polarisation values are classified and these areas are marked with colour of classified road condition (dry, ice, snow, wet or unknown).

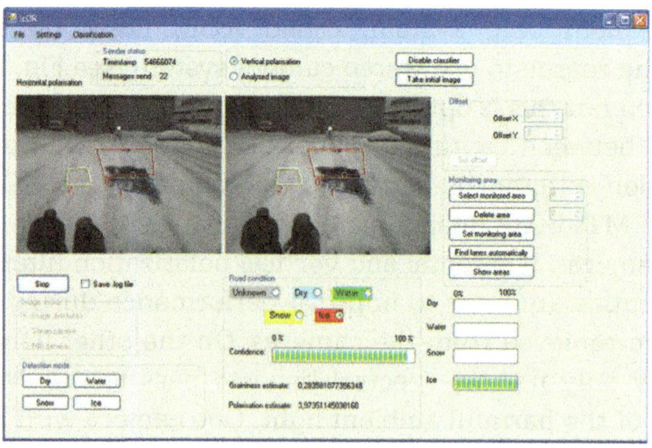


Fig. 3. User interface of road condition monitoring system

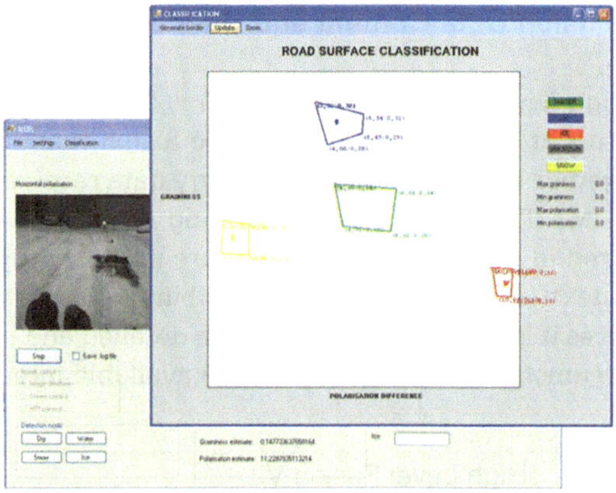


Fig. 3.1 Classification interface

Road condition monitoring system data contains time stamp in seconds, sensor location, classified road condition (dry, ice, snow, wet or unknown) and confidence level of the classified road condition. In test application maximum output frequency is 1 Hz and maximum detection range 5 meters.

## 3.2 Other Components

In the SAFESPOT project the Local Dynamic Map (LDM) concept has been introduced for detailed road and traffic related information [6]. In the local dynamic map, the road geometry from a standard digital map is integrated with the information collected by vehicles and the infrastructure. When comparing this information with the LDM specifications, conflict areas and especially unique map lanes are not included. In the SAFESPOT LDM specifications, lanes are not separate objects and their properties are linked to road elements. Other differences are e.g. the coordinate system in which paths are described.

The Laserscanner will be used in order to monitor road users in the intersection from the demonstrator vehicles and the infrastructure side. The road users (like vehicles, pedestrians, bicyclists) present at the intersection are tracked and classified based on the 3D Laserscanner measurements. The added value of the infrastructure integration is the capability to detect vulnerable road users.

## 4 High Level Fusion Data Content and User Interface

The aim of the High Level Fusion module (HLF) is to collect data coming from the different sensors at a certain moment and to fusion that data into a shared data structure together with intersection topology data received from the Local Dynamic Map (LDM). The present understanding of the principle of the HLF module is illustrated in Fig. 4. The data structure produced by the fusion process is called Context of Interest (COI). By combining the data received from the different sources it is possible to form a more detailed and improved picture of the existing circumstances depending on the available intersection data.

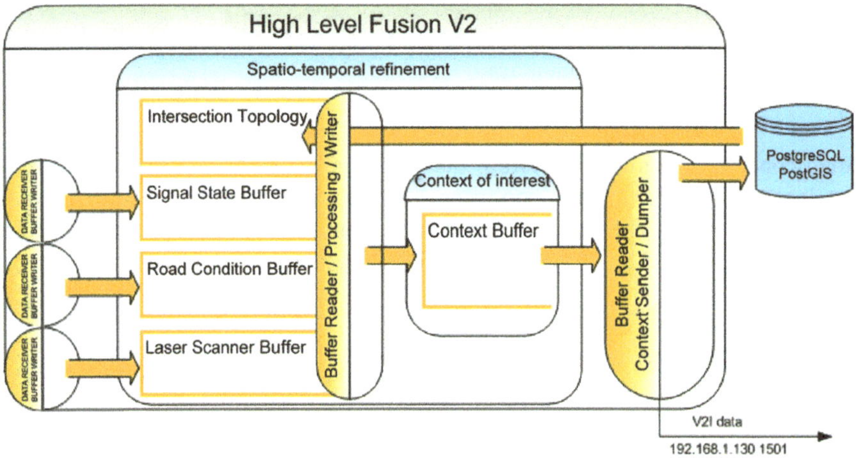


Fig. 4. Principle of High Level Fusion module version 2

Data input structures from the sensors as well as the intersection topology data from the LDM are described using the standardized ASN.1 metalanguage in accordance to the SIMTD specification [7]. The HLF module receives data from sensors through data receiver processes. Data is written into the shared memory buffers that enable data transfer and communication across process boundaries. This approach enables also the attachment of timestamps to the data packets as they are written into the shared memory buffers. This feature is used in the fusion to associate the contents of the buffers with a certain moment of time. The outlined user interface of the HLF is shown in Fig. 5 as it was at the time of writing.

Incoming data:

- **Traffic light controller data (Signal State Buffer)** data contains periodic information about the traffic light's signal state. Signal state buffer writer writes this data into the buffer approximately once in a second.
- **Road condition monitor data (Road Condition Buffer)** contains information about the road conditions in the intersection. Most critical pieces of information are the number of monitored areas, road condition type in the monitored areas and the position of the monitored areas. Road condition buffer writer writes this data into the buffer approximately once per second.
- **Laserscanner data (Laserscanner Buffer)** contains information about the detected objects in the intersection. First of all this data consists of information about age, time, classification, position, orientation, velocity and size of the detected objects. Laserscanner buffer writer writes this data into the buffer within 12,5-25 Hz frequency.
- **LDM data (Intersection Topology)** provides a formal representation of the surrounding intersection topology. The HLF module reads the intersection topology stored in an XML-file. This data has a more static character and is not read in repeatedly like the data to the other buffers.

Outgoing data:

- **Context of Interest (Context Buffer)** data structure retaining a sequence of information from the above-mentioned buffers at a certain moment of time. This data is sent to the Communication module and can be stored in the database.

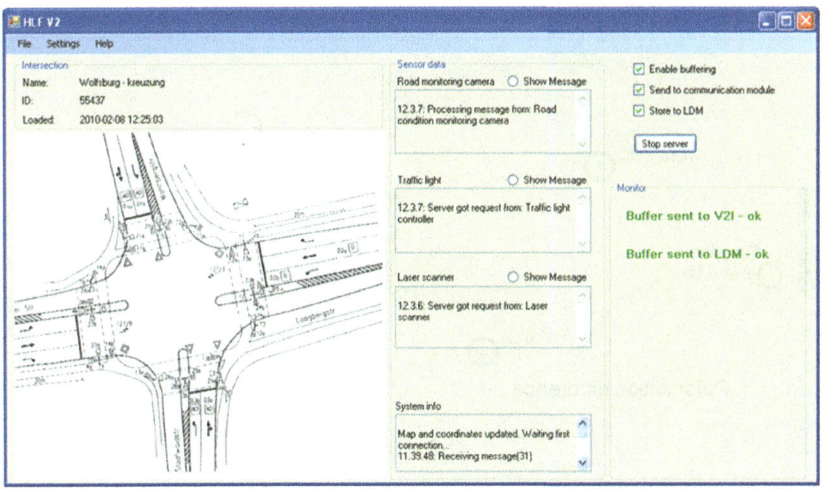


Fig. 5. User interface of High Level Fusion module version 2

## 5 Preliminary Results

The outdoor tests of the road condition monitoring system and effect of IR illumination were performed typical Finnish winter conditions in January 2010. Fig. 2 shows test situation and view from stereo camera. The test scenarios covered typical Finnish winter season alternatives dry, snowy and icy asphalt. Focus of the test was to investigate effect of the IR light source to separate different road condition types. For this purpose test sequences were captured with stereo camera with and without IR light source. The IR light source was 16 meter and monitored area 7.5 to 9.5 meter from stereo camera. Stereo camera was 3.2 meters and IR light source 2.2 meters above the ground.

In snowy area were selected 3 different areas. First area was shiny ice, second area dry asphalt and third area snow. These areas were 3 meters apart from the road monitoring camera. The IR light source was opposite site of the monitored areas and 4 meters away from the camera. Fig. 6A illustrates graininess and polarisation values without IR light source. Fig. 6B is the same situation but with IR light source. The obvious benefit of the IR light is in the polarisation difference. In Fig. 6A polarisation value for ice area of is slightly above 7. Using IR light polarisation value raise over 11 in Fig. 6B. With snowy and dry asphalt difference between polarisation values is not signifantly changed. This is mostly because ice reflects different wavelengths better than dry asphalt or snow. Also the graininess level of icy asphalt has slightly changed. Without the external light graininess is slightly above 0.1 and with the IR light about 0.15 whereas snowy and dry asphalt has not practically changed at all. The test indicates that the IR light strengths difference between road conditions.

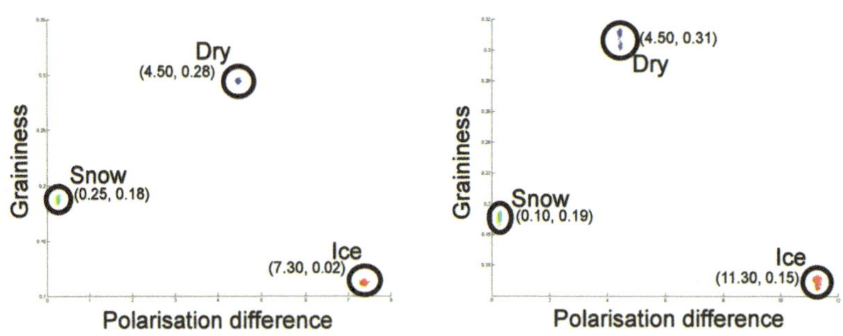


Fig. 6. Graininess and polarisation values for monitoring without IR light source (A, left side) and with IR light source (B, right side)

In Tab. 1 and Tab. 2 averages and standard deviation values of test samples are shown for different illumination conditions (with and without the IR light source). As there can be seen, the IR light increase polarization difference and graininess values for icy asphalt. In cases of snowy and dry asphalt the situation is relatively stable. In all cases standard deviation remains small indicating acceptable error for measurements. This test shows effect of the IR light source to increase chances to detect road conditions correctly.

| *Polarisation* | *Snow* | *Ice* | *Dry* |
|---|---|---|---|
| ***Average*** | *0,263* | *11,285* | *4,430* |
| ***Standard deviation*** | *0,012* | *0,053* | *0,024* |
| *Graininess* | *Snow* | *Ice* | *Dry* |
| ***Average*** | *0,192* | *0,150* | *0,310* |
| ***Standard deviation*** | *0,0016* | *0,0019* | *0,0034* |

Tab. 1. Average and standard deviation of polarisation and graininess with IR light source

| *Polarisation* | *Snow* | *Ice* | *Dry* |
|---|---|---|---|
| ***Average*** | *0,255* | *7,378* | *4,477* |
| ***Standard deviation*** | *0,010* | *0,031* | *0,018* |
| *Graininess* | *Snow* | *Ice* | *Dry* |
| ***Average*** | *0,188* | *0,113* | *0,295* |
| ***Standard deviation*** | *0,0016* | *0,0009* | *0,0007* |

Tab. 2. Average and standard deviation of polarisation and graininess without IR light source

## 6 Conclusions

The test results indicate that the road condition monitoring sub-system have taken significant steps forward compared to the system reported in [8]. The system is not just more accurate but also the adaptation capability with selecting multiple areas under analysis is improved. The adaptation capability is the most important feature when trying to improve systems performance in an intersection environment. Accuracy improvement is mostly the result of integrating active illumination to the road condition monitoring system (IcOR). The crucial future work is evaluate how much active illumination improves detection of wet asphalt for warning potential aquaplaning.

In addition to the IcOR improvements, data fusion framework is implemented for integrating heterogeneous data of the intersection infrastructure. The data from Laserscanners and local dynamic maps database is merged in the fusion level to be sent over a wireless link to the vehicles entering to the intersection area. The fusion level synchronizes the sensor measurement and increase confidence level to make data more consistent and reliable for vehicles to adapt the manoeuvring accordingly.

## Acknowledgements

We would like to express gratitude to the European Commission for funding the INTERSAFE-2. We also thank the partners of the aforementioned projects for interesting discussions related to the design and implementation of the intersection infrastructure sub-systems. Finally, we would like to assign special thanks to our colleagues Ms. Maria Jokela and Dr. Johan Scholliers due to hard work in developing the intersection safety components.

## References

[1] SafetyNet. Annual Statistical Report 2007, Deliverable No: 1.16. Data based on Community database on Accidents on the Roads in Europe – CARE. Available in http://ec.europa.eu/transport/road_safety/observatory/statistics/reports_graphics_en.htm, cited on 10th Feb 2010, 2007.

[2] Haas, T., Bian, N., Gamulescu, C., Köhler, M., Koskinen, S., Kutila, M., Jokela, M., Pesce, M., Hartweg, C., Casselgrein, J., Fusion of vehicle and environment sensing in Friction project, Proceedings of the 17th Aachen Colloquium "Automobile and Engine Technology" congress, Germany, Aachen, 6-8 Oct 2008.

[3] Kutila, M., Jokela, M., Burgoa, J., Barsi, A., Lovas, T., Zangherati, S., Optical road-state monitoring for infrastructure-side co-operative traffic safety systems, Proceedings of 2008 IEEE Intelligent Vehicles Symposium (IV'08), the Netherlands, Eindhoven. 4-6 June 2008.

[4] Casselgrein, J., Kutila, M. & Jokela, M., Road surface classification using methods of polarised light, IEEE Transaction on Intelligent Transportantion Systems, Special Issue on ITS and Road Safety, Revised on 27 Oct 2009, 2009.

[5] Vaisala Oyj web pages. www.vaisala.com, cited on 10th Feb 2010.

[6] Yuen, A., Brown, C., Zott, C., Hiller, A., Ahlers, F., Wevers, K., Dreher, S., Schendzielorz, T., Bartels, C., Papp, Z., Netten, B., Local dynamic map specification. SAFESPOT-EU-FP6 project, Deliverable 3.3.3., 2008.

[7] Hiller, A., Neumann, C., Matthess, M., Festag, A., Santos, H., Zhang, W., Sorge, C., Wiecker, M., Spezifikation der Kommunikatiosprotokolle, simTD, Deliverable 21.4., 2009

[8] Kutila, M., Jokela, M., Rössler, B., Weingart, J., Utilization of optical road surface condition detection around intersection, In: Meyer, G., et al. (Eds.), Advanced Microsystems for Automotive Applications 2009, Springer Berlin 2009.

**Matti Kutila, Pasi Pyykönen, Jussi Yliaho**
VTT Technical Research Centre of Finland
Tekniikankatu 1
33720 Tampere
Finland
matti.kutila@vtt.fi
pasi.pyykonen@vtt.fi
jussi.yliaho@vtt.fi

**Bernd Rössler**
Sick AG
Merkurring 20
22143 Hamburg
Germany
bernd.roessler@sick.de

**Keywords:** IcOR, camera, data fusion, ice, water, monitoring, laserscanner, LDM, intersection, cooperative traffic

# Components & Systems

# From Sensors to Sensor Systems

S. Günthner, B. Schmid, A. Kolbe, Continental


## Abstract

A new electric/electronic (E/E) vehicle architecture is proposed introducing a flexible sensor node (SN) concept. A SN collects signals from distributed sensors, calculates physical quantities out of the raw sensor data and provides the information to all vehicle functions connected to the node. Fusion of signals in the SN renders the E/E-architecture more structured and transparent; redundant assembly of sensors is avoided and more accurate, reliable and broadly available physical sensor information can be provided.


## 1 Current Situation

Sensors play an important role in determining the dynamic behavior of cars, monitoring the driver's intention, observing the environment, controlling the combustion engine etc. Different measurement principles detect the speed of the individual wheels, the lateral and longitudinal acceleration the car experiences and the angular velocities about the roll and yaw axis of the chassis; the steering angle and the steering torque are monitored as well as the brake and accelerator pedal positions. Environmental sensors, such as short- and long-range radar observe other traffic participants and cameras record road signs etc. Fuel delivery to the engine and in-cylinder combustion pressure are measured – just to cite two examples.

### 1.1 Description of the Current Situation

The sensor signals are collected by vehicle functions that then deduce the eventual need for action and trigger actuators, see Fig 1.

Electronic stability control (ESC) is a typical car feature. It consists of control and function algorithms that receive wheel speed signals and information about lateral acceleration, yaw rate and steering angle; based on the measured values, the algorithms decide if deceleration of a single tire is necessary in order to counter oversteer or understeer; in this case, adequate brake torque is

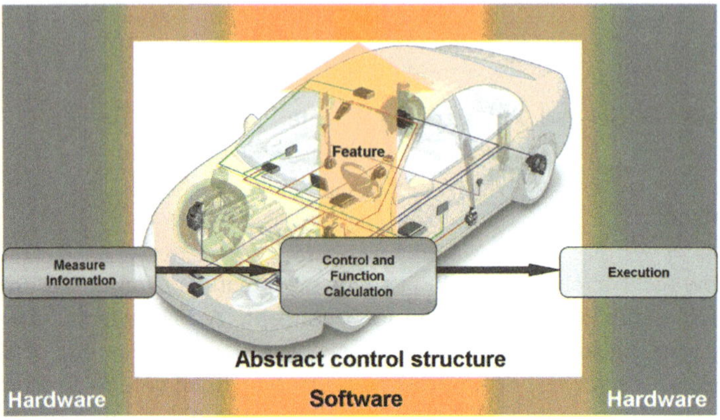


Fig. 1. Abstract control structure

applied to the corresponding tire to force the vehicle into the driver's intended track. Generally, a function consists of calculation and control algorithms (software) and the corresponding interfaces to sensors and actuators (hardware).

Fig. 2 shows a typical sensor-function structure. The sensor hardware normally consists of the sensing technology itself, e.g. a micro electro mechanical system (MEMS), and the electronics, which trigger the MEMS, process, filter, and condition the raw sensor signals, contain basic fail safe checks, and generate the interface protocol to the transport layer. The sensing element and electronics are typically contained in one package, which is why the sensor is considered part of the hardware here. The function software contains algorithms for guaranteeing a high failsafe-level and for filtering and calibrating the incoming sensor signals. Function software can also derive abstract information out of the sensor signals for further usage within the function algorithms, e.g. for cross-checks with other derived sensor information.

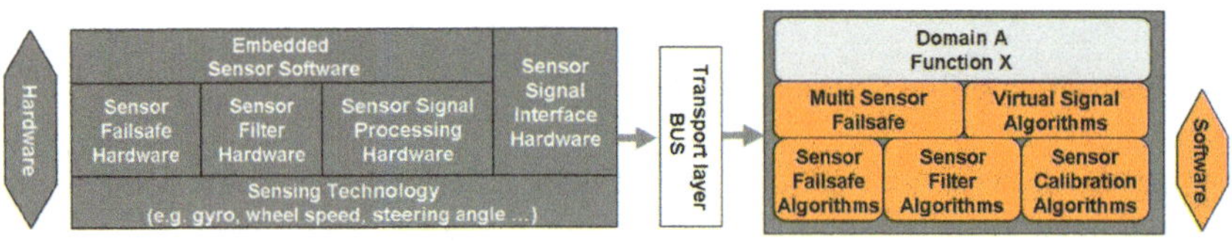


Fig. 2. Abstract sensor function structure today

In today's cars, several hundred functions perform the tasks of active and passive safety, driver assistance, chassis control, powertrain management, comfort enhancement and multimedia operability. Typically, a function is assigned to one of those domains and is connected to sensors and actuators in that domain;

the sensors are connected directly to the electronic control unit (ECU) containing the function software, particularly if sensor signals are used for safety relevant functions. The architecture is actually more complex. Fig. 3 shows the underlying principle of how several sensors deliver sensor signals to functions in different domains. Some sensors may be dedicated solely to a specific domain (sensor hardware 4, e.g. crank angle sensor in the powertrain domain); or sensors might be primarily intended for one function (sensor hardware 1, e.g. wheel speed sensor in the active safety domain) but share their signals with other functions in different domains, in which case the signals may be buffered by a gateway.

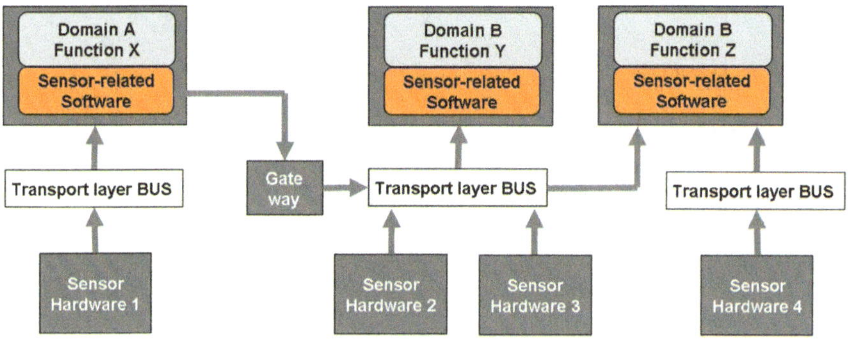


Fig. 3. Abstract sensor function architecture today

Examples for typical E/E-architectures can be found in [1,2]. Table 1 illustrates the analysis of typical sensor function architecture. Studies identify which sensor signals are proprietarily intended for certain functions and which functions make additional use of these sensor signals. The functions examined are the electronic stability program, steering assistance applications, chassis control, airbag activation, engine management, automatic start/stop, rollover detection and navigation. It can clearly be seen that certain sensor signals are used in a large number of functions, as is the case, for example, with the wheel speed signals, which are used in 54 functions developed by Continental.

## 1.2 Limitations of the Current E/E-Architectures

| *Sensor* | *Function* | | | | | | | |
|---|---|---|---|---|---|---|---|---|
| | *ESC* | *Steer-ing* | *Chas-sis* | *Air-bag* | *En-gine* | *Start/ Stop* | *Roll-over* | *Navi-gation* |
| *Yaw rate* | x | x | x | | | | | x |
| *Lateral acceleration* | x | | x | x | | | x | x |
| *Longitudinal acceleration* | x | | x | x | | | | |
| *Roll rate* | | | x | | | | x | |
| *Vertical acceleration* | | | x | | | | | |
| *Steering angle* | x | x | x | | | | | x |
| *Steering torque* | | x | | | | | | |
| *Axle acceleration* | | | x | | | | | |
| *Chassis height* | | | x | | | | x | |
| *Chassis acceleration* | | | x | | | | | |
| *Wheel speed* | x | x | x | x | x | x | | x |
| *GPS* | | x | | x | | | x | x |
| *Accelerator position* | x | | | | x | x | | |
| *Crank angle* | | | | | x | x | | |
| *CAM angle* | | | | | x | | | |
| *Front impact acceleration* | | | | x | | | | |
| *Side impact acceleration* | | | | x | | | | |

Tab. 1. Multiple usage of sensor information by different functions

There are four major issues with the current sensor-function structure:

- **Transparency/organization:**
  As can be seen in Fig. 3, there is no clear, defined architecture joining the functions and corresponding sensors. The present structure has grown historically; different interface protocols are used for certain sensors and there is no standardized procedure for the connection of sensors to functions. The resulting complexity is becoming increasingly hard to control; more effort must be invested in development; and there is a greater risk of system failures.
- **Redundancy:**
  In the present E/E architecture, it may be difficult or even impossible for a newly introduced function to access sensor signals already available, especially if the sensor and the function are situated in different vehicle domains. The same measurement is carried out more than once, e.g. if the navigation system in a car equipped with ESC has its own yaw rate gyroscope.
- **Reliability:**
  If a function uses a sensor signal emitted by a sensor device intended primarily for a different function, there is no clear measure of the reliability of that specific signal. As indicated in Fig. 3, the primary function may perform some pre-processing of certain sensor signals passed on to secondary functions. This means, for example, that latency times are no longer traceable, in particular for secondary applications requiring real-time conditions.
- **Availability:**
  Certain sensors – GPS systems, for instance – deliver no or poor signals under certain circumstances (in tunnels, city streets with tall buildings on both sides etc). Failure of an individual sensor results in the total loss of the sensor signal, although the same information could also be derived from a different device.

## 2 Sensor Node Concept

The limitations just cited for the integration of sensors into a vehicle's E/E architecture will become even more challenging for future cars as their electronics, software and functions become more complex. A modified approach with a clear and defined interface between sensors and functions is therefore proposed: It is the sensor node concept. Fig. 4 outlines the principle structure of the proposed architecture. It differs from the actual situation (s. Fig. 2 and Fig. 3) mainly in that the sensor signals' pre-processing software – like failsafe, filter and calibration algorithms – have been removed from the functions and are placed in the sensor node.

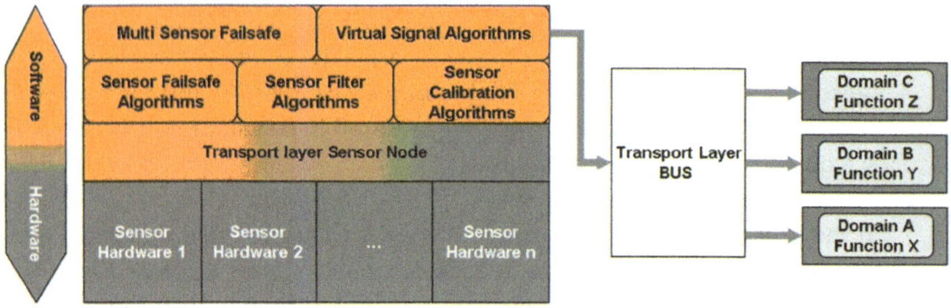


Fig. 4. Abstract sensor function architecture

The sensor node collects signals from several satellite or onboard sensors, performs the corresponding pre-processing, and outputs calibrated and validated sensor information to the vehicle functions via a defined interface. Several concepts for bundling sensor signals have been presented (e.g. [3,4]); but those sensor fusion scenarios are clearly dedicated to certain vehicle domains and the exchange of sensor signals across domain-borders remains difficult. There are two major challenges to be addressed when introducing the sensor node. Firstly, the problem of timing between different sensor signals has to be solved. For a reliable multi-sensor fusion, the individual signals must have a time-correlation, so either the latency times have to be well-known, which is sometimes difficult due to varying bus latencies, or the signals have to get a timestamp. Secondly, there must be an indication for the reliability or accuracy of the sensor information exported to the transport layer so that a function requiring a certain safety level can decide if it can rely on the information.

## 2.1 Advantage of the Sensor Node Concept

The proposed adaption of the actual E/E architecture allows for optimization of several aspects:

- **Improved sensor information accessibility:**
  Introduction of the sensor-node approach makes sensor information available beyond the function for which it is primarily intended. Other functions in different domains can also benefit from the same information.
- **Reduced Costs:**
  The sensor node concept can reduce hardware by rendering installation of similar sensors superfluous and by generating information by fus-

ing sensor signals. The integration of sensor signal pre-processing into the sensor node supersedes similar failsafe and calibration calculations performed in several functions. Standardization of sensor architecture reduces the development effort for the functions.

- **Higher reliability and availability:**
  The fusion of sensor signals across vehicle domains allows for advanced failsafe, calibration and filter algorithms so that the sensor information generated is more reliable and more widely available for all functions accessing the corresponding signal.
- **Better accuracy:**
  By relying on a larger amount of raw sensor signals, the sensor node can derive and provide sensor information with greater accuracy.

## 3 Vehicle Dynamics Sensor Node

For practical reasons it makes sense to introduce several sensor nodes that combine signals that belong together. A surrounding sensor node would thus be able to collect the information from optical devices like cameras, radar sensors or ultrasonic proximity sensors. A second possibility is a vehicle dynamics sensor node (VSN), which will be the focus of the following remarks. The basic idea behind the VSN is to fuse all sensor signals supplying information about vehicle dynamics – like inertial sensors (gyroscopes, accelerometers), sensors monitoring the driver's intention (steering angle sensor), wheel speed sensors, chassis sensors (ride-height), position sensors (global navigation satellite system - GNSS) etc.

### 3.1 Scalability

One characteristic feature of the VSN, which currently exists only as a concept, is its scalability: starting from a basic version, different configuration levels are possible up to a high-end variant. The only difference is the number of input sensor signals and therefore the accuracy, availability and reliability of the output information and the number of derived quantities. The principle E/E architecture is, however, identical for all configuration levels.

In its basic configuration, the VSN collects the signals from the sensors measuring longitudinal and lateral acceleration, angular velocity about the yaw-axis, wheel speeds and steering angle. The input interfaces are PSI5 for the inertial sensors, a 7/14 mA protocol for the wheel speed sensors, and CAN for the steering angle sensor. In the so-called core algorithm of the VSN, several software modules perform the center-of-gravity transformations, initial offset calibrations of the gyroscope and accelerometer signals, as well as the plausibility checks for all signals measured. The verified values for yaw rate, lateral and longitudinal acceleration, wheel speeds and steering angle are output via a CAN or FlexRay© interface. Other software modules compute derived quantities from the input signals for the vehicle's speed over ground. The accuracy and reliability of the output signals is also to be indicated by adding a so-called integrity measure to the output values. The detailed definition of the integrity measure is currently under discussion.

At the highest configuration level, the VSN combines wheel speed sensors, the steering angle sensor, a six-degree-of-freedom inertial measurement unit, a speed-over-ground sensor recently developed by Continental, GNSS signals, and a suspension ride height and chassis acceleration sensor. The core algorithm developed for the VSN is expanded by means of a more advanced virtual sensor algorithm that does the enhanced calibration and the validation calculations. Apart from the basic VSN configuration's output sensor information, with optimized reliability, availability and accuracy, the global position and the ride height are also communicated to the transport layer. What is more, by fusing signals, it becomes possible to derive additional physical parameters, like slip angle, tire pressure, friction and chassis accelerations.

## 3.2 Possible VSN Architectures

The abstract VSN architecture (Fig. 4) contains sensors, the transport layer from the sensors to the software algorithm and the interface for the transport layer to the functions. There is a certain latitude, but also restrictions, as regards to the placement of the individual building blocks within the E/E architecture.

The wheel speed sensors and the steering angle sensor have to be located where they perform their measurement task. There are, however, no major restrictions on the placement of the inertial sensors. Nowadays, there are three main trends for the integration of inertial sensors into existing/new control units [5]. Firstly, all inertial sensors for active and passive safety are integrated into the airbag control unit (ACU), with the sensor software being embedded in the airbag algorithm. Secondly, only the inertial sensors for active safety are placed on the hydraulic-electronic control unit (HECU), with the sensor software embedded in the ESC algorithm. Thirdly, all active and passive inertial sensors are placed on electronic control units (ECU) serving as general function controllers. The VSN concept does not run counter to these integration trends. In particular, a modular VSN concept is imaginable in which all sensors are satellites and the sensor-related software is embedded into an existing ECU. But there is also a fourth option: the introduction of a VSN ECU housing the inertial sensors and the sensors' related software. A comparison between the different scenarios is necessary to reasonably evaluate the manner in which the sensor node concept can be implemented in the vehicle hardware architecture. Apart from the integration of the inertial sensors, consideration has to be given to the matter of where to embed the sensor-related software and the function software. At the moment, the architecture of several feasible hardware configurations is under evaluation with respect to functionality and variable and fixed cost. The different architectural options are shown in Fig. 5.

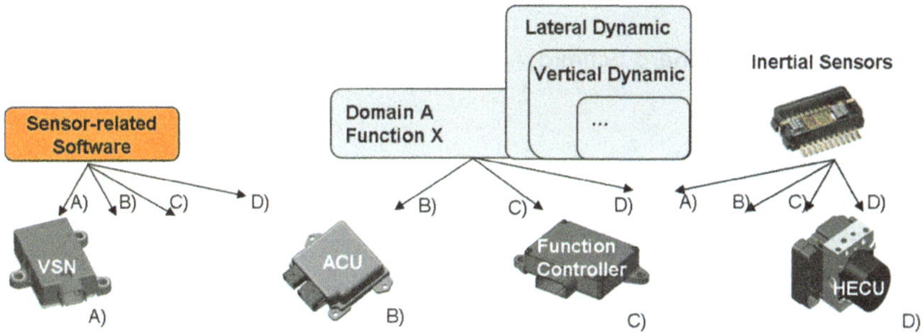


Fig. 5. Architecture options

## 4 Summary and Outlook

The concept for a vehicle dynamics sensor node has been proposed. Several advantages of the new architectural approach have been depicted: the sensor information provided would be more reliable, available and broadly accessible. What is more, structured vehicle architecture would result in lower component and development costs. The next step is to objectively benchmark the sensor node concept in the case of the VSN by means of appropriate architecture evaluation tools for realistic integration scenarios. Parallel to this, concept and feasibility studies are underway at Continental with the aim of developing a basic VSN and demonstrating its feasibility.

## References

[1] Gentner, H., Demmeler, T., Hohmann, S., Wronn, O., Architectures for Networked Active and Passive Safety, VDI-Berichte Nr. 2075, pp. 461-475, 2009.

[2] Knaup, V., Integration elektronischer Systeme bei GME, Elektronik-Systeme im Automobil, 2006.

[3] Schleuter, W., Reich, A., Kernkompetenz Diagnose - Diagnose zur Beherrschung der Elektroniksysteme, Elektronik-Systeme im Automobil, 2006.

[4] Müsch, E., Intelligente Sensor- und Aktuatorknoten, Elektronik-Systeme im Automobil, 2006.

[5] Axten, E., Schier, J., Inertial Sensor Performance for Diverse Integration Strategies in Automotive Safety, In: Valldorf, J., Gessner, W., Advanced Microsystems for Automotive Applications, pp. 251-263, Springer, Berlin 2007.

**Stefan Günthner, Bernhard Schmid, Alexander Kolbe**
Continental
Guerickestr. 7
60489 Frankfurt am Main
Germany
stefan.guenthner@continental-corporation.com
bernhard.schmid@continental-corporation.com
alexander.kolbe@continental-corporation.com

**Keywords:** sensor system, vehicle dynamics, sensor node, sensor fusion, E/E architecture

# Solutions for Safety Critical Automotive Applications

M. Osajda, Freescale Semiconductor


## Abstract

Over the last few years, automotive electronic systems have become a dominant factor in defining the driving experience of modern vehicles. Increasingly the automotive electronic systems need to fulfill functional safety requirements not only in active and passive safety systems, but also in chassis, powertrain and body applications. In this context functional safety is often considered as the part of the overall safety relating to the equipment under control (EUC) and the EUC control system which depends on the correct functioning of the electronic system. The new MPC564xL from Freescale is a controller family optimized for safety relevant applications such as electric power steering, vehicle stability control and driver assistance. It combines an industry leading functional safety architecture with new levels of performance and flexibility.


## 1 Introduction

What do electronic stability control, power steering and adaptive cruise control have in common? Designing such systems while meeting state of the art functional safety requirements can be a pretty challenging job for system designers. Application functions increase in number and complexity, development cost pressure is high and time to market is shortening. Design engineers targeting safety critical applications with complex control algorithms have a seemingly wide range of system architectures to choose from. However, most of the microcontroller solutions existing today either lack the flexibility to support varying functional safety concepts or require significant efforts in terms of safety software. Additional software again adds complexity and increases probability of systematic failures. As a consequence the following mantra has been established for development of the Freescale MPC564xL family of dual core controllers:

**Be Efficient** – provide highest level of performance, but do more with less, lower clock rates and enable intelligent peripheral coordination.

**Be Flexible** – build a dual core concept that supports multiple safety architectures and allows the user balancing of performance and safety levels.

**Be Safe** – generate a safety concept which is SIL3/ASIL D certifiable and reduces software complexity by putting key safety elements and self-tests in hardware.

## 2 Functional Safety Concept

### 2.1 Industry Trends

Driven by the introduction of higher value functions in cars and the continuous trend to vehicle electrification, safety critical functions are increasingly carried out by programmable electronic systems rather than mechanical components. The complexity of these systems makes it impossible to fully determine all potential failure modes or to test all possible behavior. Consequently, the challenge for system engineers is to architect control units in a way that dangerous failures are prevented or at least sufficiently controlled when they occur. Dangerous failures may arise from:

- Random hardware failure mechanisms
- Systematic hardware failure mechanism
- Software errors
- Common cause failures

Being a challenge for electronic control unit design, these failure modes are also specifically relevant for complex components such as microcontrollers. Therefore, industry standards such as IEC61508 and upcoming ISO26262 specify four safety integrity level, each corresponding to a range of target likelihood of failures of a safety function.

### 2.2 Freescale Safety Concept Fundamentals

Freescale can look back on more than a decade design experience in dual core controller technology for safety critical applications. Aiming for a holistic safety concept for its latest dual core processor families, third-party functional safety experts were engaged for monitoring and assessment of concept implementation as well as design processes. On this basis, an IEC61508 SIL3 compliant functional safety concept for the MPC564xL family was developed. Focus was on:

**Measures against single point faults**

Single point faults can be immediately critical for system safety functions and typically require a fast detection. Typical examples of such faults are bit flips in cores or memories induced by external influences such as radiation or electromagnetic interference. Minimum requirement is the detection of these faults within the system safety time which typically ranges between 1ms and 30ms in automotive applications. As key measure against single point faults the MPC564xL family introduces a so called 'sphere of replication' which allows the user to run key elements of the microcontroller in dual core lockstep mode.

**Measures against latent faults**

Latent faults are typically 'hidden'. Already occurred these faults do not yet compromise system safety functions. An example is a fault in the ECC logic for memory error detection/correction. It would only become critical, when a memory bit flip (e.g. in a Flash module) occurs and consequently cannot be detected/corrected anymore.

The MPC564xL controller architecture offers hardware self test (BIST) mechanisms for detection of this fault category. These tests exercise the microcontrollers logic elements with a coverage of 90% and higher. Hence, potential latent faults can be identified even when the actual application is not triggering all hardware blocks.

**Measures against common cause faults (CCF)**

Common cause faults may result from the fact that redundant elements of the MPC564xL architecture still share a common die. Typical examples are system clock or the power supply issues which can influence chip-internal blocks in a similar way and potentially cause identical failures. Consequently, in lockstep mode, where both channels of the 'sphere of redundancy' execute the same software, such faults would not be detected.

The MPC564xL family provides hardware blocks for detection of clock deviations as well as hardware monitors for main voltages, e.g. internal core voltage, Flash supply voltage etc.

## 2.3 Sphere of Replication

The 'sphere of replication' is the logical part of the MPC564xL device architecture which can be configured to run in a 'lockstep mode'. Lockstep mode means that this part of the controller runs the same set of operations at the same time in parallel. The output from lockstep operations can be compared via so called redundancy checking units (RCU). These units determine if there has been a fault. In the fault case, an error signal is forwarded to a separate hardware block, the Fault Collection and Control Unit.

In the past the principle of replication and lockstep mode was predominantly used for the cores. This allowed exceptional fast response to core faults typically in the range of a couple of clock cycles.

The MPC564xL family is going one step further and adds other key hardware blocks to the 'sphere of replication'. Main elements are:

- Crossbar including memory protection units
- Interrupt Controller
- DMA Unit
- Software Watchdog Timer

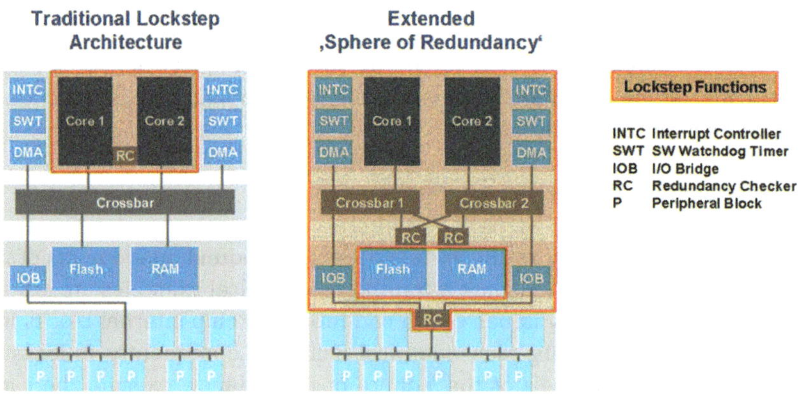


Fig. 1. MPC564xL Extended Sphere of Redundancy

## 2.4 Memory ECC

The ECC scheme implemented for the MC564xL family can correct all single-bit errors, detects all dual-bit errors and detects several faults affecting more than two bits. ECC calculation does not impact the performance of the device.

Specifically for SRAM, the address information is included in the calculation and evaluation of ECC. This allows the detection of potential addressing faults in RAM arrays. Dual or multi-bit faults are forwarded to the Fault Collection and Control unit.

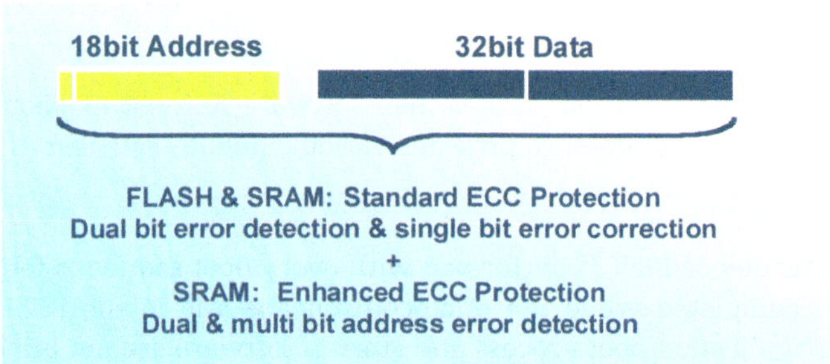


Fig. 2. MPC564xL ECC scheme with SRAM address monitoring

## 2.5 Voltage and Clock Monitoring

Voltage monitoring capabilities are an important aspect for design and implementation of safety critical systems. In order to simplify ECU power supply design and avoid additional failure sources with regard to power sequencing, the MPC564xL family builds on a 3.3V singly supply voltage concept.

A Power Management Unit (PMU) manages supply voltages for all modules on the device and provides respective on-chip monitors for low and high voltage detection. The fault indicators of these monitors are forwarded to the Fault Collection Unit and to the Reset Generation Unit.

MPC564xL voltage monitors are testable. Over/under voltage detectors for the internally generated core voltage provide a hardware-assisted self test. The test needs to be initiated by software during the boot process. Voltage monitoring during runtime is done in background without further software interaction.

For IEC61508 or ISO26262 certifiable systems according to SIL3/ASIL D, monitoring of clock signals is mandatory. The MPC564xL family uses a specific Clock Monitoring Unit (CMU) for supervision of clock source integrity.

Dedicated clock monitors are used for key elements of the microcontroller such as the 'Sphere of replication', peripherals for motor control and communication blocks such as the FlexRay module.

Clock faults are signaled to the Fault Collection Unit. In order to guarantee independence of the Fault collection unit in case other clock sources are erroneous, the module can run on a separate internal 16 MHz RC clock.

## 2.6 Built-in Selftest

Aiming at hardware solutions rather than software self test mechanisms, the MPC 564xL device architecture provides various Build-in Self Test (BIST) functions.

An automatic device BIST is performed with every boot sequence for example. The test is completed while the microcontroller is still in the RESET phase. Hence, the application boot process and startup software are not affected. The application software will only start when this initial self test finished without detection of a fault.

## 2.7 Fault Collection and Management

The Fault Collection Unit is a central element of the MPC564xL functional safety architecture. This hardware module is intended to simplify controller-level fault reporting and management in safety-critical applications. It offers a redundant hardware channel which allows controlled transitioning of the device in a safe state when a critical failure is present. No CPU intervention is requested for this operation.

The Fault Collection Unit can handle controller internal fault signals and allows the user to select how different fault signals will be treated. The Fault Collection Unit logic is checked at the start-up by a self-test procedure, immediately after the unit is active based on a default configuration. For external failure signaling the FCU provides two bidirectional signals.

In order to guarantee FCU independence in case other controller modules or the main core malfunction, the module runs on a separate internal 16 MHz RC clock. Hence, a deterministic computation of output signals and time outs is guaranteed.

## 3 Summary

The trend is alive! Specifically in the automotive chassis and safety domain application functions increase in number and complexity, development cost pressure is rising and time to market is shortening. Microcontroller features which help the designer to focus on the actual application and simplify challenges like safety concept development and certification are a clear added value for ECU architects.

Freescale has developed the MPC564xL family feature set to match these market requirements. A groundbreaking dual core concept gives the designer flexibility in the choice of the system safety architecture and allows optimal balancing of performance and safety requirements with a single controller family.

**Marc Osajda**
Freescale Semiconductor
Schatzbogen 7
81829 Munich
Germany
marc.osadja@freescale.com

**Keywords:** functional safety, ISO26262, microcontroller, multi core, redundancy, chassis & safety applications

# Secure Wireless Control Area Network

D. T. Nguyen, J. Singh, H. P. Le, B. Soh, LaTrobe University

## Abstract

Wireless Control Area Network (WCAN) is an effort to replace the wiring harness CAN system with wireless communication due to the increasing number of new electronic devices within modern vehicles. However, the WCAN design has not been researched enough in terms of security issues that a conventional wireless network has to face such as eavesdropping, malicious broadcast and Denial of Service (DoS) attacks. This paper proposes a security scheme for WCAN that provides an encryption and authentication mechanism for transmitted data. In this scheme, a shared secret key is adopted to exchange a session symmetric key every time the car owner starts the engine. The session key is selected randomly for every session to avoid the risk of being compromised. Networking nodes will use this session key for encryption and authentication purposes during the session. The security mechanism guarantees the confidentiality and integrity of the network communication from suffering severe attacks such as DoS attacks. Because the intra-vehicle network is composed of sensors and limited processing devices like electronic control units (ECUs), the security scheme must be lightweight in terms of processing.

## 1 Introduction

The CAN network otherwise known as the Control Area Network developed by Bosch, has proved its readiness for use in industrial applications and particularly for intra-vehicle networking systems. Such networks require very high real-time responses and extremely reliable communication. Recently, the need for wireless communications has increased in the domain of in-vehicle communication due to the increasing number of new electronic devices within a modern car. With the main considerations of CAN, Wireless Control Area Network (WCAN) was proposed for wireless communication [1].

There are two protocols for WCAN, Wireless Medium Access Control (WMAC) and the Remote Frame Medium Access Control (RFMAC) protocol for distributed and centric topology networks respectively. WMAC however can be further improved for use in a centric wireless sensor network [2]. Therefore, WCAN becomes more suitable for many kinds of in-vehicle networks in which sensors, ECUs, or other devices can communicate directly to a central node within a short-range distance (<10m). Although WCAN has potential for use in in-vehicle networking, research on security issues has been lacking. These kinds of sensor networks are highly susceptible to popular wireless sensor network attacks, such as eavesdropping, malicious broadcast and, in particular DoS attacks. Such attacks can cause loss of control or malfunction of the in-car network and lead to unexpected results. Therefore, before WCAN can be deployed in a car, a security mechanism must be provided to guarantee that the network works properly even in a hostile environment.

In this paper, we propose a security mechanism using a symmetric key technique to encrypt and authenticate transmitted data. Encryption provides the confidentiality for the whole network, thereby preventing eavesdropping. Similarly, authentication provides the data integrity to avoid malicious broadcast and in particular reduces the effect of DoS attacks. The security scheme employs a main secret shared key to distribute the session key. Both the main key and the session key are symmetric keys, but the main key has the longer bit-length, making it more difficult to be compromised. In contrast, the session key will be of shorter bit-length because it is used only over a short period, and the session will start every time the car owner starts the engine. Although the network has the flexibility to employ a cryptographic technique, we propose using the block cipher RC5 [3] in our protocol and hence the long-term key and the session can be 128-bits and 56-bits respectively. In addition, a Code Block Chaining mechanism with a few encryption rounds is employed to procedure Message Authentication Code tag, called the CBC-MAC. The advantage of using CBC-MAC is that it enables the reuse of the encryption module code of RC5, thereby saving memory while still achieving high speed [4]. Since the WCAN MAC layer was not designed with security functions, we have to modify the MAC packets in order to integrate security components into exchange messages. The header message will contain more index information and a message authentication code for authentication will be extended at the message tail. The mechanism also provides three flexible security functions: Encryption only (without the extend authentication), authentication only, and a combination of the encryption and authentication function. The remaining paper is organised as follows. In section 2, an overview of security issues including threats and security objectives for in-vehicle wireless networks is discussed. Then, the details of the proposed protocol are presented in section 3. In section 4 the security of the model is analysed. The last two sections 5 and 6 conclude the paper and acknowledge the sponsors.

## 2 Security Issues in Wireless Automotive Networking

Typically the In-car Wireless Network (IWN) is a WSN because it comprises a number of sensors, ECUs and a few electronic devices communicate in a wireless fashion. Therefore, most of the threats and attacks on wireless in-car networking are almost similar to their counterparts on WSNs. Nevertheless, there is an important difference in the network architecture of these two kinds of networks. Although the network architectures of both are centric oriented, the IWN is a one-hop architecture. This means that every client node can communicate directly with the central node without any routing agent between them. In other words, there is no need for routing information in an IWN. This one-hop network feature eliminates several attacks of WSNs relevant to routing data in nature [5] such as spoofing or replay attacks.

### 2.1 Attacks in Wireless Automotive Networking

There are several attacks of WSN such as sybil, blackhole, wormhole attacks etc. [6] but as mentioned above, the one-hop architecture eliminates many sorts of routing attacks in IWNs. In this section, only those attacks that may really affect on Wireless Automotive Networking are discussed.

**Attacks on Information in Transit:** In an IWN, sensors monitor the changes of specific parameters or values and report to the sink according to the requirement. While sending the report, the information in transit may be absorbed, altered, spoofed, and replayed again at other times. Thus, wrong information is provided to the base stations or sinks. The out-of-date information or improper information can cause a malfunction of the whole network, thereby leading to severe consequences for drivers.

**Denial of Service (DoS):** DoSs are attempts, through the unintentional failure of nodes or malicious action, to prevent a system from providing proper services [6]. One of the simplest DoS attacks is to exhaust the resources available to the victim node by sending extra unnecessary packets, thereby causing the network to fail or malfunction. With a DoS attack, the adversary attempts to not only subvert, disrupt, or destroy a network, but also to diminish network users' ability to access services. Several types of DoS attacks in WSNs in different layers might also be performed in IWNs. At the physical layer the DoS attacks could be jamming and tampering, at the link layer, collision, exhaustion, unfairness; at the network layer, neglect and greed and at the transport layer this attack could be performed by malicious flooding and de-synchronization. The mechanisms to prevent DoS attacks include pushback, strong authentication and identification of traffic.

### 2.2 Security Objectives

**Confidentiality:** Confidentiality means keeping information secret from unauthorized parties. An IWN should not leak sensor readings to neighbouring networks. In many applications (e.g. key distribution) nodes communicate highly sensitive data. The standard approach for keeping sensitive data secret is to encrypt the data with a secret key that only the intended receivers possess, hence achieving confidentiality. Since public-key cryptography is too expensive to be used in the resource constrained sensor networks, most of the proposed protocols use symmetric key encryption methods. According to Karlof et al. [4], the cipher block chaining (CBC) is the most appropriate encryption scheme for sensor networks and only tiny Encryption algorithms like RC5 and Skipjack are most appropriate for software implementation on embedded microcontrollers [7].

**Authenticity:** In an IWN, an adversary can easily inject messages, so the receiver needs to ensure that the data used in any decision-making process originates from the correct source. Data authentication prevents unauthorised parties from participating in the network and legitimate nodes should be able to detect messages from unauthorised nodes and reject them. In the case of communication, data authentication can be achieved through a purely symmetric mechanism: The sender and the receiver share a secret key to compute a message authentication code (MAC) of all communicated data. When a message with a correct MAC arrives, the receiver knows that it must have been sent by the sender.

**Integrity:** Data integrity ensures the receiver that the received data has not been altered in transit by an adversary. Note that Data Authentication can also provide Data Integrity.

**Freshness** implies that the data is recent, and it ensures that an adversary does not replay old messages.

## 3 Secure Wireless CAN

### 3.1 System Model and Assumptions

- An in-vehicle sensor wireless network is modelled as a star-network as shown in Fig. 1. In this model, the base station placed in the dash board acts as a central node which controls the network communication and provides cryptographic materials to other nodes.

- We assume that the underlying physical layer of WCAN is reliable enough whereby a message is transmitted properly without introducing any error. This assumption could be a reality due to the reliability of WCAN tested in an environment communication with the presence of obstacles and engine noise [8].
- Another assumption is that a long-term secret key is pre-distributed and stored securely at every node integrated into the network. This long-term key will be renewed every 6 months or at a regular car-service time.
- The network communication is one-hop communication.

### 3.2 Proposed Security Mechanism

There are two major phases in the proposed protocol:

1. The Central Node selects randomly a secret session key, encrypts it by the long-term secret key and broadcasts to all nodes. $\mapsto * : E_K [K_{session}]$
2. Client nodes use the secret session key to exchange messages. $\mapsto * : E_{Ksession} [Data] \parallel MAC$

Where $E[...]$ is encryption operation, $\parallel$ denotes the concatenation operation, MAC is the message authentication code, $K$ and $K_{session}$ are the main key and the session key, respectively. The protocol provides three security modes for the communication. The first mode, called Encryption mode as in Fig. 2, focuses on confidentiality; the whole data section is encrypted by using the session key. In this mode, the message excludes the authentication part, thus reducing the communication overhead.

The second mode is the Authentication mode where the message is authenticated using a digital signature technique and an extension of the digital signature is attached to the message. The last security mode is the combination of the Encryption and Authentication modes. The data section is encrypted and then the whole message will be authenticated by using the authentication extension part. Like other network security mechanisms, the header of every message will not be modified. This helps the receiver to process incoming message quickly.

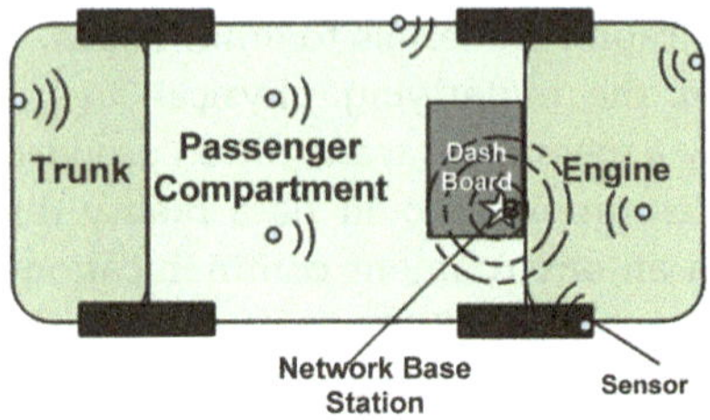


Fig. 1. In-vehicle Wireless Star-Network Topology

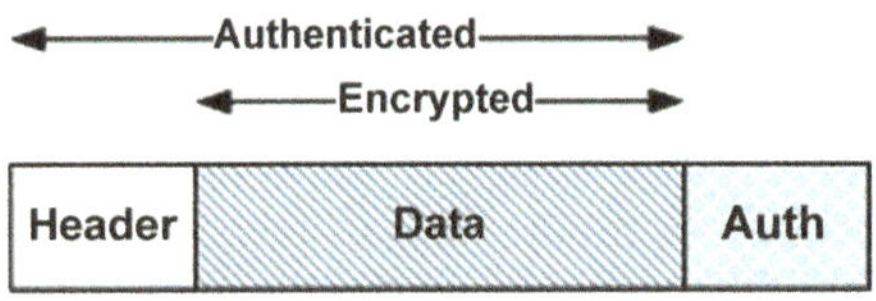


Fig. 2. Authentication and Encryption Mode

### 3.3 Cryptographic Primitives

**Encryption** Due to the stringent resource constraints of the sensor nodes, the candidate cipher to be implemented into the IWN must be lightweight in terms of computation cost. Apparently, a symmetric cryptosystem is a better choice than is an asymmetric one. According to Perrig et al. [9], the symmetric block cipher RC5 [3] is the most suitable for a WSN because of its small code size but sufficiently high efficiency. Compared to RC5, AES [10] using over 800 bytes of lookup tables and DES [11] using 512-entry SBox table are more secure but demand far higher memory for a sensor. Furthermore, the other tiny code size cipher TEA [12] and TREYFER [13] are not matured enough. RC5 has a variable block size (16, 32, 64 or 128 bits), key size (0 to 2040 bits) and number of rounds (0 to 255). Therefore, RC5 is identified by a tube of three indexes $(w, r, b)$ where $w$ = *word-size* in bits, $r$ = *rotation-round-number* and $b$ = *byte-length* of secret key, respectively. For example, *RC5(32,12,8)* means the word size is 32 bits, number of rotation round is 12 and the secret key length is $8*10 = 80$ bits. Obviously, the choices of $w$, $r$ and $b$ affect both encryption speed and security. The original suggested choice of parameters was a block size of 64 bits, a 128 bit key and 12 rounds. There are two types of encryption in our model. The first uses the long-term secret key to encrypt the session key and distribute to client nodes. A minimum long-term key for RC5 could be 128 bit due to sensor ability (see table 1 [3, 14]). The second encryption is to encrypt data in a common message, using a 56 bit session key. Therefore, we use *RC5(32,12,16)* to convey the session key. The $w$ should be 16 or 32 bits to fit the sensor transfer demand and processing ability. In an IWN, most of data transfer content are 8 bits in length or less (see the Benchmark Signals Table in [8]), only few message signals are 32 bits. Since the output blocks are $2w$, there are two options for the $w$ index, 16 bit if the data content is less than or equal to 32 bits, otherwise $w$ is 32 bits. As a result, the data field of a transferred message in our network will be 32 bits or 64 bits. The last feature is the number of rotation rounds. Although Rivest [3] suggested $r$ could be 6 within the given processing speed constraint, the recommended $r$ is still 12 to avoid differential attack [15]. So far, there are two RC5 algorithms in our model to exchange messages: *RC5(16,12,7)* and *RC5(32,12,7)*, this means $w$ = 16 or 32 bits, $r$ = 12 rounds and the session key is $b$ = 7*8 bits = 56 bits.

**Authentication**

There are several methods for authenticating a message; using digital signature and message authentication code (MAC) are two of those. The use of MAC, however, is preferable due to its lightweight computing cryptographic operations which are suitable for sensors. A MAC algorithm accepts as input a secret key and an arbitrary-length message to be authenticated, and outputs a MAC. The MAC value protects both a message's data integrity as well as its authenticity, by allowing verifiers (who also possess the secret key) to detect any changes to the message content. MAC algorithms can be constructed from other cryptographic primitives, such as cryptographic hash functions (as in the case of HMAC) or from block cipher algorithms. Although techniques to procedure MAC by using hash functions have proven to be advantageous over their high speed cryptographic operations, we still prefer using a block cipher algorithm, called Code Block Cipher Message Authentication Code or CBC-MAC, to reuse the code modules of RC5 implemented for the encryption function. Furthermore, applying CBC with a few rounds could achieve high speed and security similar to sophisticated hash functions [16]. A block diagram is shown in Fig. 3. In this CBC-MAC model, the message is divided into several word-size blocks $m_1$, $m_2$, ..., $m_n$, and the last block $m_n$ will be padded to equal a word-size if needed. The session key serves as another input at each round and the MAC tag is the result at the last round. According to Perrig et al. [9], a 30 bytes message needs 8 bytes MAC; in our model, a normal and a maximum transfer message are 16 bytes and 20 bytes respectively (see in section 3.4) and hence, we need only a 4 byte MAC [4]. Thus we use 16 bits word-size RC5 block with 8 or 10 rounds. The MAC digest is 32 bits and is attached at the end of the message.

| *Security* | *RC5* | *AES/DES* | *ECC* | *RSA* |
|---|---|---|---|---|
| *Short-term* | *56 bits* | *64 bits* | *128 bits* | *700 bits* |
| *Middle-term* | *80 bits* | *80 bits* | *160 bits* | *1024 bits* |
| *Long-term* | $\geq$ *128 bits* | *128 bits* | *256 bits* | *4096 bits* |
| | $\leq$ *2040 bits* | | | |

Tab. 1. Recommended key length for public-key and symmetric-key cryptographies

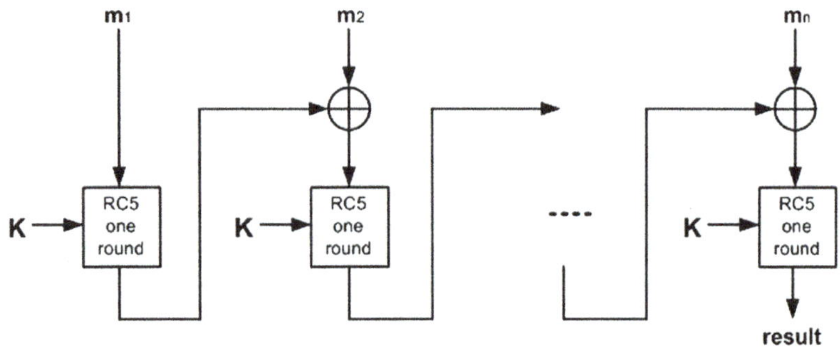


Fig. 3. CBC-MAC

### 3.4 Medium Access Control

There are four types of messages in the previous version WCAN [1] as shown in table 2: Data Packet, Request To Send (RTS), Clear To Send (CTS) and Acknowledgment (ACK). In the RTS, CTS and ACK frames, the MAC-addresses are not used at all, and they will be simply replaced by the 29 bit CAN ID, within the arbitration field. The 2 bits ACK field within the CAN frame is replaced by an ACK frame. In our proposed model, all messages are also added with three security fields: Sequence Number (*SN*), Security Mode (*SM*) and Authentication Extension (*Auth*). The 3 bits *SM* field informs the security modes: Encryption only, Authentication or the combination mode. The first of the three bit indicates the mode; the other two bits indicate cryptographic options. For example, the first 0 bit of the *SM* compound 001 shows the message is encrypted and the other two show the cipher choice is RC5. If the first bit is 1 that means the message is in Authentication or Combination mode, and the other two bits show the chosen cipher in case Combination mode, otherwise they are 00. Because we use only RC5 in our proposed model, the other spare combinations of SM can be kept as a reserve. The *SN* field keeps the information fresh for security purposes. If an attacker wants to send a malicious data packet, the *SN* must be in sequencer to persuade the receiver; otherwise the message will be discarded. However, this will be very difficult if data is sent in authentication mode. In this case, the 32 bits *Auth* field will be extended at the end of the message. Another modification is the flexible length for the Data Packet. In the previous version of WCAN [17], the data field in Data Frame was fixed to 64 bits. This can cause wastage if data is less than 64 bits. We provide a more flexible data field length in our WCAN protocol. This length ranges from 0 to 8 bytes and the length is indicated by the Data Length Code (*DLC*) field. This field is 4 bits (Table 2(a)). Although the maximum length of the Data Frame is 192 bits in length, the actual message length is around 136 bits in length. This is because most of automotive networking data messages are 8 bits in length [8]. Therefore, our protocol somehow provides a shorter transferred message than do those in the previous version [2], thereby guaranteeing that the network latency and real-time requirements are met.

WCAN Data Frame (a)

| *Field name* | *Length (bits)* | *Purpose* |
|---|---|---|
| *MAC header: control frame* | *16* | |
| *MAC header: duration ID* | *16* | |
| *Start-of-frame (SOF)* | *1* | *Denotes the start of frame* |
| *Arbitration field* | *29* | *CAN ID* |
| *Sequence number (SN)* | *12* | *Indicate message order* |
| *Security Mode (SM)* | *3* | *Denotes Security mode, and Cryptographic type Number of bytes of data* |
| *Data length code (DLC)* | *4* | |
| *Data field* | *0-8 bytes* | *Data to be transmitted (length dictated by DLC field)* |
| *Frame Check Sum (FCS)* | *15* | |
| *Authentication extension (Auth)* | *32* | *Message authentication code* |

| | RTS Frame (b) | CTS/ACK Frame (c) |
|---|---|---|
| *Field name* | *Length (bits)* | *Length (bits)* |
| *Control Frame* | *16* | *16* |
| *Duration ID* | *16* | *16* |
| *RA* | *29* | *29* |
| *SN* | *12* | *12* |
| *SM* | *3* | *3* |
| *TA* | *29* | |
| *FCS* | *15* | *15* |
| *Auth* | *32* | *32* |

Tab. 2. Wireless CAN Parameters

## 4 Security Analysis

We aim to achieve two main security goals in this model: the Confidentiality and the Integrity combined with Authenticity. The security level of the proposed mechanism is analysed in the following sections.

### 4.1 Confidentiality

The confidentiality of the mechanism depends heavily on the security of the RC5 which is challenged by three most powerful types of attack including exhaustive search, differential cryptanalysis, and linear cryptanalysis [18]. In general, attacks on RC5 focus on finding either the original secret key or the expanded key table. Clearly, if the latter approach is used, the attack is independent of the length of the secret key. The long-term secret key used in the proposed mode is 128 bits with *RC5(32,12,16)* model. Hence, both the length of the secret key and the number of rounds are sufficiently large to achieve security against the above attacks [15, 18]. The remaining problem is the security while using *RC5(16,12,7)* and *RC5(32,12,7)* in exchange messages. In this section, we discuss only the security of *RC5(16,12,7)*, because the *RC5(32,12,7)* is apparently safe if *RC5(16,12,7)* is safe.

Differential and linear cryptanalysis are two powerful techniques developed in recent years for analysing the security of block ciphers. For differential cryptanalysis, the basic idea is that two chosen plaintexts $P$ and $P^*$ with a certain difference $P' = P \geq P^*$ provide two ciphertexts $C$ and $C^*$ such that $C' = C \geq C^*$ has a specific value with non-negligible probability; such a characteristic ($P'$, $C'$) is useful in deriving certain bits of the key. For linear cryptanalysis, the basic idea is to find linear approximations (parity relations among certain bits of plaintext, ciphertext, and key) which hold probability of $p \neq ½$ (i.e., $bias = |p - ½|$); such approximations can be used to obtain information about the key. Both differential and linear cryptanalysis principles are based on the number of chosen plaintexts, meaning that the attacker must be able to obtain encrypted ciphertexts for some set of chosen plaintexts. Therefore, the two above attacks will be impractical if the number of plaintext requirements is large enough. According to Kaliski et al. [18], the plaintext requirement for differential attacks is determined by $w$ and $r$ as follows: $8w/p^{\Omega_{2r}}$ if $r \geq 12$ where $1/p^{\Omega_{2r}}$ is the maximum expected number of plaintext pairs to obtain a single good pair. Thus, the number of chosen plaintext pairs with $r = 12$, $w = 16$ is $\sim 2^{50}$. Similarly, the number of chosen plaintexts for linear cryptanalysis is $4 * w^{2*(r-1)} = 2^{90}$ with $w = 16$, $r = 12$. These two numbers of chosen plaintexts, $2^{50}$ and $2^{90}$, seem too large for current computers to compute within the few hours of our proposed model session.

Exhaustive search attack is another severe attack on RC5 but it is related more to the key size rather than the word size and number of rotation rounds. The effort for this such a brute-force attack on *RC5(w,r,b)* is $min\{2^{8b}, 2^{w(2r+2)}\}$ [18]. Again, the amount of effort for *RC5(16,12,7)* is too large to be processed in a few hours. Thus, the RC5 encryption method in our model is really safe against cryptanalysis in the context of IWNs.

### 4.2 Message Integrity and Authenticity

The security of CBC-MAC is directly related to the length of the MAC. Conventional security protocols use MACs of 8 or 16 bytes, again erring on the side of caution. We show here that our choice of a 4 byte MAC is not detrimental in the context of IWNs.

Our CBC-MAC model produces 4 bytes of output. Given a 4 byte MAC, then, an adversary has a 1 in $2^{32}$ chance in blindly forging a valid MAC for a particular message. If an adversary repeatedly attempts blind forgeries, she should succeed after about $2^{31}$ tries. Note that adversaries cannot determine *off-line* whether or not a forgery will be successful or not; the adversary can only test the validity of an attempted forgery by sending it to an authorized receiver. This implies the adversary must send about $2^{31}$ packets before being successful in forging the MAC for a single malicious packet. In conventional networks, this number is not large enough for security. However, in IWNs, this may provide an adequate level of security. Adversaries can try to flood the channel with forgeries, but in IWN, most sensors can send only 10 packets per second, so sending $2^{31}$ packets at this rate would take well over a year. Therefore, trying to fool one node by broadcasting malicious MAC message is not really effective. Since the Dos attacks work on the same principles, it is both desirable and feasible to detect when these attacks are underway. The shortness of the MAC tag and the high speed of CBC-MAC processing reduce the consequences of DoS attacks. We leave the solution once DoS attacks have been detected to another time.

## 5 Conclusion

The paper proposes a security mechanism for IWNs which is similar to WSNs. In order to achieve security, the proposed model provides both encryption and authentication function in exchange messages. A secret shared key is pre-distributed so that the network nodes can use this long-term key to deliver a session key at the beginning of each session. Both the long-term key and session key are symmetric cryptographic RC5 keys. A CBC-MAC is employed to produce a MAC tag for message authentication. The use of RC5 and CBC-MAC guarantees that the requirements of tiny sensors are met while still achieving a high security level in the context of IWNs.

## Acknowledgement

This research is supported by AutoCRC, and the Commonwealth of Australia, through the Cooperative Research Centre for Advanced Automotive Technology.

## References

[1] Kutlu, A., Ekiz, H., Powner, E., Wireless control area network, Networking Aspects of Radio Communication Systems, IEE Colloquium, pp. 3/1–3/4, 1996.
[2] BenGouissem, B., Dridi, S., Data centric communication using the wireless control area networks, IEEE Conference of Industrial Technology, pp. 1654–1658, 2006.
[3] Rivest, R. L., The rc5 encryption algorithm, Springer-Verlag, pp. 86–96, 1995.
[4] Karlof, C., Sastry, N., Wagner, D., TinySec: A link layer security architecture for wireless sensor networks, Proceedings of the 2nd international conference on Embedded Networked Sensor Systems, pp. 162-175, 2004.
[5] Zhu, S., Xu, S., Setia, S., Jajodia, S., LHAP: A lightweight hop-by-hop authentication protocol for ad-hoc networks, in the 23rd ICDCSW, 2003.
[6] Perrig, A., Canetti, R., Song, D. X., Tygar, J. D., Efficient and secure source authentication for multicast, in NDSS, 2001.
[7] Saraogi, M., Security in wireless sensor networks, Tennessee University, 2003.
[8] Kutlu, A., Ekiz, H., Powner, E., Performance analysis of MAC protocols for wireless control area network, in Parallel Architectures, Algorithms, and Networks, 1996.
[9] Perrig, A., Szewczyk, R., Tygar, J. D., Wen, V., Culler, D. E., SPINS: security protocols for sensor networks, Wirel. Netw, vol. 8, no. 5, pp. 521–534, 2002.
[10] Daemen, J., Rijmen, V., AES proposal: Rijndael, 1998.
[11] Data Encryption Standard (DES): Draft Federal Information Processing Standards Publication, U. S. National Institute of Standards and Tech (NIST), 1999.
[12] Wheeler, D., Needham, R., Tea, A tiny encryption algorithm, http://www.ftp.cl.cam.ac.uk/ftp/papers/djw-rmn/djw-rmn-tea.html, 1994.
[13] Yuval, G., Reinventing the travois: Encryption/MAC in 30-ROM bytes, Springer-Verlag, 1997.
[14] Marko Wolf, A. W., Wollinger, T., State of the art: Embedding security in vehicles, EURASIP Journal on Embedded Systems, Volume 2007, 2007.

[15] Biryukov, A., Kushilevitz, E., Improved cryptanalysis of RC5, Springer-Verlag, 1998.

[16] Bellare, M., Kilian, J., Rogaway, P., The security of the cipher block chaining message authentication code, Journal of Computer and System Sciences, Volume 61, pp. 362-399, 2001.

[17] Dridi, S., Gouissem, B., Hasnaoui, S., Rezig, H., Coupling latency time to the throughput performance analysis on wireless can networks, in ICCGI '06, p. 39, 2006.

[18] Kaliski, B. S., Jr., Yin, Y. L., On the security of the RC5 encryption algorithm, RSA Laboratories Technical Report TR-602, Version 1.0, 1998.

**Dat Tien Nguyen, Jack Singh, Hai Phuong Le, Ben Soh**
LaTrobe University
Melbourne
Australia, VIC 3086
d2nguyen@students.latrobe.edu.au
Jack.Singh@latrobe.edu.au
hai.le@latrobe.edu.au
B.Soh@latrobe.edu.au

**Keywords:** intra-vehicle network, wireless communication, wireless CAN, security, encryption

# Design and Control of a Linear Electromagnetic Actuator for Active Vehicle Suspension

J. Wang, W. Wang, University of Sheffield
S. Tuplin, M. C. Best, Loughborough University

## Abstract

Linear electromagnetic actuator (LEA) based active suspension has superior controllability and bandwidth, provides shock load isolation between vehicle chassis and wheel due to absence of any mechanical transmission, and therefore has a much great potential in terms of improving ride performance and comfort, vehicle safety, and manoeuvrability. It also has the ability to recover energy that is dissipated in the shock absorber in the passive systems, and results in a much more energy efficient suspension. This paper discusses the issues pertinent to the design, integration and control of a LEA based active vehicle suspension unit, including the choice of linear electromagnetic actuator technologies and actuator topologies, design optimisation, integration with passive components and control strategies for improving riding comfort.

## 1 Introduction

Traditionally, automotive suspension designs with passive components have been a compromise between the three conflicting demands of road holding, load carrying, and passenger comfort. In contrast to passive systems, active suspension systems can adjust their dynamic characteristics in response to varying road conditions, and offer superior handling, road feel, responsiveness, and roll stability and safety without compromising ride quality. Although electro-hydraulic active suspension has been extensively researched and developed [1 - 3], it does not possess the bandwidth of a fully electric system, and thus does not achieve the anticipated performance improvements.

Linear electromagnetic motor based active suspension has superior controllability and bandwidth, provides shock load isolation between vehicle chassis and wheel due to absence of any mechanical transmission, and therefore has a much great potential [4]. It also has the ability to recover energy that is dissipated in the shock absorber in the passive systems [5 - 6], and results in a much more energy efficient suspension system. There are various linear machine technologies and numerous topologies [7] which might be employed in active vehicle suspension applications. Of the possible topologies, tubular configurations are compatible with the packaging/integration requirements of active suspension system-since they have no zero net radial force between the armature and stator, no ending windings, and they are volumetric efficient [8 - 11]. However, due to their significantly higher efficiency, only permanent magnet excited machines are deemed to be appropriate for the proposed active suspension applications. These may be classified as moving-coil, moving magnet or moving-iron.

The performance of moving-coil machines is limited by their specific thrust capability and the flying leads to the current-carrying coils. The moving coil motor topology has a relatively large air-gap, and requires a significant volume of magnet material to produce a high force, which clearly has cost implications. In addition, as the motor power increases, its dynamic capability, i.e. thrust force per moving mass, becomes progressively less due to the increase in efficiency requirement and difficulties in thermal dissipation.

Moving-magnet motor topologies have been shown to have a high force capability. These may be radially [8], axially [9], or Halbach magnetized [10], and either slotted or slotless, [11]. Although a Halbach magnetization distribution yields desirable features, such as a virtually "self-shielding" armature, an essentially sinusoidal electromagnetic force (back-emf) waveform, a high force capability, and a low force ripple, it remains relatively difficult to manufacture magnets with an ideal Halbach magnetization. A simpler form, referred to as a quasi-Halbach magnetization [12 - 13], is, therefore, often employed.

Recently, a tubular, 3-phase, moving-iron, flux switching permanent magnet machine has been developed [14]. It is based on the flux-switching principle, which is also being exploited in rotary machines. The mover is similar to that of a linear switched reluctance (SR) machine, and, therefore, has a simple and robust structure. Thus, the machine combines salient features from SR and PM machines. In addition, it has no end-windings and zero net radial force, and has the potential for operation in harsh environment. Further, since the armature reaction field is essentially perpendicular to the axis of magnetisation, the risk of irreversibly demagnetising the magnets is low. However, detailed design studies have been have shown that while the flux switching machine has a robust moving-armature structure it suffers from the low force density and high material cost compared with tubular PM machine counterparts [15].

A moving-magnet motor topology with quasi-Halbach magnetization and modular winding configuration [12-13] is particularly attractive for high force density and ease of manufacturing. Each phase of the machines comprises a number of concentrated coils which are disposed adjacent to each other. This results in a high fundamental winding factor and a small number of stator slots for a given number of poles. This is not only conducive to a lower manufacturing cost, but also results in a fractional number of slots per pole. Thus, the cogging force due to stator slotting can be very small without employing skew. This machine topology is, therefore, considered as the most suitable candidate for active vehicle suspension and the issues pertinent to its design will be described.

## 2 Design Consideration

Figure 1 shows a 9-slot, 10-pole quasi Halbach magnetised tubular machine with modular winding. Each pole of the magnetic armature comprises of one radially and one axially magnetised ring magnet. The salient feature of the quasi-Halbach magnetisation is that the axially magnetized magnets essentially provide a return path for the radial air-gap flux, and hence, the flux in the inner bore is relatively small. As a result, the use of a very thin ferromagnetic tube or even a non-magnetic tube on which to mount the magnets will not significantly compromise the thrust force capability. This is conducive to reducing unsprung mass and hence enhancing the dynamic capability of active suspension.

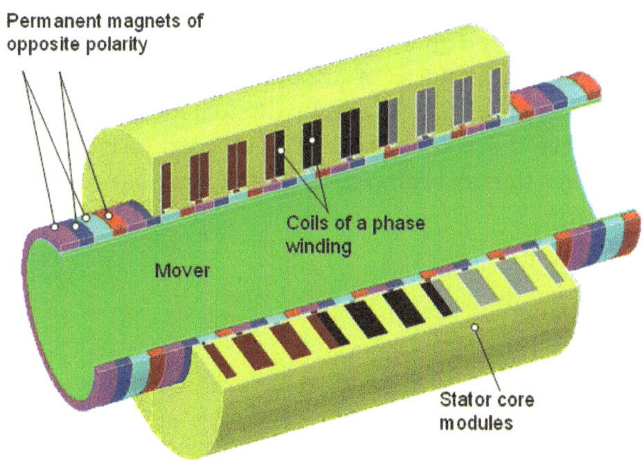


Fig. 1. Schematic of 9-slot, 10-active pole tubular PM machine with quasi-Halbach magnetization

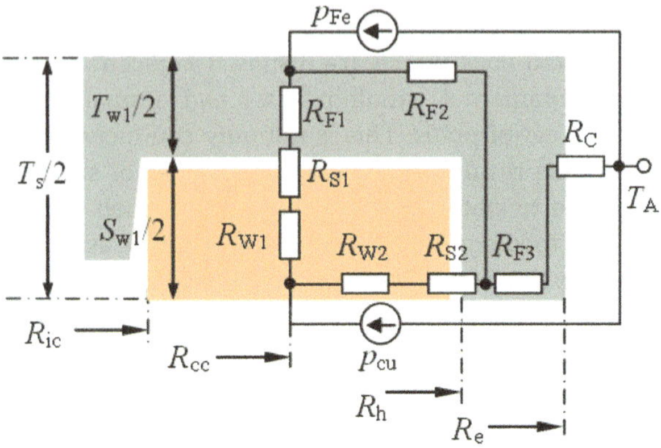


Fig. 2. Schematic of thermal model for tubular PM machines

An analytical framework for predicting the magnetic field distribution in the tubular brushless PM machine with modular windings has been established in [12], and the performance of the machine can be subsequently evaluated. However, in order to optimise machine design for operation in the harsh environment, it is essential that an integrated, computationally efficient electromagnetic and thermal analysis can be performed.

## 2.1 Thermal Modeling

Assume that the thermal dissipation in the machine is predominant in the radial direction, the heat conduction in the machine may be represented by the thermal network shown in Fig. 2, where a half of the tooth pitch is modelled. At the axial symmetrical surfaces of a tooth or a slot, the heat flows tangentially. The copper loss and iron loss can be dissipated both radially and axially via the representative thermal network. The thermal resistances of the model can be derived using the governing principle of the heat conduction, and are given by:

$$R_{F1} = \frac{T_{w1}}{4K_{Fe}\pi(R_h^2 - R_{ic}^2)} \;;\; R_{F2} = \frac{\ln(R_h/R_{cc})}{\pi T_s K_{Fe}} \;;\; R_{F3} = \frac{\ln(R_e/R_h)}{\pi T_s K_{Fe}} \;;\; R_{S1} = \frac{\delta_s}{\pi(R_h^2 - R_{ic}^2)K_{line}}$$
$$R_{S2} = \frac{\delta_s}{\pi R_h S_{w1} K_{line}} \;;\; R_{W1} = \frac{S_{w1}}{4K_w\pi(R_h^2 - R_{ic}^2)} \;;\; R_{W2} = \frac{\ln(R_h/R_{cc})}{\pi S_{w1} K_w} \tag{1}$$

where the geometric parameters are shown in Fig. 2. $K_{\mathrm{Fe}}$ and $K_{\mathrm{line}}$ are the thermal conductivities of the stator core and slot liner, respectively. $\delta_s$ is the slot liner thickness. $K_w$ is the equivalent thermal conductivity which combines the effect of the copper and resin, and is given by:

$$K_w = \frac{K_{co}K_r}{P_f K_r + (1 - P_f)K_{co}} \tag{2}$$

$K_{co}$ and $K_r$ are the thermal conductivities of copper and resin used in the vacuum impregnation, respectively. $P_f$ is the packing factor. $p_{cu}$ and $p_{\mathrm{Fe}}$ are the copper loss and iron loss of the half slot-pitch stator.

The thermal resistance which represents the heat convection at the stator surface is given by:

$$R_C = 1/(\pi R_e T_s h_c) \tag{3}$$

where $h_c$ is the thermal convection coefficient. The thermal network in Fig. 2 can be further simplified to that shown in Fig. 3.

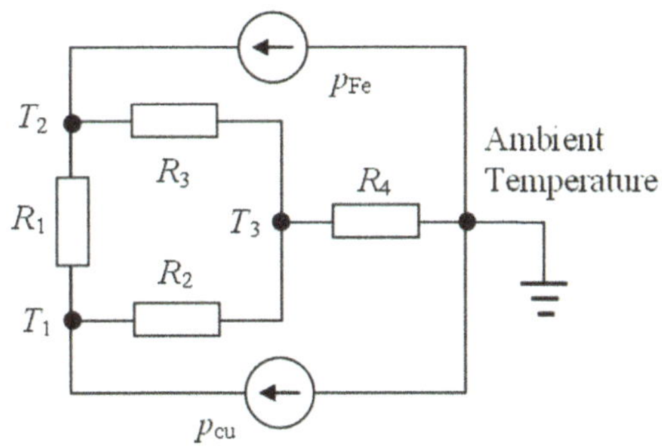


Fig. 3. Simplified thermal model

| | |
|---|---|
| ***Outer stator diameter*** | *0.180 m* |
| ***Active length*** | *0.200 m* |
| ***Air-gap length*** | *0.001 m* |
| ***Stroke*** | *± 0.065 m* |
| ***Maximum temperature rise*** | *120 $^0$C* |
| ***Ambient temperature*** | *60 $^0$C* |
| ***Thermal convection coefficient*** | *40 W/$^0$Cm$^2$* |
| ***Remanence of magnets*** | *1.15 T* |

Tab. 1. Fixed design parameters and thermal conditions

where

$$R_1 = R_{F1} + R_{S1} + R_{W1} \ ; \ R_2 = R_{W2} + R_{S2} \ ; \ R_3 = R_{F2} \ ; \ R_4 = R_{F3} + R_C$$

Thus the temperature rises in the centres of the coil, the tooth and back-iron can be predicted by:

$$\begin{bmatrix} T_1 \\ T_2 \\ T_3 \end{bmatrix} = \begin{bmatrix} G_1 + G_2 & -G_1 & -G_2 \\ -G_1 & G_1 + G_3 & -G_3 \\ -G_2 & -G_3 & G_2 + G_3 + G_4 \end{bmatrix}^{-1} \begin{bmatrix} p_{cu} \\ p_{Fe} \\ 0 \end{bmatrix} \tag{4}$$

where $G_1$, $G_2$, $G_3$ and $G_4$ are the inverses of $R_1$, $R_2$, $R_3$ and $R_4$, respectively. Assume that $IGA$ is the inverse of the conductance matrix, the temperature rise, $T_1$, at the centre of the coil can be determined by:

$$T_1 = IGA(1,1)p_{cu} + IGA(1,2)p_{Fe} \tag{5}$$

The copper loss in the half slot can be calculated by:

$$p_{cu} = \rho_{cu} P_f (V_s / 2) J_{rms}^2 \tag{6}$$

where, $\rho cu$ is the resistivity of the copper at the operating temperature, and $V_s$ is the slot volume. Thus for a given permissible temperature rise $DT$, the permissible rms current density, $J_{rms}$, can be determined by:

$$J_{rms} = \sqrt{\frac{2(DT - IGA(1,2)p_{Fe}}{\rho_{cu} P_f V_s IGA(1,1)}} \tag{7}$$

## 2.2 Design Optimisation

Equation (7) together with the analytical method for predicting electromagnetic performance of the tubular PM machine reported in [12] provides a computationally efficient design tool for machines operating in harsh environment under thermal limits. The design of the quasi-Halbach magnetized moving-magnet machine can be optimized with respect to three leading dimensional ratios, as shown in Fig. 4, viz. the ratio of the outer mover radius, $R_m$, to the outer stator radius, $R_s$, the ratio of the axially length, $\tau_{mr}$, of the radially magnetized magnets to the pole-pitch, $T_p$, and the ratio of the pole-pitch, $T_p$, to the outer radius of the stator, $R_s$, for the given specification in table 1. By way of example, Fig. 5 shows the variation of the thrust force with $\tau_{mr}$ to $T_p$ ratio for the motor. It is evident there is an optimal $\tau_{mr}$ to $T_p$ ratio that yield the maximum thrust force. Table 2 lists the optimised design parameters of the machine under the thermal constraints given in table 1.

Fig. 6 shows the open-circuit flux distribution of the quasi-Halbach magnetized PM machine. It is evident that the axially magnetized magnets essentially provide flux return paths for the radial flux in the air-gap. Hence the flux path through the back-iron of the mover is relatively low. This is conducive to a light moving armature.

Fig. 7 compares analytically and finite element (FE) predicted phase A back-emf waveforms of the machine at an armature velocity of 1.0 m/s. A small difference in phase shift is visible due to the finite length effect which is modelled in the FE analysis, but is neglected in the analytical prediction.

Figure 8 shows the thrust force capability as a function of rms current, where the minimum current on the curve corresponds to the rated (permissible) continuous value under the thermal condition given in table 1. As can be seen, the thrust force capability of the machine is virtually linearly proportional to the current up to 220A. However, the maximum thrust force will be limited by demagnetisation effect as reported in [16].

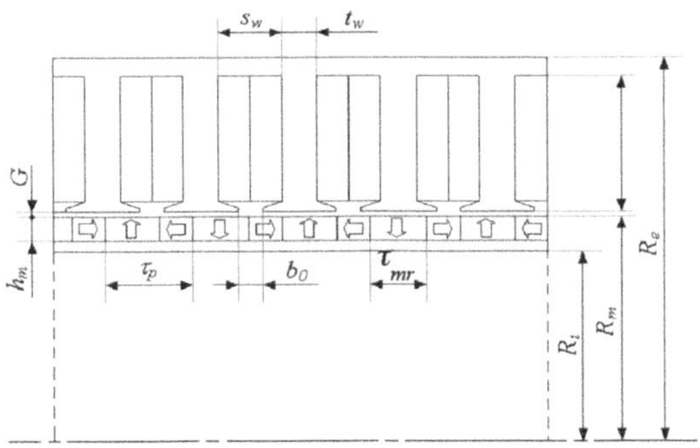


Fig. 4. Leading design parameters of 3-phase tubular modular PM machine

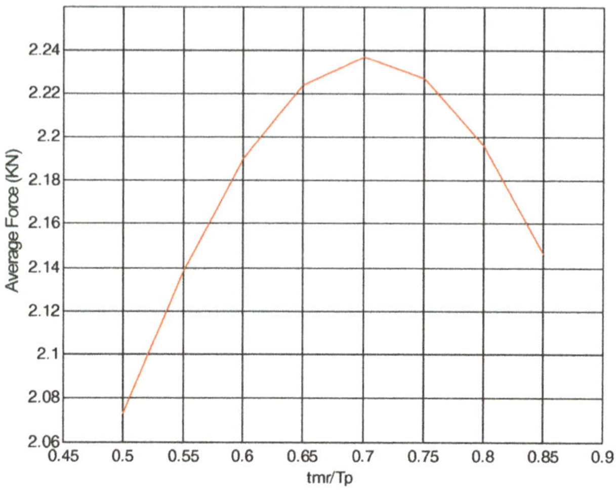


Fig. 5. Variation of thrust force with $\tau_{mr}/T_p$ for quasi-Halbach magnetized PM machine

| | |
|---|---|
| ***$R_m/R_s$*** | *0.55* |
| ***$\tau_{mr}/T_p$*** | *0.70* |
| ***Number of poles*** | *10* |
| ***Pole-pitch, $T_p$*** | *20 mm* |
| ***Number of slots*** | *9* |
| ***Slot-pitch*** | *22.2 mm* |
| ***Magnet thickness*** | *5.0 mm* |

Tab. 2. Optimised dimensions of candidate machine

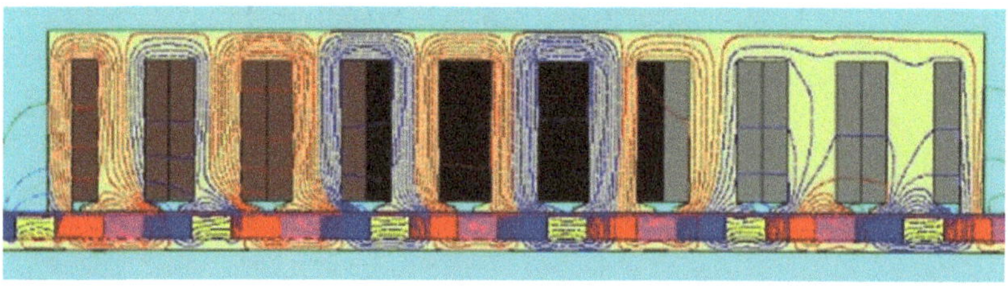

Fig. 6. Open-circuit flux distribution at the armature position of $z_d$ = 0.005 m

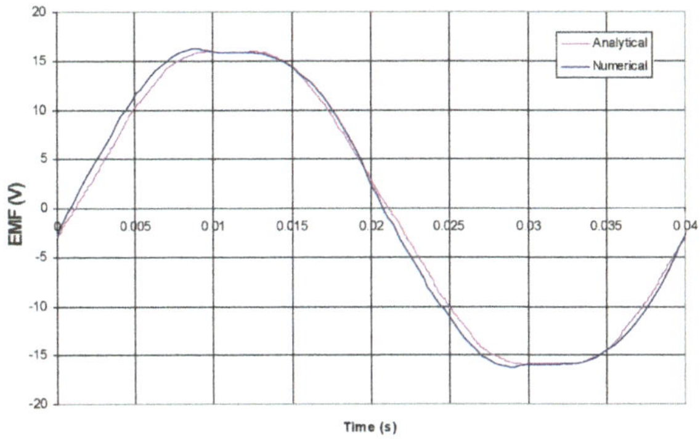


Fig. 7. Comparison of analytically and FE predicted phase A back-emf waveform

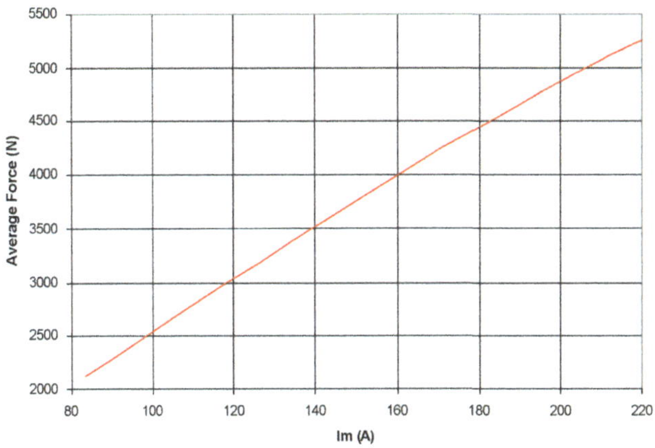


Fig. 8. Variations of thrust force with rms current density

## 3 Integration of LEA Into Strut Design

A compact strut design has been realised based on the actuator topology in Fig. 1, and incorporates an integral coil spring inside the mover to support the vehicle weight, thus offering an LEA package of similar dimensions to that of air spring / damper units currently used on production vehicles. The solution has the obvious advantage of no redesign of the chassis for the vehicle manufacturer as the load paths will be identical, with reduced load through the LEA as the coil spring will take the static load and a self contained unit with no dependency on other vehicle systems (e.g. cooling). Figure 9 shows the assembly drawing, with the lower clevis and vehicle chassis top mount omitted. Thick wall plastic bush bearings have been employed to minimise eddy current losses in the bearing housing through which longitudinal slots are cut to prevent eddy current from circulation in the circumferential direction. The plastic of choice is PTFE with 25% "glass fibre" infill. This has a very low dynamic coefficient of friction (~0.07) and several orders of magnitude increase in wear life over virgin PTFE as well as improved dimensional stability.

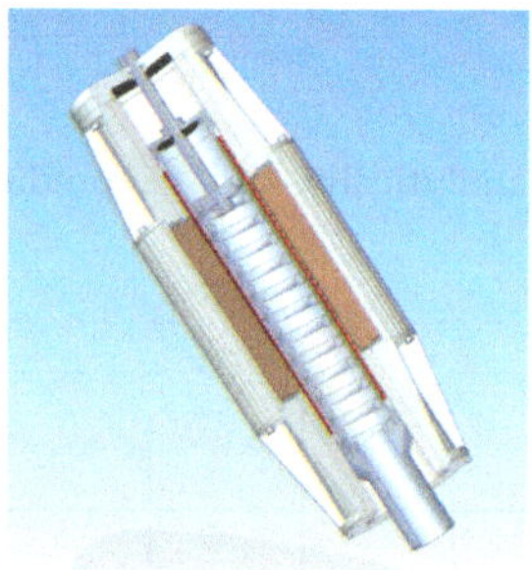

Fig. 9. Schematic of integrated design of LEA based active vehicle suspension unit.

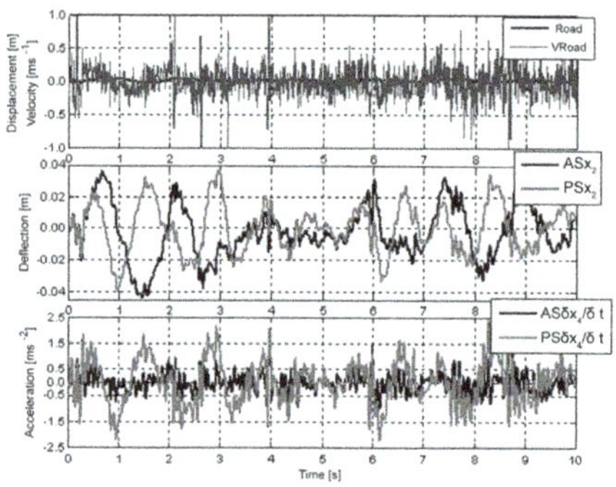


Fig. 10. Ride control performance with quarter car model and real road data

The apparent over-sizing of the bush bearings is a conservative design to prevent locking. The spring top mount is connected to the vehicle chassis through a 14 mm diameter stainless steel "damper tube" which is rated to 20 KN in bump.

## 4 Control of Active Suspension

The control of the suspension system can be divided into three distinct areas; ride control, roll control and handling control. However, only ride control is considered in this paper with the objective to minimise vehicle body acceleration $\dot{x}_4$ whilst maintaining the tyre deflection ($x_1$) and suspension deflection ($x_2$) at passive levels. The linear quadratic regulator (LQR) design technique is employed to determine the state feedback gains $K$ of the ride controller which outputs the force demand $F_C$ for the linear actuator:

$$Fc = -KX$$

where $X = [x_1\ x_2\ x_3\ x_4]^T$, $x_3$ and $x_4$ denotes wheel velocity and vehicle body velocity respectively. The control gains $K$ are determined via LQR design such that the control force minimises the infinite time integral cost function given in equation (9):

$$J = \int_0^\infty (\alpha x_1^2 + \beta x_2^2 + \dot{x}_4^2)dt$$

The costing of body acceleration is incorporated in (9) with $\alpha$ and $\beta$ turned to maintain the passive system rms responses of the tyre deflection and suspension travel. The full state feedback LQR control is realised by employing a Kalman filter. The passive spring is included and 100% of the system damping is realised by the LEA in order to maximise the regenerative capability. The control is then applied in conjunction with the passive support components.

Using the selected spring rate of 30 N/mm and feedback gains $K = [-10700, -29700, 925, -530]$, the ride control performance realised by the active suspension unit was evaluated by simulation with real world road information on a quarter car model. A significant improvement of body acceleration for the active system has been observed, as shown in Fig. 10, with RMS suspension deflection, $x_2$, Fig. 10 (b), and RMS tyre deflection (not shown) being maintained at passive levels. The RMS body acceleration is reduced from 1.00 $ms^{-2}$ (passive) to 0.62 $ms^{-2}$ (active), Fig 10 (c). The peak force observed was 1354 N, with the RMS force being 518 N. Inspection of the body acceleration suggests that body bounce is well controlled albeit the wheel hop less so. Further performance evaluation with full vehicle model has also been carried out and will be reported in future together with experimental results on a quarter car rig.

## 5 Conclusion

The issues pertinent to the design, integration and control of a high force density tubular permanent magnet machine for active suspension have been described. A thermal model representing the heat dissipation in the machine has been established, and utilized for design optimization against a given set of volumetric and thermal constraints. The performance of the active suspension system has been demonstrated by simulation with a quarter car model. It has been shown that a significant improvement of body acceleration for the active system can be achieved

## References

[1] Sharp, R. S., Crolla, D.A., Road vehicle suspension system design - a review, Vehicle System Dynamics, Vol. 16, No. 3, pp. 167-192, 1987.

[2] Rakheja, S., Hong S., Sankar, T.S., Analysis of a passive sequential hydraulic damper for vehicle suspension, Vehicle System Dynamics, Vol. 19, No. 5, pp. 289 – 312, 1990.

[3] Sims, N.D., Stanway, R., Semi-active vehicle suspension using smart fluid dampers: a modelling and control study, International Journal of Vehicle Design, Vol. 33, No. 1-3, pp. 76 – 102, 2003.

[4] Martins, Esteves, J., Marques, G.D., da Silva, F.P., Permanent magnet linear actuators applicability in automobile active suspensions, IEEE Trans. on Vehicular Technology, vol. 55, no. 1, pp. 86-94, 2006.

[5] Martins, Esteves, J., da Silva, F. P., Energy recovery from an electrical suspension for EV", Proc. Electric Vehicle Symposium (EVS-15), Brussels, Belgium, 1998.

[6] Graves, K.E., Loventti, P.G., Toncich, D., Electromagnetic regenerative damping in vehicle suspension systems, International Journal of Vehicle Design, Vol. 24, No.2-3, pp. 182-197, 2000.

[7] Howe, D., Zhu, Z.Q., Clark, R.E., Status of linear drives in Europe, Invited Paper at Linear Drives for Industry Applications (LDIA 2001), Nagano, pp.468-473, 2001.

[8] Wang, J., Howe, D., Design optimisation of radially magnetised, iron-cored, tubular permanent magnet machines and drive systems, IEEE Trans on Magnetics, vol. 40, no. 5, pp. 3262-3277, 2004.

[9] Wang, J., Howe, D., Analysis of Axially Magnetised, Iron-Cored, Tubular Permanent Magnet Machines, IEE Proceedings on Electric Power Applications, vol. 151, no. 2, pp. 144-150, 2004.

[10] Wang, J., Jewell, G. W., Howe, D., A general framework for the analysis and design of tubular linear permanent magnet machines, IEEE Trans. on Magnetics, vol. 35, No. 3, pp. 1986-2000, 1999.

[11] Wang, J., Jewell, G. W., Howe, D., Design optimisation and comparison of tubular permanent magnet machine topologies, IEE Proc. Pt B. Electric Power Applications, 148 (5), pp. 456-464, 2001.
[12] Wang, J., Howe,D., Tubular modular permanent magnet machine equipped with quasi-Halbach magnetized magnets ---- Part I: Magnetic field distribution, emf and thrust force, IEEE Trans. on Magnetics, vol. 41, no. 9, pp. 2470-2478, 2005.
[13] Wang, J., Howe, D., Tubular modular permanent magnet machine equipped with quasi-Halbach magnetized magnets ---- Part II: Armature reaction and design optimisation, IEEE Trans. on Magnetics, vol. 41, no. 9, pp. 2479-2489, 2005.
[14] Wang, J., Wang, W., Clark, R. E., Atallah, K., Howe, D., A tubular flux-switching permanent magnet machine, Journal Applied Physics, vol. 101, 2008.
[15] Wang, J., Wang, W., Atallah, K., Howe, D., Comparative studies of linear permanent magnet motor topologies for active vehicle suspension, Proc. 2008 IEEE Vehicle Power and Propulsion Conference (VPPC2008), Paper ID, H08428, Harbin, China, 2008.
[16] Wang, J., Wang, W., Atallah, K., Howe, D., Demagnetization Assessment for 3-phase Tubular Brushless Permanent Magnet Machines, IEEE Trans. on Magnetics, vol. 44, no. 9, pp. 2195-2203, 2008.

**Jiabin Wang, Weiya Wang**
The University of Sheffield
Mappin Street
Sheffield S1 3JD
UK
j.b.wang@sheffield.ac.uk
w.wang@magnomatics.com

**Simon Tuplin, Matthew Best**
Loughborough University
Loughborough LE11 3TU
UK
S.Tuplin@lboro.ac.uk
M.C.Best@lboro.ac.uk

**Keywords:** linear motor, linear actuator, permanent magnet machines, active vehicle suspension

# Optimising Efficiency using Hardware Co-Processing for the CORDIC Algorithm

A. Glascott-Jones, P. Kuntz, T. Masson, P.A. Pinconcely, B. Diasparra, A. Tatat, F. Berny, F. Salvi, M. Fadlallah, e2v
D. Kerr-Munslow, Cortus

## Abstract

The CORDIC algorithm is a computationally efficient method of calculating trigonometric functions. The algorithm is put to good use in angle measurements systems used extensively in automobiles for measuring throttle and also steering wheel position for stability control applications. This article covers the use of a high performance 32 bit processor to implement this technique and illustrates an application of the algorithm in angle measurement using magneto-resistive sensors. Particularly, this article will cover the basics of the CORDIC algorithm and its implementation and applicability for the application is explained. An introduction to angle measurement using magneto-resistive sensors is given and the requirements of this type of sensor in an automotive application are discussed.

## 1 Introduction

CORDIC stands for Coordinate Rotation Digital Computer. Its first application was in very accurate navigation systems for aircraft and an early application for the algorithm was for trigonometric function calculation in early calculators such as the H.P. 35 [1]. It was highly useful in those days where every transistor counted in a design. Other applications for this algorithm have been found in radar processing, Direct Digital Conversion, numerically controlled oscillators, Tone Generation, angle measurement, electric motor control and sunlight angle measurement. The partitioning of the algorithm between a high performance 32 bit processor and hard coded digital logic will be described and simulation results presented. The requirement for high speed and low power placed difficult constraints on this partitioning. Real world additions such as algorithm adjust facilities and output angle linearization tables are described.

## 2 The CORDIC Algorithm

The CORDIC algorithm emerged in the 1950's however the techniques had been known since ancient times [1]. Jack Volder produced the algorithm based on applications in real-time navigation, the real advance was the calculation of complex functions using simple add and shift operations. [2, 3]. The algorithm was developed to compute the transcendental functions. It does this by decomposing each rotation operation into successive basic rotations.

For example, considering the diagram below, the rotation from $x$, $y$ to $x'$ $y'$ can be described as:

$$x' = x\cos\beta - y\sin\beta \tag{1}$$

$$y' = y\cos\beta + x\sin\beta \tag{2}$$

This can be rearranged so that :

$$x' = \cos\beta\left[x - y\tan\beta\right] \tag{3}$$

$$y' = \cos\beta\left[y + x\tan\beta\right] \tag{4}$$

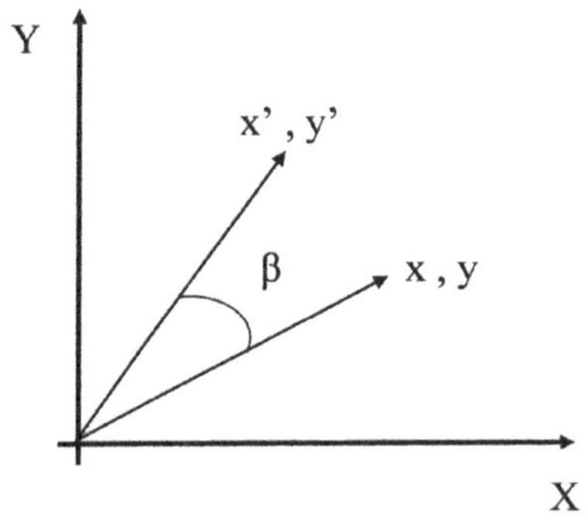


Fig. 1. The CORDIC Principle

If now the rotation angles are restricted such that $tan(\beta) = +/- 2^{-i}$. then the multiplication is no longer required and can be replaced by a simple shift. Where $i$ is an integer.

The Cosine term becomes a constant because $\cos d = \cos -d$ ($d$ is the small increment of angle) [4]:

$$x_{i+1} = K_i\left[x_i - y_i d_i\, 2^{-i}\right] \tag{5}$$

and

$$y_{i+1} = K_i\left[y_i + x_i d_i\, 2^{-i}\right] \tag{6}$$

where

$$K_i = \cos\left(\tan^{-1} 2^{-i}\right) \tag{7}$$

and

$$\left[d_i = +/-1\right] \tag{8}$$

So each angle of rotation can be made up of additions or subtractions of these basic series of angles. There are two different modes of calculation that use this basic approach.

### 2.1 Rotation Mode

This operates the algorithm in a mode where a specific angle of rotation is required. The goal is to reduce the magnitude of the residual angle between the input vector and the successive approximation vector. Sine and Cosine can be computed using this mode

### 2.2 Vector Mode

In this mode the goal of the algorithm is to diminish the $y$ vector until the remaining vector is aligned with the $x$ axis. If the vector magnitude is required there is a need to take into account the CORDIC gain for magnitude so $x$ needs to be reduced by 1.608. ArcTangent is directly computed using the vectoring mode, the argument is presented as a ratio and the angle is deduced from the angle accumulator. No adjustment for gain is required. To calculate the phase during this process it is just necessary to accumulate all the angle data required to get back to zero. For example, to obtain Arctangent, the function rotates the vector with a successive approximation approach until the final vector reaches 0 degrees. Each successive rotation uses predefine angles whose tangent value is 1/ $2^{-i}$.

| *tan-1(2-i)* | Angle ° |
|---:|---:|
| 1 | 45 |
| 0.5 | 26.5650512 |
| 0.25 | 14.0362435 |
| 0.125 | 7.12501635 |
| 0.0625 | 3.57633437 |
| 0.03125 | 1.78991061 |
| 0.015625 | 0.89517371 |
| 0.0078125 | 0.44761417 |
| 0.00390625 | 0.2238105 |
| 0.00195313 | 0.11190568 |
| 0.00097656 | 0.05595289 |

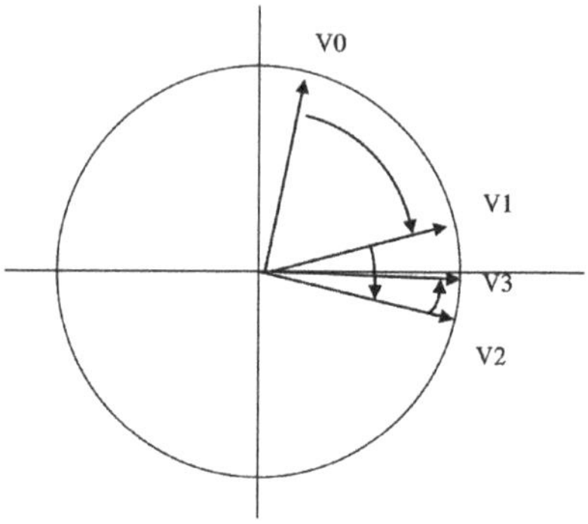


Fig. 2. Using the CORDIC to calculate ArcTangent

The algorithm can also be used for computing Discrete Fourier and Discrete Cosine functions [3].

## 3 A CORDIC Algorithm: xMR Angle Measurement

The Cordic algorithm is widely used in systems for angle measurement. These include systems based on inductive techniques where a sinusoidal signal emitted from a rotor is picked up and rectified by stationary coils; in systems using Hall effect sensors where a rotating permanent magnet causes voltage changes in an array of Hall sensors and in systems using Magneto-resistive sensors where the rotating magnetic field causes changes in the resistance of the sensor array. In magneto-resistive sensors the two main technologies are An-isotropic Magneto-Resistance (AMR) and Giant Magneto Resistance (GMR).

Magneto resistive technologies provide advantages in their higher sensitivity, faster response and more tolerant of mechanical variations than Hall effect techniques. The AMR effect was discovered by W. Thomson in 1857 where the resistivity of some ferromagnetic materials was found to vary as a function of the angle of the incident magnetic field. The resistance change is of the order of 3%. The GMR effect was discovered more recently in 1988 and occurs in materials which have been manufactured to have a stack of ferromagnetic and nonmagnetic materials. In this case the resistance change is of the order of 5 -15%.

AMR sensors are typically arranged in a bridge structure since this reduces temperature effects and resistance offsets in addition to providing extra sensitivity. The resistors are arranged so that two will increase in value with magnetic field and the other two decrease in value [5].
The theoretical equations governing the change in resistance are :

For a GMR sensor $R= R0\ (\ 1+ k\ sin(\Theta))$

For an AMR sensor $R=R0\ (1+ k\ cos(\Theta)2)$

Note that because of the squared term in the AMR equation the output repeats around +90° -> -90° whereas the GMR sensor allows an angle range of -180° to 180°

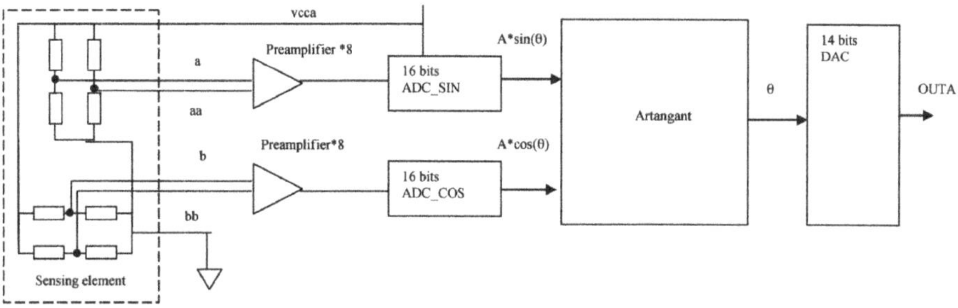


Fig. 3. Typical magneto-resistive readout system

A typical readout component for this type of system is shown in Fig. 3. This will contain 2 channels of low noise, low offset preamplifiers followed by 16bit ADCs. The CORDIC block follows the ADCs and the calculated angle is output using a DAC or digital interface.

### 3.1 Angle Measurement System Requirements

Depending on the application the output data rate or the data refresh rate required can vary from 1000 Hz for slowly varying measurements such as valve angle applications to 50 kHz for electric motor control applications. Accuracy is an important factor and values of better than 0.3° accuracy over the full temperature and angle range are required.

Although the use of this component is not usual in situations requiring ultra low power consumption, there is a requirement for a power down mode and current consumption which is as expected by a analogue industrial sensor (of the order of 5mA). Low power is also important for intrinsically safe applications. Since the typical application for this component is part of a safety critical system; self test and verification functions are essential. These include test signals, watchdog functions, sensing element disconnection and signal level alarms. The ability to function as part of a redundant double sensor system is very important. In addition to the CORDIC algorithm some post linearization algorithms are required; these can take the form of a look up table prior to the output interface where the magnet imperfections and even mechanical effects can be 'calibrated out'. Also the use of emerging interfaces such as SENT and LIN is an important advantage in particular applications.

## 4 Choice of Processor

The processor choice is critical, the requirements for the processor are:

- small silicon footprint
- low power consumption
- high precision and power to perform complex algorithm calculations.

All these factors were taken into account in the choice of the APS3 32 bit core from Cortus. The APS3 processor is a modern RISC processor with a load/store architecture. Sixteen 32 bit general purpose registers are available alongside an optimised 32 bit ALU. A 32 bit uniform address space permitted the application to be developed in a high level language such as C or C++ without any constraints. The core has a Harvard architecture, but a crossbar switch is used to provide a single bus interface. The instruction set features 16 bit long instructions, with an optional 16 bit operand, which ensures excellent code density and efficient interfacing with 16 bit memories. The pipeline design ensures that most instructions are single cycle, including load and store. Memory latency is hidden through out-of-order instruction completion. Fully vectored interrupts allow the low latency required in real time applications. A configurable, programmable priority interrupt controller is available to manage interrupts. A simple co-processor interface enables the easy extension of the instruction set to further optimise the processor for the application, for example increasing the speed of signal processing algorithms. Typical coprocessors are barrel shifters, multiplier, or the CORDIC co-processor described in this paper.

### 4.1 Hardware/ Software Partitioning

A full software implementation of the CORDIC algorithm allows the precise evaluation of the number of CPU clock cycles required to perform the various calculations, and which parts of the algorithm take up most cycles. These corresponding functions can then be implemented in hardware in order to optimise the cycle time. Thanks to the coprocessor interface, which provides direct access to processor resources and arbitration, communication between the executing software and functions implemented as coprocessors, occur without latency. From the software point of view, the coprocessor accesses use dedicated instructions, directly supported in the development toolchain. The developer can then directly use high level CORDIC functions without needing to deal with low level functions or specific management of peripheral functions that would otherwise be needed.

### 4.2 Profiling Results for CORDIC Software Implementation

Along with the CORDIC software algorithm itself, a testsuite calling the different usable functions of CORDIC algorithm on a large set of input values; allow profiling of the execution of the corresponding functions.

(It also allows a trace of the precision ranges of the functions depending on data sizes and algorithm execution regions. The precision can then be known for each implementation, and checked against Matlab error tracing program for further validation).

| | *Cordic 16 bits* | *Cordic 32 bits* |
|---|---|---|
| ***Resolution*** | *$4.88 \times 10^{-4}$* | *$7.45 \times 10^{-9}$* |
| ***Cycles for CORDIC circular function*** | *11426* | *19954* |
| ***Cycles for algo core (main iteration loop)*** | *11324* | *19876* |
| ***Cycles for extension function*** | *45* | *26* |
| ***Cycles for linear function*** | *570* | *1216* |
| ***Memory Size*** | *1.848 KBytes* | *2.32 KBytes* |

Tab. 1. Profiling results software implementation of the CORDIC Algorithm

The figures shows the CORDIC algorithm efficiency, as the complete set of functions only takes up to 2,32kbytes of code, which remains quite low. They also show that a significant number of cycles are required for processing, in both the circular and core functions. To reduce the cycles needed, the most cycle consuming functions can be implemented as optimised hardware blocks. Note, these figures are for a complete CORDIC solution which includes hyperbolic, exponential and linear functions a reduced feature set implementation is described later.

### 4.3 Coprocessor Architecture

This schematic represents a full hardware implementation. A parametrisable generator builds the VHDL files corresponding to a required architecture. The architecture itself depends upon the required functions, hence some blocks (flagged with a star on the schematics) are optional, and only components and functions specified in the parametrization list are actually generated.

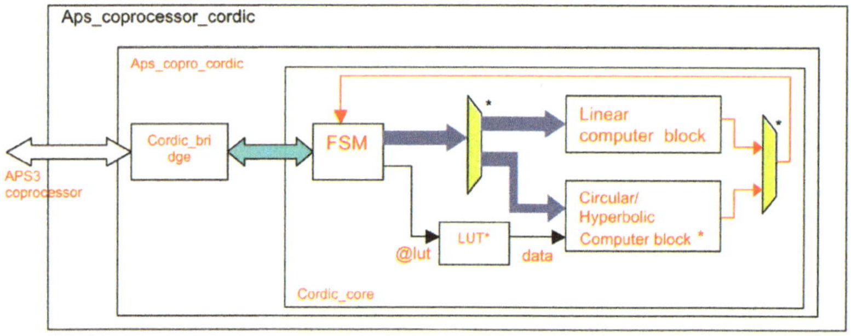


Fig. 4. The CORDIC coprocessor

### 4.4 Simulation

The test suite is run using the hardware coprocessor, and can be compared to the results of the full software implementation, for a full 32 bit CORDIC algorithm.

| | *CORDIC full SW* | *CORDICcoprocessor* |
|---|---|---|
| ***Cycles Nb algo core*** | *19870* | *80* |
| ***Average cycle number per function*** | *19950* | *150* |
| ***HW block size*** | - | *70 000 $nm^2$ (0,18um process)* |
| ***Memory code size*** | *2.32 KBytes* | *1.1 KBytes* |
| ***Equivalent code size*** | *2.32 KBytes* | *2,25KBytes* |

Tab. 2. Comparison between SW and CORDIC coprocessor performance

The figures show the clear performance improvement of 133 (150 cycles versus 19950). The area required by the coprocessor is equivalent to 1.15Kbytes of ROM. Thus the total coprocessor implementation area is equivalent to 2.25Kbytes of code, which is better than the 2,32Kbytes of code required by the full software implementation. Note, again this is for a complete CORDIC solution. Each individual function can be used standalone which reduces further the cycle times and footprint. For example, to illustrate a solution where Sine, Cosine, and ArcTangent modules are implemented with 16bit precision:

| | | |
|---|---|---|
| Functions : | CORDIC 16bits ; ATAN() ; SINCOS() | |
| Footprint : | Hardware: 2800 gates - Software : 292bytes (ArcTangent 112bytes, SINCOS 180) | |
| Execution time: | ATAN -HW only | : 34 cycles ( 16 iterations ) |
| | all SW+HW | : 68 cycles |
| | SINCOS -HW only | : 34 cycles ( 16 iterations ) |
| | all SW+HW | : 81 cycles |
| Accuracy : | +/- 4 LSB on 16bits | |

In this example, maximum angle error due to the Cordic calculations (rounding, limited number of iterations etc.) is estimated at 0.02 degrees.

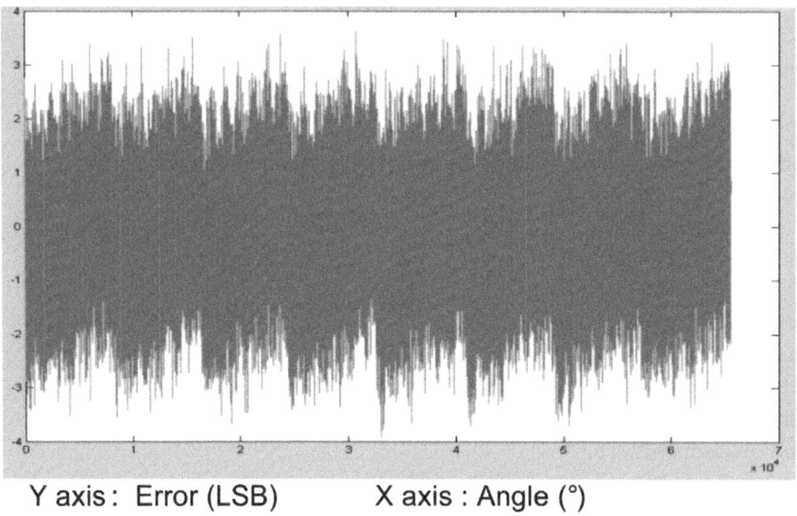


Fig. 5. Simulation of CORDIC error performance

## 5 Conclusions

To obtain the best performance in small footprint but high processing applications the use of a co-processor interface was found to be the best solution. The APS3 processor, which already has a small silicon footprint and excellent performance, offers an ideal and simple co-processor interface to achieve this. In this way competitive solutions for angle measurement can be implemented in silicon. e2v will be producing standard products based on this technology, these will offer excellent performance in terms of accuracy, stability and speed. Even higher speed products are expected in due course.

## References

[1] Website http://www.jacques-laporte.org/cordic_for_dummies.htm

[2] Volder, J., The CORDIC Trigonometric Computing Technique IRE Trans Electron Comput, 1959

[3] Welther, J. S., A unified algorithm for elementary functions, Hewlet Packard, Computer Conference, 1971

[4] Andraka, R., A survey of CORDIC algorithms for FPGA based computers, Andraka Consulting Website. http://www.andraka.com/

[5] Mason, H., A basic Introduction to the use of magnetoresistive sensors, Zetex Applications Note 37, Sept., 03.

[6] APS3 Short form, Cortus website http://www.cortus.com

**Andrew Glascott-Jones, Philippe Kuntz, Thierry Masson, Franck Berny**
e2v
Mixed Signal ASICs Business Unit
Ave. de Rochepleine
BP 123
38521 Saint Egrève
France
andrew.glascott-jones@e2v.com
philippe.kuntz@e2v.com
thierry.masson@e2v.com
franck.berny@e2v.com

**David Kerr-Munslow**
Cortus S.A.
Le Génésis
97 rue de Freyr
34000 Montpellier
France
david.kerr-munslow@cortus.com

**Keywords:** CORDIC, 32 bit processor, magneto-resistive sensors, angle measurement

# Microsensor Based 3D Inertial Measurement System for Motion Tracking in Crash Tests

F. Niestroj, J. Melbert, Ruhr-Universität Bochum

## Abstract

Novel inertial measurement units render possible for the first time the observation of the trajectories of arbitrary and especially hidden objects in vehicle crash tests. These units employ standard micro-electro-mechanical system inertial sensors that need a non-recurring comprehensive calibration to meet precision requirements. Inertial measurement units are the complement of today's photogrammetry and permit the improvement of passive vehicle safety.

## 1 Introduction

Simulation tools are becoming increasingly important for the development of new automobiles. Today's methods and models used for the improvement of passive vehicle safety still necessitate a multitude of time consuming and expensive crash tests. Enhancements of these tools demand observability of highly dynamic motions of arbitrary and especially hidden objects. Photogrammetry alone is inherently not capable of performing this task. Only visible objects can be observed. Thus a new sensor technology is required that provides for observability of high dynamic motions without external reference during the observation period. Various approaches based on ultrasound, static and dynamic electric or magnetic fields, microwaves, X-radiation and inertial measurement have been considered. Taking complexity, expenses, performance and robustness into account, inertial measurement turns out to be the best trade-off for this application.

The newly developed compact inertial measurement units (IMUs) use standard micro-electro-mechanical system (MEMS) inertial sensors for acceleration and yaw rate sensing. For motion reconstruction extraordinary measurement performance is required, which is not fulfilled by the sensors inherently. Due to their long term stability, a non-recurring complex multi-position and centrifuge calibration together with error correction algorithms can overcome this restriction.

The first IMU version is applied in dummy heads for motion tracking [1, 2]. This discloses for the first time the position of the dummy head, even in the non visible phase while the head is plunging into the airbag, and allows the determination of the remaining safety distance to the steering wheel or chassis. This IMU version proved its performance in several crash tests. The deviation of the trajectories is less than 5mm. Figure 1 shows the latest IMUs. On the left side are the third and second version sensor systems that are additionally capable of tracking motions with high dynamics, which occur, for example, at the vehicle's body or power train. On the right side is the data recorder.

Fig. 1. Third and Second Version Sensor Systems with Data Recorder

## 2 Inertial Motion Tracking

MEMS inertial sensors utilize mass inertia to sense accelerations and yaw rates in all three dimensions of space and thus need no external reference. In order to calculate the absolute and not only the relative position, initial velocity $\vec{q}_0$ position $\vec{s}_0$ and orientation $\vec{q}_0$ need to be known. If the initial orientation is not available it can also be derived from given positions. The first integration of the acceleration $\vec{a}(t)$ yields the velocity $\vec{v}(t)$ and the second integration the position $\vec{s}(t)$ . $\vec{g}(\vec{s})$ is the gravity at the crash test site and is subtracted from the accelerations.

$$\vec{v}(t) = \int \left(\vec{a}(t) - \vec{g}(\vec{s})\right) dt + \vec{v}_0 \quad \text{and} \quad \vec{s}(t) = \int \vec{v}(t)\, dt + \vec{s}_0 \qquad (1)$$

Since the IMUs are strapped down to the vehicle, accelerations are measured in the vehicle frame, which rotates with respect to the fixed site frame when the car moves. In order to calculate trajectories referenced to the site frame, the acceleration vectors are rotated from the vehicle into the site frame by quaternion mathematics. Quaternions constitute a four-dimensional vector space, which characteristics are well suited for computing rotations in three-dimensional space [3]. A normalized quaternion contains all information needed about rotation axis and angle and does not change the Euclidean norm of the vector to be rotated. They are recursively computed starting with the initial quaternion $\vec{q}_0$ and are updated by the yaw rates $\vec{\omega}(t)$ . Double quaternion multiplication carries out the frame conversion. Since data is recorded only over a short period of time, earth's angular velocity of $\vec{\omega}_V^e = 0.004° / s$ can be neglected. Figure 2 depicts the tracking principle.

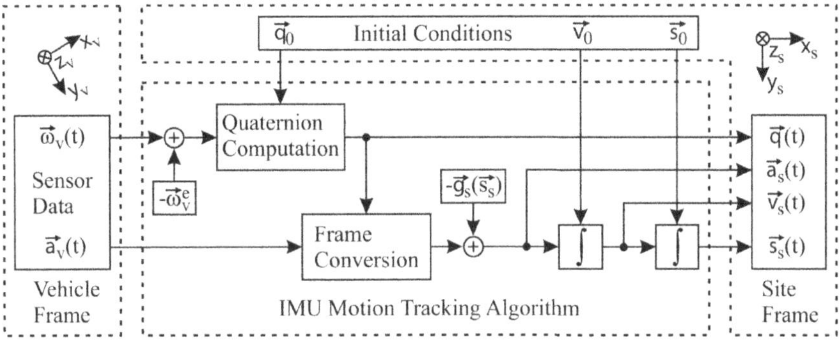


Fig. 2. IMU Motion Tracking

## 3 Sensor Systems

Space constraints demand small sensor systems that limit the sensor technologies in question to MEMS inertial sensors for acceleration and yaw rate sensing in all three dimensions. Accelerations at a vehicles body top out at more than 1000g at frequencies in the kilohertz range. The suitable sensor technology unfortunately suffers from offset voltage drift and the high range comes at the price of reduced sensitivity. The employed high-g sensors have a range of 2000g, a sensitivity of less than 100µV/g and a bandwidth from 0Hz through beyond 5kHz. An offset drift due to temperature variation of 100µV is not uncommon and yields a position error of 5cm after 100ms. This problem is overcome by introducing an additional sensor technology in the second sensor system version (see middle of figure 1). It provides better offset stability and higher sensitivity, but lower range and bandwidth. These mid-g sensors have a range of 120g, a sensitivity of 20mV/g and a bandwidth of 400Hz. Both data is combined in the analysis after a crash test. The yaw rate sensors' range is 1800°/s and their bandwidth 2.4kHz. A sensor system thus comprises nine sensors whose sensor signals are conditioned, anti-alias filtered and analog-to-digital converted in the IMU itself with 16bit resolution and at 50kHz sampling rate. The digitized data is then transferred via a serial interface to a miniaturized recorder in the vehicle for intermediate storage until it is finally transmitted to a PC using an USB link. For best mechanical coupling between sensors and vehicle the back sides of the printed circuit boards (PCBs) are glued to the aluminum sensor carrier. The units are 49mm wide and deep and 48mm high, weight 175g and can sustain accelerations up to 2000g.

The results and experience of the crash tests conducted with second version IMUs led to the development of the third version IMUs, see left side in figure 1. Analog signal conditioning and alias filtering are improved. This and careful layout permit the readout of the high-g acceleration sensors at 18bit resolution as remedy for their lower sensitivity and a reduction of the sampling rate down to 40kHz. The increased resolution of the high-g exempts the mid-g sensors from motion tracking. Their function is offset correction of the high-g sensors and orientation estimation in the pre-impact phase. The latter is to relax precision requirements of the photogrammetry. Thus they are replaced by sensors with a range of 5g and sensitivity of 200mV/g. The PCBs are now mounted with self-centering screws to the carrier. A consequence of the new mounting principle are smaller PCBs, which allow a size reduction of the units to a width of 43mm, depth of 41mm, and height of 36mm. The space consumption is reduced by 45% and the weight is 135g.

The sensors are located neither in the center nor on an axis through the center of the carrier. If a sensor system is exposed to a rotation, a centrifugal acceleration is imposed on the sensors in parallel and / or orthogonal to their sensitive axes. An additional acceleration is acting on the sensors if the rotation is accelerated. Both effects are compensated utilizing the sensed yaw rates and the sensor positions with respect to the center of the carrier.

Up to four sensor systems can be readout at a user-defined sampling rate by a single data recorder (see right side in figure 1). A single system can be recorded for 44s and four systems accordingly for 11s at 40kHz sampling rate. The recorder is powered by its own Li-Ion battery and supplies four connected sensor systems up to six hours. Consequently the IMUs are electrically autarkic. Recording can be started by either a radio or galvanic isolated electrical trigger.

## 4 Calibration

Motion reconstruction necessitates extraordinary measurement performance that is not met by the sensors inherently. Due to their long term stability, a non-recurring complex multi-position and centrifuge calibration, together with error correction algorithms, can overcome this restriction. The low-g sensors are calibrated using the automatic multi-positioner shown on the left side of figure 3. The mid-g, high-g and yaw rate sensors are calibrated using the centrifuge shown on the right side in the same figure.

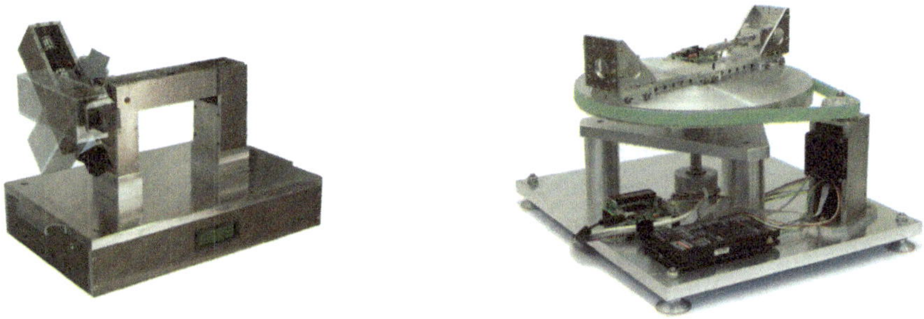

Fig. 3. Multi positioner and centrifuge

Obvious parameters the calibration has to find are the sensors' offsets and sensitivities. Unfortunately the sensing principle, a proof mass in a suspension, introduces cross axis sensitivities and nonlinearities. Misalignments of the proof mass in the chip carrier, of the chip carrier on the PCB and of the PCB in the sensor carrier, add to the inherent cross axis sensitivity and cause non-orthogonal alignment of the three spatial axes made up by the 9 sensors of an IMU. An acceleration sensor whose sensitive axis is parallel to the x-axis, for example, is misaligned by 1° around the y-axis. This causes the sensor to output a spurious acceleration of sin(1°)*10 m/s$^2$ = 0.17 m/s$^2$ at just 1g acceleration, which results in a position error of 1.7cm in a mere 100ms.

A calibration has thus to populate this equation:

$$\vec{a} = M \cdot (\vec{u} - \vec{u}_{\text{off}}) \Leftrightarrow \begin{pmatrix} a_{\text{x}} \\ a_{\text{y}} \\ a_{\text{z}} \end{pmatrix} = \begin{pmatrix} m_{11} & m_{12} & m_{13} \\ m_{21} & m_{22} & m_{23} \\ m_{31} & m_{32} & m_{33} \end{pmatrix} \begin{pmatrix} u_{\text{x}} - u_{\text{x,off}} \\ u_{\text{y}} - u_{\text{y,off}} \\ u_{\text{z}} - u_{\text{z,off}} \end{pmatrix} \tag{2}$$

$\vec{a}$ is the acceleration or yaw rate vector fed into the motion tracking algorithm (see figure 2). $M$ is a (3,3)-matrix that translates the sensors' voltages into accelerations or yaw rates and compensates cross axis sensitivity. Each element of $M = (m_{i,j})$ is a polynomial that compensates nonlinearities. $\vec{u}$ is the voltage vector sensed by the ADCs in the sensor system and $\vec{u}_{\text{off}}$ takes the offsets into account.

The calibration of the low-g acceleration sensors is conducted with the multi-positioner exploiting gravity. A sensor system to be calibrated is mounted on the lever of the positioner and can be rotated around two axes. The lever is actuated by two micro stepping motors and the orientation of the sensor system is detected by two high resolution encoders. The whole measurement process and parameter extraction is automated and needs no user interaction. A calibration takes a minimum of 12 measurements to generate the parameters and more measurements increase precision.

High-g and mid-g as well as yaw rate sensors are calibrated on the centrifuge. A brushless dc motor drives the disk by a flat belt. A gear belt would introduce mechanical noise. The rotation speed is adjusted by a proportional plus integral controller and measured with two encoders, one on the shaft of the motor and one high resolution on the shaft of the disk. Two sensor systems of the same version are always mounted on sleds on opposite ends of the rail on the disk for best balance. Since the sensors are not in the center of the carrier and thus not coaxial to the centripetal acceleration, the effective acceleration parallel and orthogonal to each individual sensor's sensitive axis is calculated by the rotational speed and its exact position and orientation with respect to the center of the centrifuge. At a given acceleration the sensor system is sampled several hundred times. The sampling period is always the rotation period divided by the number of samples. Thus a measurement is always taken over a whole revolution, which cancels out mechanical noise and vibrations that are rotation-symmetric. The sensor system is readout by a small recorder in the center of the disk and the data is then wirelessly transferred to the PC that is running the calibration program. The calibration of the acceleration sensors in one unit involves six different orientations and a multitude of accelerations up to 300g. In contrast, the calibration of the yaw rate sensors comprises only three different orientations and therefore the centrifuge spins clockwise and counterclockwise up to the limit of the sensors. The measurement process and parameter extraction, except the re-orientation of the sensor system on the centrifuge, is controlled by a PC and needs no user interaction. Figure 4 depicts exemplarily the measured sensor voltages of a mid- and high-g acceleration sensor and a yaw rate sensor versus acceleration and yaw rate respectively.

Although the three calibrations are spread over almost a year with a crash test between the second and third calibration and including dis- and reassembly of the sensor system's mounting on the centrifuge, the variation of parameters is less than 0.5%. After the second calibration the centrifuge has been updated to the latest evolution with stiffer sleds. The high reproducibility of the yaw rate sensor measurements does not only show the performance of the sensors, but also the performance of the rotation rate measurement of the centrifuge, and accordingly the acceleration sensor measurements show the quality of geometrical accuracy. The figure shows also the offset drift of the high-g sensors, which can excess 1mV and translates into accelerations of more than 14g. Typical cumulative cross axis sensitivities translate into misalignments of up to 1.3°.

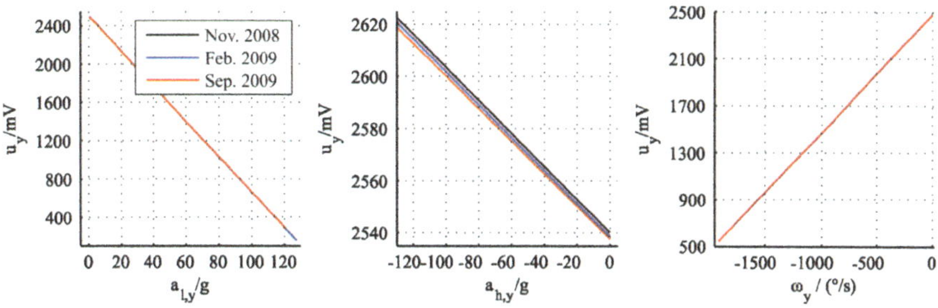


Fig. 4. Output voltages of mid-$g$ and high-$g$ acceleration and yaw rate sensors vs. acceleration and yaw rate

## 5 Motion Reconstruction in Crash Tests

Photogrammetry is inherently not capable of tracking neither high dynamic motions nor motions of hidden objects. But a combination of both, photogrammetry and inertial measurement, can overcome this limitation. The photogrammetry can provide the initial conditions and additional reference data during low dynamic phases of the crash test, that is before the impact and after the high dynamic motions are decayed. Since the vehicle's body can be considered rigid during the low dynamic phases, the positions of the hidden sensor systems can be linked to markers that are visible to the motion cameras. This referencing needs to be carried out before and after the crash test, because the vehicle is deformed plastically, and can be done using still cameras. Several crash tests have been conducted to date with second version IMUs and two of them are mentioned below. Four sensor systems are mounted in the passenger compartment in different locations at the vehicle's body and are not visible to the motion cameras. They are connected to one data recorder, which is triggered in the moment the vehicle touches the barrier.

The first crash test setup is a frontal crash against a deformable barrier with offset. For the motion reconstruction the initial conditions velocity $\vec{v}_0$ , position $\vec{s}_0$ and orientation $\vec{q}_0$ are provided by the photogrammetry. The orientation is most difficult to measure of all initial conditions and most crucial for the motion reconstruction. An analysis showed that the precision of the orientation measurement may not be sufficient for the aspired accuracy. This problem is resolved by utilizing additional position data provided by the photogrammetry for time intervals, where the deformation of the vehicle's body is negligible. An error function is formulated that benchmarks the initial conditions against the spatial difference between photogrammetry and inertial measurement and varies the initial conditions if necessary. Figure 5 illustrates the optimization procedure.

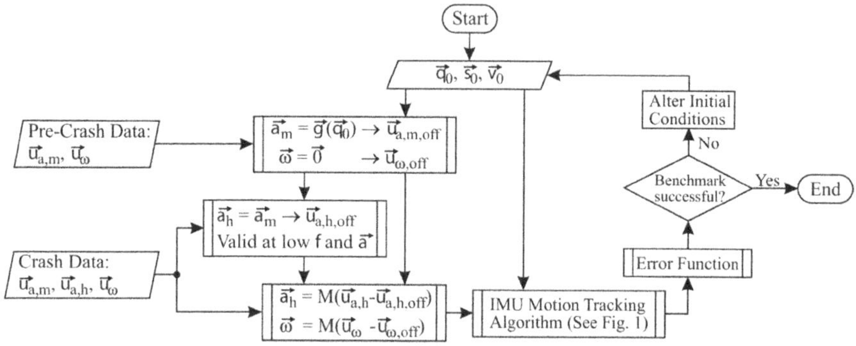


Fig. 5. Optimization Procedure

It starts with the initial conditions provided by the photogrammetry. A recording as close to the crash as possible is conducted while the vehicle is motionless to compensate the $1/f$ noise in the mid-$g$ acceleration and yaw rate sensors' offsets. Periods during the crash, where the accelerations are within the range and bandwidth of the mid-g sensors, are utilized to find the high-$g$ sensors' offsets. Using the derived offsets of the high-$g$ and yaw rate sensors, the accelerations and yaw rates are computed in the vehicle's frame. Only the high-g sensors' data are used for motion reconstruction but not the mid-$g$ sensors' data. Accelerations and yaw rates are fed into the IMU tracking algorithm (see figure 1), and its output is then fed into the error function. If the error is sufficiently small, the optimization terminates, otherwise the initial conditions are altered. The initial conditions are optimized independently for each sensor system. Figure 6 displays on the left side the trajectories of the four sensor systems during the first 200ms after the impact. The maximum distance between inertial measurement and photogrammetry is less than 5mm in the intervals where the photogrammetry is capable of providing a reference.

On the right side of the figure the elastic differential deformations parallel to the y-axis are shown. This proves the assumption that there are time intervals where the deformation is negligible. As a benchmark, the initial orientations have been substituted by position data provided by photogrammetry without adverse effects on accuracy.

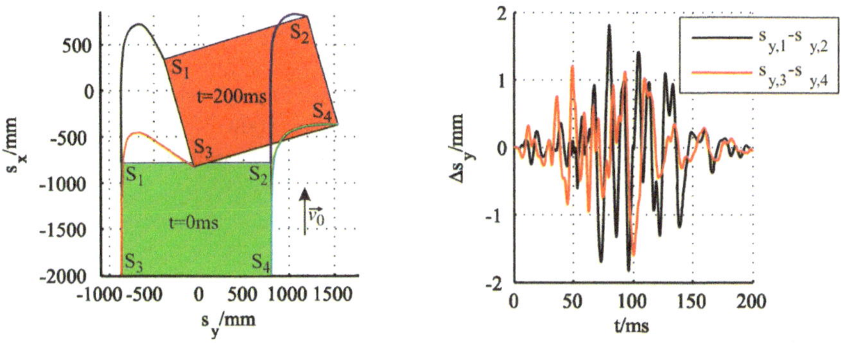


Fig. 6. Trajectories and elastic differential deformations

The second test setup is a frontal crash against a non-deformable barrier with no offset. For the motion reconstruction only the initial conditions velocity $\vec{v}_0$, position $\vec{s}_0$ and orientation $\vec{q}_0$ are utilized. The additional position data available in the first crash test is substituted for two boundary constraints. The first is that the car is motionless at the end of the crash test, which means that there is no acceleration or velocity or yaw rate acting on the vehicle except gravity. The second constraint is that each sensor system's position is fixed with respect to the other three except for the time periods where the vehicle is deformed plastically and elastically. Using these constraints an error function is formulated that benchmarks the initial conditions, and changes them in order to meet these constraints. The reached accuracy is similar to the first scenario.

## 6. Conclusion

The introduced novel IMUs are capable of recording accelerations and yaw rates at unprecedented precision and sampling rate. They permit for the first time the observation of trajectories of arbitrary and hidden objects in crash tests and the analysis of elastic and plastic deformation in the millimeter range and below. Utilization of this information can verify and improve CAD models and methods like crash test simulation and thus reduce the number of time consuming and expensive crash tests. Standard sensors can be employed after they have undergone an extensive calibration procedure with the multi-positioner and centrifuge. Standard photogrammetry is capable of providing the initial conditions and additional reference data. The latest version IMUs increases precision further, reduces space consumption and requires less information from photogrammetry.

## References

[1] Schönebeck, K., Melbert, J., Weiser, F., Motion Tracking in Crash Test Applications with Inertial Measurement Units, SAE International, 2009.

[2] Schönebeck, K., Melbert, J., et. al, Präzise Bahnkurvenbestimmung im Crashtest durch Inertialmesstechnik, crash.tech, 2010.

[3] Titterton, D., Weston, J., Strapdown Inertial Navigation Technology, American Institute of Aeronautics and Astronautics, 2004.

**Fabian Niestroj, Joachim Melbert**
Ruhr-Universität Bochum
Universitätsstr. 150
44780 Bochum
Germany
fabian.niestroj@lems.rub.de
joachim.melbert@lems.rub.de

**Keywords:** inertial measurement system, IMU, crash test, crash test simulation, motion tracking, passenger safety, photogrammetry

# Virtual Reality and Hardware In-the-Loop Testing Methods in the eCall In-Vehicle Module Research and Verification

J. Merkisz, Politechnika Poznańska
R. Grzeszczyk, Automex sp. z.o.o.

## Abstract

The main feature and purpose of the eCall in-vehicle unit is to initiate autonomously an emergency call, to extend it with a data message containing information of current position of the vehicle, its prior-to-crash speed, type of the vehicle, VIN, VRN, number of passengers travelling, and other available information which can be of substantial value to the operator of the Public Safety Answering Point (PSAP), allowing the officer responsible for receiving incoming 112 emergency calls despatching paramedic squads in appropriate numbers and suitable equipped. The main purpose of the project is to design a testing methodology and set-up a testing bench for the research and certification of the in-vehicle e-Call system units. The test stand should allow reproduction of precise and pre-programmed testing sequences of conditions, to excite the device-under-test sensors and its signal inputs, and to relate their logged data and observed control outputs to the reference set of values used during simulation.

## 1 Introduction

Within the European Union initiative referenced to as eSafety programme [3, 4, 5], eCall system is currently being implemented. When the eCall system is finally fully operational, it will significantly improve the notification of road accidents and emergency actions within the whole EU territory. A few years ago, a need to introduce a standardized emergency call number 112 across the EU territory was recognised by European legislation. The 112 number, usually existing in parallel with prior emergency services, is supposed to be operated with the same efficiency and effectiveness in each member state, and the personnel of the emergency call centres should be able to process incoming emergency calls not only in local language, but also in other languages, making the Europe territory friendlier and safer for all EU citizens, also when travelling abroad their homelands [3, 4, 5]. The following technical progress has lead to further modifications of the emergency system scheme, the emergency-call number 112, became E112 (location-Enhanced 112), which will enable the emergency centre to automatically determine the geographical location of the caller. This new development is particularly useful for people travelling abroad, since the victims of the accidents very often cannot inform precisely on their current location, and the timely disposal of the ambulance in emergency situation often means the difference between life and death of the injured. The eCall system is further development in E112 technology, the vehicles will be equipped during manufacturing with crash sensors and communication unit, which, if the predefined crash conditions are met, will automatically dial the emergency centre number and provide exact information on the location of the accident, vehicle identification, its type, and initiate the voice connection allowing the call centre personnel to communicate with the passengers in the car, and to gather additional information on the accident, resulting in optimal decision as to the parameters of the rescue operation and allocated resources. The in-vehicle unit consists of the following sub-systems:

- measurement (accident detection),
- GSM-based, communication link able to establish the emergency call,
- positioning sub-system (GNSS receiver),
- user-interface sub-system (voice communication, manual activation and de-activation).

All those sub-systems need to be thoroughly surveyed before issuing the device with the formal type approval from the notified laboratory. Normally the certification is to be carried out during the type approval of the car, as performance of the unit is strongly correlated with the chassis design, its rigidity, dynamical parameters of the vehicle body and number and locations of sensors used to detect accident. In the case of modern vehicle, when the eCall unit will be developed and integrated by the vehicle manufacturer, it is to be implemented as an integral part with other vehicle safety systems and components, such as airbag control unit/module (ACU/ACM), or restrain control module (RCM), using all the data available in the vehicle information network.

## 2 The eCall System Functional Framework

The eCall system consists of four main subsystems, which should be adequately designed and compatible in order to allow their interoperability and functionality. The first main element of the system in-vehicle module – a device containing a set of inertial sensors which detect excessive linear and angular acceleration values, caused during an impact into the fixed obstacle or into another vehicle, which allow detection of the accident. The GPS module is used to supply geographical position of the vehicle, as well as its direction and speed shortly before the accident. And finally, the GSM module is utilised for automatically or manually initiated connection to the Public Safety Answering Point (PSAP).

Other necessary parts of the eCall system are the ground telecommunication infrastructure of the GSM operator and the ground and space infrastructure of the satellite navigation systems (GNSS). The infrastructure of GSM operator should allow dialling of emergency number installed in the vehicle module, also outside the native mobile operator network. In general, the roaming should be available not only outside the country in which the vehicle is registered and where the SIM card was issued, but also in the territory of that country because the connecting may be limited when the accident takes place outside the coverage of the native mobile operator or when the position of the vehicle after the accident does not allow connection to the native operator network whose signal in the given location can be weak, or when the vehicle is turned over and the GSM antenna is shielded by the vehicle body and/or local topographical conditions. This would make possible the use of another mobile network, that provides stronger signal. The PSAP centres must be equipped with appropriate technical means for reception of emergency information (i.e. incoming calls generated automatically by eCall modules or those initiated manually), its efficient verification and processing, as well as effective management of rescue resources.

## 3 Performance Related Issues of the GNSS Sensor Used in eCall Rover Modules

The Global Positioning System (GPS) is the only fully functional global system of satellite navigation, which was developed, implemented and is run by the US Department of Defence in the framework of NAVSTAR programme. Every satellite navigation system consists of three segments: the ground segment, the orbiting vehicles, and receivers which position is determined by receiver firmware. Currently, all of the systems in operation and in development support passive receivers which means the user's receiver can calculate its own position, time, and direction and velocity of the movement, based on the signals obtained from satellites found in the line of sight (i.e. when in direct, unobstructed optical contact between the rover sensor and the orbiting satellite). In order to produce a valid fix it is necessary to receive navigational messages from at least four satellites. The quality of the obtained solution depends on many factors, among others, the number of available satellites, their orbital positions and signal interferences between the satellite and the receiver (these distortions are result of refraction in the ionosphere, troposphere and multipath of signal – i.e. receiving signal indirectly from satellites, or when it is reflected from various obstacles and objects located near to the receiver). All of the currently existing and planned satellite navigation systems provide free of charge and normally unlimited access to publicly available services. In the case of the augmented navigation systems, when the calculation of the position of the given rover receiver is done externally and is then sent back to the user, or corrective data indispensable for differential GPS (DGPS) is sent, such services may be payable. The satellite navigation system enables its operators to selectively turn off public availability of the service (designed to be used especially in the area of warfare), while providing unchanged accuracy of dedicated services to authorised users (i.e. its own and allied military users).

| *Parameter* | *Satellite Navigation Systems* | | |
|---|---|---|---|
| | *GPS* | *GLONASS* | *Galileo* |
| ***Accuracy of horizontal positioning (95%)*** | *100m (with SA distortion) 13-36m (without SA distortion) 3,7m (after modernisation - GPSIII)* | *30-40 m (50-70m according to spec.)* | *15/4 m ≤ 0,8-7 m* |
| ***Accuracy of vertical positioning (95%)*** | *300m (with SA distortion) 22-77m (without SA distortion) 7m (after modernisation - GPSIII)* | *60-80 m (70m according to spec.)* | *35/8 m ≤ 1-15 m* |
| ***Accuracy of determining speed (95%)*** | *≤ 2m/s* | - | *20cm/s -50cm/s* |
| ***Time (95%)*** | *340ns* | *1 μs* | *50/30 ns* |

Tabl. 1. Comparison of satellite navigation systems accuracy [6, 9].

The following features of the existing GNSS solutions seem fundamental for their use in eCall application: general availability of service in the territory in question (how often and for how long the service may be unavailable) regarding both the geographical location and time, accuracy of determining location, speed and time, minimal sensitivity of the receivers rendering them usable in difficult environmental conditions: in urban areas, in mountains and forests, when the visibility of satellites is limited and so is the possibility of receiving the signal. Further technological development of the space segment of the navigation system, i.e. increase of radiated power of broadcasted navigational signals, more active satellites in the orbit, improved orbit models, as well as development of the receivers (new designs of the antennas, increase of their sensitivity, improved algorithms for elimination of signal distortions, rejecting multipath effects, etc.) will allow in the near future to improve functional parameters of receivers, and more accurate positioning also inside buildings, in the areas of intense vegetation or diverse topographical features.

In the case of the eCall performance, the sensitivity of the receiver is of utter importance for determining the position of the vehicle after the accident, considering that the vehicle after the accident can be turned over or stuck under construction elements such as a bridge or a flyover. Additionally, the system principal assumption is that the receiver is working continuously, and it is required that the last three correctly calculated positions and the direction and speed, are sent in the assembled after-crash data message. Therefore, even in the case of the unfavourable conditions when obtaining a new, valid location fix is not possible, it should be still enough information available to be sent to determine where the accident had happened, and to find the vehicle and the victims of the accident. Another condition to be met necessary for the message to be successfully delivered to the PSAP is the functional GSM link, which depends on the GSM network signal current availability.

## 4 HIL – Hardware-in-the-Loop Testing

Over the last few years tremendous advancements in virtual reality (VR) tools and technology have been achieved. Many of those projects have been developed as open source tools, such as OpenGL, ODE, Open Scene Graph, OpenAL, and Delta3D suite to name but a few [11]. This area of computer science, once perceived only as of interest to the entertaining and gaming industry, is now almost omnipresent in all areas of serious research in the simulation, testing and verification stages of every technologically advanced project. This statement seems to be particularly true for motor industry research laboratories and design centres.

Hardware-in-the-loop testing (HIL) allows real hardware to be tested (i.e. physically existing modules wired to the PC system running the VR simulation software) in a VR environment in near-real-world conditions, taking advantage of the high level of complexity modelled into today's VR systems. Graphic engines, when combined with physics engines, sound effect engines, etc., provide an extraordinary tool to be used at the design and testing stage. This approach also involves a human operator (driver of the simulated vehicle) in the control loop of the system, as the driver is directly interacting with the system and operating all the available controls, and the evolution of the state of the system in time evolves in response to his / her actions. This approach is also sometimes referred to as H2IL testing, i.e. Hardware-and-Human-In-the-Loop systems.

Testing of the eCall unit in a conventional way would be difficult, time-consuming and prohibitively expensive [1, 2, 7, 10]. Although the VR / HIL testing is probably one of the best tools you can use, it can be used only for partial testing, and it will never become a substitute for an ultimate full-scale experiment and testing, as all mathematical models describe reality with ever better, but still limited precision and will never reflect the real world complexity and imperfections in every possible detail. Using the HIL allows thorough testing in all possible (predictable) working conditions, with varying values of motion parameters, road characteristics such as surface smoothness, friction parameters, road slope and road lateral profile.

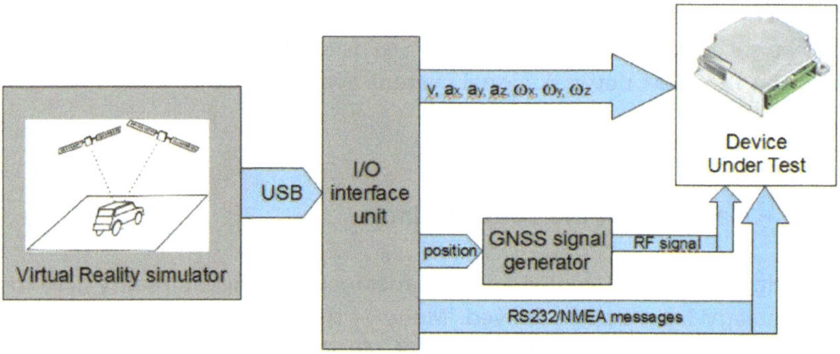


Fig. 1. Block diagram of the testing bench.

All simulation can be repeated several times with different selected parameters to assess how susceptible the device might be to these, and how the simulated situation would evolve in time for different parameters. Using scripting languages to programmatically control the simulated car and other traffic participants (agents) allows the researcher to control them as robotic cars (and other agents) in the virtual world.

All the estimated parameters of the current state of the VR system, such as vehicle speed, vehicle body spatial attitude, angular and linear accelerations, etc., can be converted to electrical signals, then wired to the signal processing inputs of the device under test (DUT).

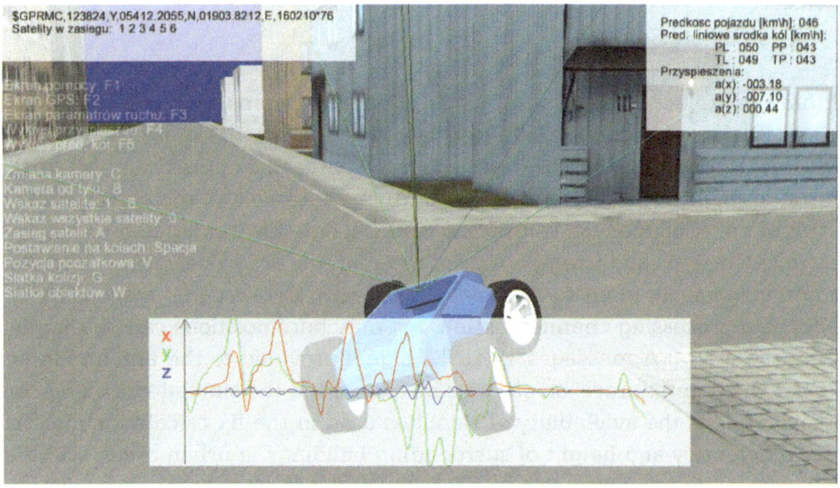


Fig. 2. Screenshot from the developed software, with the linear accelerations of the simulated vehicle drawn against time, and the auxiliary test lines testing direct availability of the satellite signals.

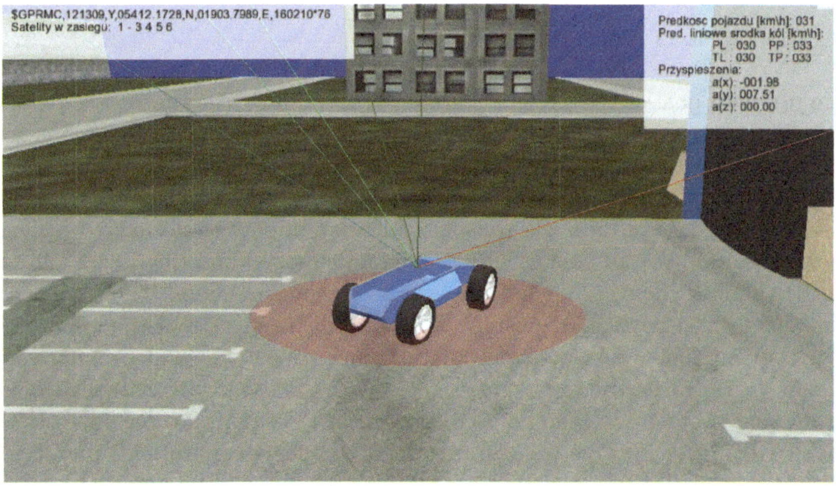


Fig. 3. Screenshot from the developed software, with the calculated position error range, as a function of simulated satellites availability and geometry, depictured in the simulation window using different colour.

## 5 The Bench Setup

Within the project scope generally, two main subsystems of the eCall on board unit are being tested: the GNSS receiver and the inertial measurement unit (IMU). In the case of the IMU testing, this approach requires removal of the sensor, and wiring the output of the hardware interface to the corresponding circuit of the eCall module. During the tests not only ideal estimated values, but also imperfections and irregularities of the real sensor device, can be modelled and provided as an input to the device. In the software module responsible for GNSS data processing channel testing, not only pure positions can be supplied as both, the NMEA messages, or GNSS signal provided to the antenna input, but the position fix precision can also be adapted to the simulated driving conditions, such as the availability of satellites used in the fix calculation (derived from the density and height of surrounding buildings in urban areas, width of the streets, and intensity of vegetation in simulated rural sites), the geometry of the satellite constellation effecting accuracy of the fix, with all effects being characterized by different values of DOP (Dilution Of Precision) parameters and the actual distortion applied to the position produced on the control output of the HIL system. In the case of using a GNSS signal simulator wired directly to the antenna input requested precision of the position fix can be controlled.

## 6 Conclusions

The ultimate goal of the project is to develop not only the methodology, but and also the complete hardware and software solution of a testing bench for the research and certification of the in-vehicle e-Call units. The test stand must allow replicable reproduction of precise and pre-programmed testing sequences of conditions, to excite the sensors and signal inputs of the tested device, so the resulting output control signals, and data logged into the device can be verified against the commanded values used during simulation. The developed methods proved particularly useful and efficient, as testing of the eCall unit in a conventional way would be difficult, time-consuming and prohibitively expensive. However, the testing carried out with the use of virtual reality tools, or any model-based tool, can be used only to certain extent, for partial testing, and it will never become a substitute for a full-scale experiment, prototype testing, and finally, for collecting data during the actual, real-world operation test. This is because of the inherent feature of any mathematical model, which describes reality with only limited precision, thus no model will ever reflect the real world complexity. All simulations can be repeated several times with selected parameters modified, making possible observations of performance of the device under test for different parameters and requested simulated conditions.

## References


[1] Bogus, P., Merkisz, J., Mazurek, S., Grzeszczyk, R., Application of GPS and other ORD sensors to detection of the vehicle state, Proceedings of FISITA 2008 World Automotive Congress, World Atomotive Congress 2008, Munich, Germany, p. 247, 2008.

[2] Bogus, P., Merkisz, J., Mazurek, S., Grzeszczyk, R., Detection Of Critical Situations In Rail Transport Using Ord Motion Parameters, Proceedings of the 10th International Conference on Application of Advanced Technologies in Transportation, Athens, Greece, 2008.

[3] Bouler, Y., et. al., Clarification Paper EG1-eCall Performance Criteria, Draft V14, 11 April 2006.

[4] Clarification Paper – EG.2. High level requirements for eCall In-vehicle system. Supplier perspective. Ver. 1.0.

[5] European e-Call functional specifications In Vehicle System, Working Document ver. 1.1., Vehicle Functionality Working Group (ECIV).

[6] Grewal, M., H., Weill, L. R., Andrews, A.P., Global Positioning Systems, Inertial Navigation, and Integration, Wiley, New Jersey, 2007.

[7] Grzeszczyk, R., Merkisz, J., Bogus, P., Kaminski, T., Methods and Procedures for Testing the eCall In-Vehicle Unit for the Purpose of Its Performance Assessment and Certification, 21st International Technical Conference on the Enhanced Safety of Vehicles ESV 2009, Stuttgart, Germany, 2009.

[8] Huang, M., Vehicle Crash Mechanics, SAE International CRC Press, 2002.

[9] Januszewski, J., Systemy satelitarne GPS, Galileo i inne, Wydawnictwo Naukowe PWN, Warszawa, 2006.

[10] Merkisz, J., Bogus, P., Wrona, A., Using Signals of Registered Linear and Angular Acceleration to Reconstruct the Trajectory of Vehicle's Movement, European Automobile Engineers Cooperation 2003, 9th EAEC International Congress European Automotive Industry Driving Global Changes, Paris, 2003.

[11] www.delta3D.org



**Jerzy Merkisz**
Politechnika Poznańska
Piotrowo 3
Poznań
Poland
jerzy.merkisz@put.poznan.pl

**Rafal Grzeszczyk**
Automex sp. z.o.o.
Marynarki Polskiej 55D
Gdansk
Poland
rafal.grzeszczyk@automex.pl

**Keywords:** eCall, hardware-in-the-loop , virtual reality, simulation, modelling

# 2020 Auto Sensor Vision

R. Dixon, J. Bouchaud, iSuppli Corporation


## Abstract

iSuppli examines the major trends affecting the automotive market over the next decade and assesses the potential impact for the sensor supply chain. Along with a historical perspective of the development of sensor implementation, fresh examples are given for the major domain subsectors—in powertrain management, for safety systems like ESC and in the cabin (body domain) to illustrate the next decade of sensor development in the vehicle.


## 1 Introduction

The automotive practice of iSuppli has recently undertaken a long-term examination of electronics trends as a result of changing architectures. In this article, iSuppli dusts off its crystal ball to look at the 2020 car and what it will mean for MEMS and related sensors. Here are some of the highlights with a focus on sensors. Three major driving forces will dramatically drive innovation in automobiles for at least the next two decades:

- "Green" issues will force the car industry to improve the efficiency of combustion engines (less $CO_2$ and less fuel used per mile). Related to this is the emergence in the public consciousness of the electric vehicle, for which the 2009 Frankfurt Automotive Show played no small part, with a host of new vehicle announcements coming from many major OEMs.
- The high cost of automotive-related accidents — both in terms of lives lost and overall economic costs — are driving new systems for accident mitigation, causing a refinement and improvement in existing driver assist systems.
- Automotive mobile connectivity (i.e. communication links) will foster huge innovation in new services and applications, benefiting customers through engine and chassis software updates, new maps, vehicle-to-vehicle communication, etc.

Much of this innovation will be handled by embedded software as hardware (ECU and associated electronic components) becomes more generic and its functionality can be programmed. The next generation of ECU system configuration has been developed by several European automotive suppliers (AUTOSAR), and will help speed this revolution across OEMs. As a result, existing systems will become more intelligent and have the potential to be updated during the life of the car. Hardware, in general, will be changed less and less.

## 2 Sensor Overview

But how will this trend impact sensor hardware? Today, as many as 150 different sensors — at least 30 of which are MEMS — can be found in high-end vehicles, distributed in as many as 70 different ECUs containing other electronic ICs like interface drivers and microprocessors. Having reached its peak, the number of ECUs is now set to decline. In comparison, the number of sensors has not crested, but growth in the number of new sensor-related applications has slowed down considerably. With hindsight, a large number of sensor applications — estimated to be around 30 to 35 — were already realized in production prior to 2000. In fact, the use of sensors started almost two decades ago — as early as the 1980s in the powertrain as air pressure and exhaust gas sensors were introduced to manage the combustion feedback loop, and in the 1990s with accelerometers for airbag systems. Other systems, such as radar, also originated around this time (1995) in luxury-class vehicles like the Mercedes S-Class. New solutions are now sought to lower costs to below $100 (for example, using SiGe radar chips) and allow wider adoption in this sector. An overview of sensor adoption — past, present, and future — is given in Table 1. A further uptake of sensors has taken place during the current decade, especially in chassis and safety applications like electronic stability control systems (ESC), tire pressure monitoring systems (TPMS), and active suspension (in the luxury segment).

ESC and TPMS are mandated technologies in the North American (U.S. and Canada) and European markets from 2012, and ESC continues to grow organically worldwide as a must-have safety feature. Both TPMS and ESC systems employ multiple sensors:

- ESC = typically a 2-axis accelerometer and single-axis yaw rate sensor, 1.2 pressure sensors in the brake modulator (on average), 4 wheel-speed sensors and one precision angular sensor in the steering (magnetic, optical, etc).
- An average of 4.5 pressure sensors is used in direct TPMS systems. For historical market reasons direct TPMS make up close to 100% of the U.S. market today. Due to the tougher environmentally-driven specifications for TPMS in Europe this market will also amount to close to 100% direct systems [1] when these systems are phased in on new 2012 series and 2014 on all new vehicles in Europe.

| *Epoch* | *Estimated number of sensors** | *Comments* |
|---|---|---|
| ***Pre-2000*** | *25 – 90* | *Rapid uptake of new sensors in luxury segment and beginning of key powertrain sensors like manifold air pressure, ABS, oxygen sensors, crank- and camshaft sensors, and airbags in the mid-range segment.* |
| ***2000 to 2009*** | *70 – 150* | *The number of sensors increases rapidly in mid-range vehicles, although overall, growth of new sensor applications has shown the first signs of slowing down. This is particularly true for MEMS-sensor-related applications.* |
| ***2010 to 2020*** | *110 –160* | *Sensor fusion of active and passive safety systems will dominate, (e.g. combining data from a number of sources such as cameras, radar, and ESC systems). No reduction in number of sensors, but no addition as well for these functions. Fast bus networks to provide more intelligent safety systems that further protect the car occupants by mitigating a crash or, if unavoidable, preparing occupants for the event.* |

Tab. 1. Evolution of automotive sensors, i.e. ultrasound, radar, cameras, acceleration, yaw rate, temperature, pressure, rotation, angular and linear position, flow, gas, infrared, sound, torque. * The ranges given assume a mid-class vehicle like the VW Golf and the S-Class Mercedes for the high end.

Powertrain applications — spurred by reduced emission requirements on vehicles mostly in developed markets — represent a strong driver for the insertion of many kinds of rotational and angular position sensors, not to mention pressure and $CO_2$ sensors as part of the closed loop engine management system. iSuppli analysis shows the number of sensors in the average powertrain will increase from 2.6 to 3.8 from 2008 to 2013, respectively [2].

In the body domain, many functions have moved from mechanical to electronic (e.g. central door locks, wipers, electric windows, sunroof) — all of which come with magnetic sensors, switches, and other similar controls. Other comfort and convenience applications — including rain sensors, door mirror switches, or advanced climate control systems — which started in high-end cars, have rapidly percolated down into lower-priced automobiles, and show no evidence of slowing down either, growing from an average of 6.7 per car in 2008 to 9.4 in 2013 [2].

## 3 System Overview

At the system level, a distributed modular architecture is evolving, with several Electronic Control Units (ECUs) managing several functions via a high degree of networking. The ECUs are increasingly complex, comprising selftest and calibration, data reduction and bus interface(s) which possess an increasingly higher functional integration. Micro-machined (MEMS) devices – especially sensors – are well suited to meet requirements for small size and functional integration.

Nonetheless, the number of ECUs will fall considerably in the next decade to just a handful of so-called "domain controller" ECUs in order to reduce weight and complexity. These domain controllers, for example, collect all powertrain sensors in a "powertrain domain" ECU or gather relevant safety sensors in a large "safety domain" ECU, with several microcontrollers to handle the ever-increasing demand on processing power.
Another scenario is the "zone controller," which collects unrelated sensors together (e.g. non-safety critical devices) that are in a similar location in the vehicle.

## 4 Impact on System Suppliers

What are the implications for sensor requirements and sensor suppliers? The following table highlights some of the potential changes for sensors ahead.

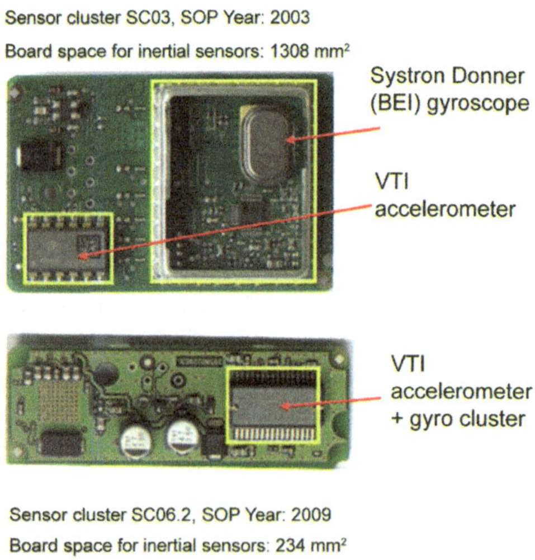


Fig. 1. Sensor consolidation: Bosch and Continental are introducing new generations of sensor clusters for ESC. This example of a cluster—from Continental using inertial sensors from VTI—is 5x times smaller than its 2003 model (CAS)

| *System trend* | *Impact on sensors* |
| --- | --- |
| **Fewer ECUs in future:**<br>*Peak today with 70+ ECUs moving to architecture with a handful of "domain ECU controllers"* | **Major sensor changes:**<br>*Premium on footprint, power consumption. Solutions include...*<br>*• Co-location of sensors in single packages (happening today at Bosch and Continental with "combination or multi-sensor" packages). Advantage: smaller footprint, e.g. see Fig. 1 comparing old and new ESC inertial sensor cluster solutions from Continental.*<br>*• Co-locate sensors on same piece of silicon, monolithic-integrated inertial sensors for ESC, sensors with shared ASIC.*<br>*• More intelligence in sensors (but still limited compared to ECU intelligence, which features millions of lines of S/W code). No general trend for distributed intelligence with 8-bit microcontrollers in each sensor.* |
| **Extension of system functionality** | **No impact for sensors:**<br>*More features at system level, leveraging existing sensors mostly with software adjustments. As early as 2004, ESC systems were extended by Bosch, Conti, and Delphi to include brake assist, roll detection and active steering intervention and other refinements of the brake controller.* |
| **Increasing bandwidth needed for communications among modules:**<br>*Applications in powertrain and chassis, and increasing data required for ADAS applications (examples are combination of ESC with ACC, airbags with radar and camera data)* | **Impact for sensors:**<br>*Sensors must accommodate several bus speeds (e.g. CAN/Flexray in powertrain/chassis). Depending on functionality (air bag/ESC) the bandwidth requirements differ and need to be accommodated in the integration process.* |
| **Sensor fusion:**<br>*Before moving to full domain controllers, more intelligent electronic systems will emerge as a result of sensor fusion. Examples are vehicle dynamics sensors combined with camera and radar data with airbag systems. This information provides additional time to avoid a crash or to prepare the occupants (seat positioning, pre-tensioning of belts, optimizing firing sequences for curtain bags in event of roll, etc)* | **Major sensor changes:**<br>*Does not mean fewer sensors, as software will combine existing sensors. Fusion requires a great deal of difficult systems engineering, however....*<br>*• It is unlikely that a safety sensor cluster will be available that does everything due to the requirements of different vehicle subsystems. Example: some applications have different specifications, such as roll and electronic parking brake accelerometers. Perhaps more important, an airbag accelerometer should not suffer latency due to bandwidth issues as a result of connection with other systems and sensors.*<br>*• Compounded issues with fault tolerance and level of redundancy must be considered (e.g. if IMU cluster fails, then potential vital operations could be lost in several locations).*<br>*• Component suppliers will need increasing systems appreciation. This could benefit an integrated company like Bosch, but many Tier 2s already talk regularly with OEMs.* |

Tab. 2. System trends and implications for sensors to come

Finally, what about the potential disruption of mandates and new propulsion systems on sensors? There are three major forces acting on the market:

- Externally imposed regulations, especially mandates for TPMS, ESC, and regulations on $CO_2$ and other gas emissions are a force for good as they drive the deployment of sensors. On the other hand, mandates act like glaciers, inexorably bringing change through commoditization, supply chain shifts, and — ultimately — innovation as companies fight to retain margins.
- Electric vehicles are, on the one hand, bad news for the suppliers of existing sensors for internal combustion engines (e.g. pressure, flow and gas sensors). On the other hand, there will be other opportunities for compact Hall-type current sensors (targeting replacement of bulky transformers and shunt resistors) in electrical current monitoring as well as for encoders and angular-position sensors used in all manner of electric motors and pumps that will replace inefficient belt-driven versions. Despite recent interest in EV, iSuppli believes only a small portion of cars will be fully electric by 2020.
- Hybrid cars will be much more common up to 2020, and have no effect on powertrain sensors. In fact, the future for sensors looks good in hybrid-electric vehicles, which will combine electric motors with downsized, turbocharged ICE engines for greater range. These ICE and electric propulsion systems ultimately need position sensors (e.g. for turbocharger, EGR vane, etc.) and better angular measurement (new variable valve systems).

## 5 Conclusions

The situation looks bright in the coming decade for automotive MEMS sensors as safety mandates continue to stimulate markets for existing sensors. Inevitably, these mandates will extend to advanced driver assist systems, further entrenching the important role of sensors.

Major changes at the system level will also be important in driving sensor design innovation in the future. In addition, emerging markets like China and India will play a key role for sensor companies as they accommodate different system requirements.

Data fusion is not a threat to existing sensors, but fewer or even no major new sensor category will emerge. However, Tier 1's in the developed markets continue to innovate, and particularly, the body domain of the car will continue to be a fertile area for differentiation for all kinds of sensors and switches. While the number of ECUs has peaked, the number of sensors has not done so in similar fashion and will grow much more slowly in the next decade.

## References

[1] Dixon, R., Bouchaud, J., iSuppli, MEMS Automotive Market Tracker H1, 2010.

[2] Dixon, R., iSuppli, Special Report Magnetic Sensors H2, 2009.

**Richard Dixon, Jérémie Bouchaud**
iSuppli Corporation
Spiegel Str. 2
81241 Munich
Germany
rdixon@isuppli.com
jbouchaud@isuppli.com

**Keywords:** MEMS, magnetic sensors, sensors, automotive architecture, ECU, ADAS, sensor fusion, safety mandates, hybrid electric vehicle, multi-sensor package, TPMS, ESC

# Traffic Management

# Towards an Active Approach to Guaranteed Arrival Times Based on Traffic Shaping

D. Marinescu, M. Bouroche, V. Cahill, Trinity College Dublin

## Abstract

To address the goal of providing drivers on highways with guaranteed arrival times, we propose a traffic management system that combines virtual slots with semi-autonomous driving to shape traffic and prevent congestion. A possible application of this system to address efficient merging from three to two lanes is outlined here. Finally, the technical requirements for traffic shaping using virtual slots and semi-autonomous driving are presented.

## 1 Introduction

Increasing road usage has led to congestion on highways as well as on rural and urban roads. In the past, congestion was commonly addressed by building new roads. Due to its financial, social and environmental implications this approach is becoming increasingly infeasible [1]. Starting from the early 1980s [2], an alternative solution to this problem has been investigated by designing and deploying intelligent transportation systems (ITS). Besides reducing congestion, ITS also addresses problems like safety and cost reduction. In this context, our work focuses in general on how ITS can be used to reduce highway congestion by increasing traffic efficiency, and more specifically on providing guaranteed arrival times to drivers on highways.

Most previous work has been focused on avoiding congestion [3, 4], increasing the capacity of the highway [5, 6] and so reducing average travel time. It has been observed that drivers would prefer to have predictable travel times in preference to on-average shorter, but unpredictable, travel times [7]. The motivation for this ranges from commuters wanting to make sure they arrive at their work place in time to businesses wanting to make sure that their deliveries arrive in time. Despite this, to the extent of our knowledge, little attention has been paid to providing guaranteed arrival times on highways. Current state-of-the-art approaches to obtaining reliable travel time estimates are passive, in the sense that they attempt to predict travel times using both historic traffic data and real-time traffic observations [8]. However, these approaches fail in the presence of unexpected events such as accidents and road works, which cause sudden bottlenecks and may lead to prolonged congestion [9].

To address the uncertainties arising from the occurrence of such events, we propose an active approach to providing guaranteed arrival times. Unlike passive travel-time prediction, an active approach implies some degree of actuation on vehicles so that initial arrival time guarantees can be maintained. In the case of bottlenecks, actuation aims to maintain the same throughput in the area of the bottleneck as elsewhere on the highway, thus maintaining the arrival-time guarantee. The problem of maintaining the travel-time guarantees in the presence of unexpected bottlenecks is thus reduced to the problem of maintaining the same throughput. We address this problem by designing a system that actuates directly upon the vehicle by combining virtual slots with semi-autonomous driving, and show how this concept can be applied on a three-lane to two-lane merging scenario.

The rest of this paper is organised as follows: section 2 defines and exemplifies our concept of traffic shaping using virtual slots. The technical requirements of our approach are presented in section 3. Finally, our conclusion as well a perspective on how these concepts could be put into practice are presented in section 4.

## 2 Traffic Shaping Using Virtual Slots

Traffic shaping is a general term that encompasses several techniques used in computer networks to optimise traffic and guarantee delivery times. One technique in particular, called leaky bucket [10], is used to control the rate at which packets enter a network, similarly to the way in which ramp metering is used to control access to the highway [11]. We adapt these techniques to highway traffic by mapping vehicles to virtual slots and show how they can be used to provide guaranteed arrival times.

### 2.1 Virtual Slots

In 2005, Morla proposed that congestion-free travel can be guaranteed by assigning slots to vehicles [12]. In Morla's vision, a slot represents a time-space corridor negotiated among vehicles. The speed of slots on the same lane is the same and vehicles are driven by human drivers, supported by a driver assistance system.

The same concept has been advocated two years later by Ravi et al., in a more business-centric vision [13]. Their approach permits the coexistence of a slot-based system with the currently available driving patterns by building a high-priority lane. In similar work [14], Cahill et al. proposed a time division multiple access (TDMA) inspired approach to allocate slots to vehicles on the highway.

Our traffic shaping approach builds on this idea of a virtual slot. We extend the previous notion of a lane-tied slot moving at a constant speed to a more flexible one, where a slot can accelerate, decelerate and change lanes. A vehicle driving in a slot shall replicate the behaviour of its slot. For the scope of this paper we propose the use of virtual slots with predefined behaviour. By predefined behaviour we understand an a priori known speed profile and trajectory. As such, we define a slots $S = \{z,p,t,b\}$ where $z$ represents the size of the slot, $p$ represents the position at time $t$ and $b$ the predefined behaviour. Possible behaviours might describe merging patterns, allowing emergency vehicles to pass and others.

## 2.2 The Advantages of Virtual Slots Against the Current Driving Pattern

The nature of human driving is inherently competitive [15]. When driving on highways, this competitive nature is somehow limited by the set of driving rules. Unfortunately, these rules were chiefly designed to address safety issues only. Too often we find drivers driving on the fast lane instead of the slow lane, not making space for faster cars or cars that intend to exit the highway. No legally binding driving rules exist forbidding such behaviour or, where they exist, they are seldom enforced. The result is the same: an inefficient form of driving, that is competitive rather than cooperative.

Unlike current driving behaviour, the predefined nature of a virtual slot means that one can design a set of behaviours that, when assigned to different slots, work together in a cooperative manner. Our hypothesis is that such behaviour is not only predictable, but can also be more efficient that the current driving pattern. To show this we chose an example where road works force a three-lane highway to merge into a two-lane highway. This causes a bottleneck and, under sufficiently high traffic volumes, can cause congestion.

One obvious reason for the occurrence of congestion in such scenarios is the drop in capacity experienced when a three-lane highway merges into a two-lane one. However, we believe that another cause for congestion is the inefficient driving behaviour of humans. To test this hypothesis, we have run a set of experiments using the VISSIM traffic simulator. We compared the total travel time of vehicles driving on a two-lane highway against the total travel time of vehicles driving on a three-lane highway that experiences a drop to two lanes. We used the same traffic volume of 4000 vehicles per hour and total travel distance of approximately 2715km. We ran each of the two experiments 10 times and we obtained an average total travel time of 199 seconds for the two-lane experiment and an average of 230 seconds for the three-lane to two-lane. Assuming that the Wiedemann driver model used by VISSIM correctly simulates human driving behaviour, these results indicate that merging from three to two lanes, as performed by human drivers, is inefficient and causes congestion that results in an increase in the average travel time.

We propose a set of predefined virtual slot behaviours to perform a more efficient merging procedure. As shown in figure 1, the behaviours are classified in two main categories: A and B. Category A slots are immediately followed – within the same lane – by category B slots. Each of the two categories has three subcategories, numbered 1, 2 and 3 after the lane in which the slots are initially moving. This results in six different slot behaviours. Note how at the beginning the slots are aligned and then B-slots reduce the distance to A-slots, leaving an empty space between these newly formed pairs. This space is then used by A2 slots to move to lane 1. Similarly, A3 and B3 slots use the remaining space to change to lane 2, leaving no slot and thus no car on the lane 3. While slots in different positions have different behaviour, their combined emergent behaviour is a merging from three to two lanes that does not require slots to slow down and thus maintains the throughput through the bottleneck at the same level as elsewhere on the highway. In general, the predefined behaviour of a slot shapes the traffic in a predictable way. Strong guarantees can therefore be provided with regard of the travel time of a slot and implicitly of a vehicle.

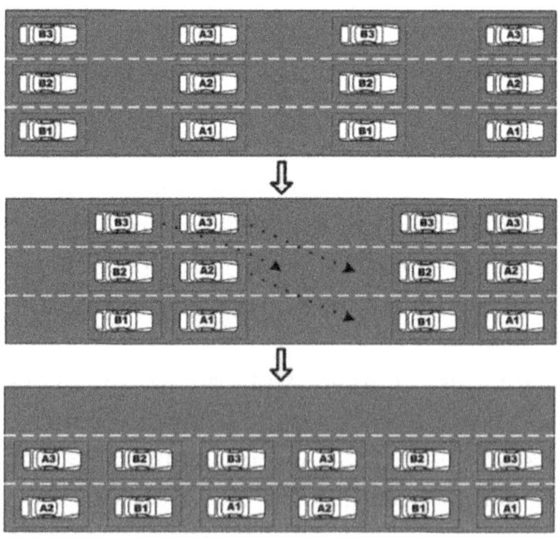

Fig. 1. Lane merging procedure using virtual slots.

## 2.3 Driving Within a Virtual Slot

Before a vehicle can drive within a slot, it needs to be aligned with the slot. The respective slot can either be assigned by the infrastructure or the vehicle can find the empty slot that suits it best. Furthermore, this mapping can be performed either when entering the highway or on the fly when the vehicles are informed that they need to get into slots. Since driving in a slot reduces the freedom of choice of the driver, this could be an impediment to the acceptance of the system. As such we believe that virtual slots should not be used in free traffic flow conditions, but only when this is mandatory to prevent congestion and so maintain the arrival time guarantees. For the lane-merging scenario, this means vehicles are informed on the fly that they need to move into slots and then choose their own slots. We are currently working on designing an efficient algorithm that vehicles can use to decide which slot to join. The algorithm is based on the idea that slots are generated with a headway equal to the average headway between vehicles driving on the highway. Some vehicles will need to slow down and some will need to speed up, but overall the throughput is the same and so the guarantees are maintained.

Once vehicles are in slots they shall replicate their slot's behaviour. To achieve this, the driver needs to be assisted by the vehicle. Such assistance can range from fully autonomous driving to simply providing the driver with the information necessary to replicate the behaviour of the slot. While fully autonomous driving mode has been demonstrated on previous occasions [5, 6], it is now commonly believed that this form of driving will not be widely deployed in the near future. On the other hand, relying on drivers following the advice of the vehicle to perfection is also unrealistic. As such, a balance needs to be obtained between autonomous driving and manual driving. In general, we refer to this as semi-autonomous driving. From the perspective of the longitudinal (throttle) and lateral (steering) control, several forms of semi-autonomous driving can be designed, such as:

- Assisted longitudinal control and manual lateral control
- Assisted longitudinal and lateral control
- Autonomous longitudinal control and assisted lateral control

Transferring the longitudinal control of the vehicle from the human driver to the vehicle poses a great advantage. Currently, the minimal distance between two vehicles is limited by slow human reaction times, which varies between 1.2 and 1.7 seconds [16]. When the human driver is removed from the control loop the reaction times are significantly reduced. This means that slots can be packed tighter together on the highway and so the capacity of the highway increases. For the lane-merging scenario, we propose a combination of automated longitudinal control and assisted lateral control

## 3 Technical Requirements

Shaping traffic by means of virtual slots poses some requirements on the underlying technologies. We discuss here the technical requirements of virtual slots.

### 3.1 Communication, Time Synchronizations and the Role of the Infrastructure

For a vehicle to be able to replicate the behaviour of a slot it needs to be able to know the current position, speed, acceleration, trajectory as well as the set of manoeuvres the slot is going to perform in the future. Since the behaviour of a slot is predefined, the current position can be determined by knowing this behaviour and the position of the slot at a certain point in time. For that, the clocks of all vehicles and infrastructure components need to be synchronized, which can be easily achieved nowadays for example using GPS clocks. The slot information needs to be communicated by the infrastructure to the vehicles. This form of communication is generally referred to as vehicle-to-infrastructure (V2I) communication.

Depending on the scenario, V2I communication might have stringent real-time requirements or not. For example, for a merging scenario that was caused by road works, the slot information could be provided in advance using the already widely deployed 3G networks or 4G technologies such as WiMAX that are in the process of being deployed on a wide scale. For other scenarios, the information might need to be communicated in real-time and so real-time vehicular communication would be required. A considerable research effort is currently targeting the development of real-time vehicular communications [17].

### 3.2 Absolute Positioning, Digital Maps and Sensing

When a vehicle drives within a virtual slot, its speed and trajectory need to replicate the speed and trajectory of the respective slot. To achieve this, the vehicle must be aware of both its own position and the position of the slot. This information can be augmented with digital maps of the highway for better accuracy (e.g. determining the lane a vehicle currently drives in). While the absolute position of the slot can be determined at each point in time using dead reckoning and the slot information communicated by the infrastructure, the absolute position of a vehicle must be determined using a global navigation satellite system (GNSS).

The accuracy of the slot positioning depends on the accuracy of the initial slot position, the accuracy of the digital maps and the accuracy of the dead reckoning method used. Since the slot can be generated at a well-known position and has a predefined behaviour, this accuracy should be very good and the eventual error range neglectable. The size of a slot is directly proportional to the error-range of the GNSS technique used by the vehicle. For example, a positioning error of up to one meter means that the slot needs to be at least two meters larger in length than the car. The error-range is not relevant for the width of the car if we assume that a vehicle can determine that lane on which it's currently driving, possibly using a combination of stereo vision and digital maps.

The highway capacity of the slot-based system is inversely proportional to the size of the slot. As such, the error-range of the GNSS in the vehicle has a direct effect on the capacity of the highway: the bigger the error range, the smaller the capacity and the throughput. While widely deployed GNSS systems such as GPS have ranges of up to 20 m, others such as differential GPS (DGPS), with a standard deviation of typically 0.2 - 0.5 m, and Galileo in the future, as well as the use of dead reckoning provide the required accuracy and are likely to be deployed in vehicles within the next few years.

Finally, for a vehicle to decide whether it can get into a slot it first needs to determine whether the slot is empty or not. This can be achieved using radar- or lidar-based sensing technologies such as the ones used by adaptive cruise control (ACC) systems [18] already available in some cars.

### 3.3 Driving Assistance and Vehicle Control

A semi-autonomous form of driving implies some driving assistance from the vehicle. Depending on the precise scenario, this can be achieved by means of voice, haptic feedback or by using a heads-up display. Semi-automated vehicle control, in particular longitudinal [19] but also lateral control [20] has been designed, implemented and proven to work. However, except for ACC systems, such technologies have not yet been deployed in production cars. This is mostly due to the liability implications of using (semi-) automated control. Nevertheless, with time this technology will mature and will eventually be deployed in production vehicles.

### 3.4 Failure Modes

One very important issue that needs to be addressed when the driving task is shifted from the human driver towards the machine is the failure model. Errors need to be detected and failure modes activated. Redundancy is the elementary technique used when fault tolerance must be provided. Where absolute positioning might fail, relative positioning using radar or lidar should assure that vehicles do not collide. Manual override from the human driver should also be available.

Obviously safety takes priority over efficiency, and failing back to a mode that ensures safety but does not necessarily keep the arrival times guarantees is an acceptable solution. A switch back to fully manual driving is a feasible option as long as the transition from automated to manual can be implemented taking into account the limitations of the human driver such as the slow reaction times.

## 4 Conclusions and Future Work

This paper introduced the concept of providing guaranteed arrival times by means of traffic shaping. Our approach of vehicles driving in virtual slots was presented and the technical requirements of implementing this were discussed. In the future we plan to thoroughly prove our hypothesis, and we will start by implementing and evaluating our proposed solution for the three-to-two lane-merging scenario.

The technical requirements of virtual slots have already been addressed by research. Prototypes were developed and proven to work, and some of this technology is already deployed in some production vehicles. The emergence of electric vehicles is a good opportunity for bringing more of these technologies into production cars.

## References

[1] Ioannou, P., Wang, Y., Chang, H., Integrated Roadway / Adaptive Cruise Control System: Safety, Performance, Environmental and Near Term Deployment Considerations, California PATH Research Report, 2007.

[2] Figueiredo, L., et al., Towards the development of intelligent transportation systems, Proceedings of Intelligent Transportation Systems Conference, pp. 1206 - 1211, 2001.

[3] Ghods, A., et al., Adaptive freeway ramp metering and variable speed limit control: a genetic-fuzzy approach, Intelligent Transportation Systems Magazine, pp. 27 - 36, 2009.

[4] de Bruin, D., et al., M., Design and test of a cooperative adaptive cruise control system, Proceedings of the Intelligent Vehicles Symposium, pp. 392 - 396, 2004.

[5] Horowitz, R., Varaiya, P., Control design of an automated highway system, Proceedings of the IEEE, 7, pp. 913 - 925, 2000.

[6] Kato, S., et al., Vehicle control algorithms for cooperative driving with automated vehicles and intervehicle communications, IEEE Transactions on Intelligent Transportation Systems, 3, pp. 155 - 161, 2002.

[7] Recker, W., et al., Development of Methods and Tools for Managing Traffic Congestion in Freeway Corridors, Proceedings of Intelligent Transportation Systems Conference, pp. 30 - 37, 2006.

[8] Margulici, J.D., Ban, X., Benchmarking travel time estimates, IET Intelligent Transport Systems, 3, pp. 228 - 237, 2008.

[9] Yulong, P., Leilei, D., Study on Intelligent Lane Merge Control System for Freeway Work Zones, Proceedings of Intelligent Transportation Systems Conference, pp. 586—591, 2007.

[10] Turner, J.S., New directions in communications (or which way to the information age?). IEEE Communications Magagazine, 24, pp. 8 – 15, 1986.

[11] Papageorgiou, M., Kotsialos, A., Freeway ramp metering: an overview, Proceedings of Intelligent Transportation Systems Conference, pp. 228 - 239, 2002.

[12] Morla, R., Vision of Congestion-Free Road Traffic and Cooperating Objects, Technical Report, 2005.

[13] Ravi, N., et al., Lane Reservation for Highways (Position Paper), Proceedings of Intelligent Transportation Systems Conference, pp. 795 - 800, 2007.

[14] Cahill, V., et al., The managed motorway: real-time vehicle scheduling: a research agenda, HotMobile '08: Proceedings of the 9th workshop on Mobile computing systems and applications, pp. 43 - 48, 2008.

[15] Vanderbilt, T., Traffic: Why We Drive the Way We Do (and What It Says About Us), Alfred A. Knopf, New York, 2009.

[16] Maciuca, D.B., Hedrick, K.J., Brake Dynamics Effect on Ahs Lane Capacity, Proceedings of the Future Transportation Technology Conference and Exposition, 1995.

[17] Willke, T.L., et al., A survey of inter-vehicle communication protocols and their applications, IEEE Communications Surveys and Tutorials, 11, pp. 3 - 20, 2009.

[18] Abou-Jaoude, R., ACC radar sensor technology, test requirements, and test solution, IEEE Transactions on Intelligent Transportation Systems, 3, pp. 115 - 122, 2003.

[19] Vahidi, A., Eskandarian, A., Research advances in intelligent collision avoidance and adaptive cruise control, IEEE Transactions on Intelligent Transportation Systems, 4, pp. 143 - 153, 2003.

[20] Kumarawadu, S., Lee., Tsu T., Neuroadaptive Combined Lateral and Longitudinal Control of Highway Vehicles Using RBF Networks, IEEE Transactions on Intelligent Transportation Systems, 7, pp. 500 - 512, 2006.

**Dan Marinescu, Mélanie Bouroche, Vinny Cahill**
Trinity College Dublin
Dublin
Ireland
dan.marinescu@cs.tcd.ie
melanie.bouroche@cs.tcd.ie
vinny.cahill@cs.tcd.ie

**Keywords:** virtual slots, traffic shaping, semi-autonomous driving, guaranteed arrival times

# Simulation Process for Vehicle Applications Depending on Alternative Driving Routes Between Real-World Locations

A. Lamprecht, Technische Universität München
T. Ganslmeier, Audi Electronics Venture GmbH

## Abstract

State-of-the-art vehicle navigation systems not only offer the option of calculating the fastest or shortest route between two known locations on a digital road network, they also offer energy optimized routes as an alternative to the driver. For being able to simulate this problem, it is necessary to not only utilize a high-precision vehicle model, but also to use an exact model of the routes which have to be investigated. The environment in which the virtual vehicle is driven has to be a precise model of the real road network. In order to minimize the effort of simulation, it is essential to automate the generation process of the virtual environment. A simulation is developed in which a virtual vehicle can drive along a model of real roads. All available environment information such as speed limits, lane information, presence and position of stop signs, road curvature and inclination can be used in the simulation as if the vehicle would drive along those routes in the real world.

## 1 Introduction

Virtual product development plays an important role in the development of many automotive applications. During the past year, dynamic vehicle models were designed that are able to compute all possible vehicle parameters in real-time. Combined with simulation databases that correspond to the real world, this approach leads to a lot of advantages in the development- and testing stage over real test drives. Functional applications can be used early in the development stage and ideal sensor information can be provided while testing, a highest possible parameter variation can be achieved and all conditions during a simulation are totally reproducible. At the end, physical prototypes with a more mature development status can be built.

This paper describes a simulation process to validate alternative driving routes between two locations in the real world. The routes which have to be analyzed are modeled into an environmental simulation. The comparison to a real test drive is made possible because of the simulation's correlation to real world coordinates. Thus, conventional vehicle routing algorithms can be analyzed. The objective of the project is to verify whether a real test drive can be replaced with simulative tools. A simulation is being used which allows to employ predictive applications in a closed-loop scenario and is able to compute all necessary sensor information in real-time. The usage of virtual methods enables the reduction of development costs despite of the rising complexity, the interconnectedness and the used perception of the environment. Real test drives can be reduced and the development can be done without relying on physical prototypes. The used environmental simulation supports four different types, such that an integrated development process is possible [1].

The simulation type **Hardware in the loop** (HiL) enables testing of real electronic control units (ECU). The vehicle application is implemented on the target ECU and supplied with sensor information from the simulation. In a HiL simulation, all sensor data must be computed in real-time to synchronize it with the connected hardware components. One possible simulation configuration consists of a target ECU and a development PC which runs the environment simulation.

**Driver in the loop** (DiL) is a static simulator where a driving dynamics model is interconnected with an environmental simulation. A user interface consisting of steering wheel, accelerator- and braking pedals, other operating elements and a visual representation of the current state of the simulation takes the test participants into a closed-loop scenario. Applications relevant for safety can be simulated without any threat to humans or material.

**Vehicle in the loop** (ViL) is a variant of DiL where the function which has to be analyzed is implemented in an actual vehicle. Sensor information like cars driving in front of the ViL-simulator are delivered to the system as virtual objects from the simulation and visually represented to the test participants via a head mounted display. This simulation variant in particular has high importance for applications that are relevant to safety. Critical scenarios like rear-end collisions can be simulated in a safe and efficient way.

Within the scope of this project, the simulation variant **Software in the loop** (SiL) was widely used. Predictive vehicle applications can be tested in an early development stage without the need for physical prototypes or ECUs. The environmental simulation generates the necessary sensor information and provides it to the vehicle model. In this study, a mathematical vehicle model of an Audi Q5 was used to model the physical properties and the demand for energy of a real vehicle. As a reference for validation purposes, an identical vehicle was available for test drives.

## 2 Method

For a test route between the cities of Manching and Gaimersheim alternative driving routes have been created by a navigation algorithm as described by Barth [2] and Bartsch [3]. The innovative algorithm not only computes the fastest and shortest path between two known geographical locations on a road network, it also determines the path with the least amount of energy necessary to drive on the route. Using an environmental simulation, the associated simulation database was recreated with the procedure described in the following paragraph. The virtual vehicle then drives along those routes in the simulation where all interesting internal and external data is stored. Afterwards, the corresponding routes were driven with a real test vehicle where again all necessary data was stored for validation purposes. The underlying concept of this method is to reduce the complexity of validation by introducing a simulation strategy consisting of three stages. The first stage aims at reaching the highest possible number of random samples with the involved consequences of low preciseness and accuracy. The final stage of a real test drive on the other hand delivers results of real-world tests, but is very complex and expensive to conduct. The introduced simulation method aims at installing an intermediate stage between the simple simulation and the real-world test drive where a high precision model of driver, vehicle and environment deliver results close to the real world with the costs and effort of only a software simulation.

## 3 Simulation

In an environmental simulation it is necessary to recreate driving routes from the real world in high quality to validate the experiments in real test drives. The following paragraph describes an innovative way of how the simulation database can be generated automatically and then be used in a virtual test drive.

### 3.1 Generation of the Simulation Database

The simulation database is usually created with a software editor that allows building of arbitrary road-segments and their attributes such as speed limits, the lane width, the number of lanes or the type of road. In general it is possible to define the simulation database manually, but this procedure is not feasible regarding the scope of the project with the extensive amount of road kilometers that would have to be created. The goal of the project therefore is to utilize available map data to automate the process of generating a simulation database. Within the scope of this project, map data provided by the company NAVTEQ was widely used. Nearly all necessary attributes are available in their software suite ADASRP. Since so far only highways are digitalized with altitude information in the used map database, not all possible roads could be reconstructed as a simulation database. The initial database was therefore expanded to another source of data. Altitude information from the ASTER mission was used to enrich NAVTEQ's map data with altitude information. ASTER is a high-spatial resolution, multispectral imaging system flying aboard TERRA, a satellite launched in December 1999 as part of NASA's Earth Observing System (EOS). According to Al-Hani [4] the digital elevation model (DEM) of ASTER with 15 m spatial resolution in general is better than 20 m. The best potential accuracy for the DEM from ASTER using the root mean square error (RMSE) is about $\pm 9.42$ m. In urban areas as well as woodland, the satellite data does not necessarily represent the earth surface but could also refer to the altitude of houses and trees. As discussed later on, this phenomenon raises some problems.

The simulation database is defined by a logical and a graphical part with the format conventions of OpenDRIVE and OpenFLIGHT [5] respectively. OpenDRIVE describes the course of the road consisting of straight lines, circular arcs and clothoids. Turning restrictions and junctions are defined by the logical links between the segments. Additional information such as the number of lanes, the width of the lane or traffic lights is defined by a separate attribute of each segment. The basic mechanism for generating an OpenDRIVE database can be seen in figure 1. For lack of interest, the graphical representation of the simulation was not pursued within this project because it has no influence in the energetic analysis of alternative driving routes.

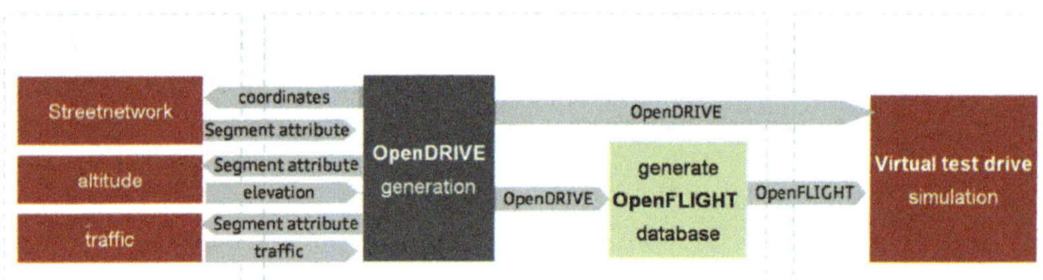


Fig. 1. Pre-processing the OpenDRIVE database

The creation of the virtual database is a necessary step before the simulation can take place. In this preprocessing step, bases for the road curvature are being calculated out of the NAVTEQ map data. The road curvature is fragmented into straight lines, circular arcs and clothoids. The elevation of the road segments is defined by cubic polynoms. Each road segment is sampled with a constant step size. The corresponding altitude information from the ASTER database is calculated. Due to the low spatial resolution of the satellite data, not every data point from ASTER is located on a road. The needed information is interpolated out of the four surrounding elevation points like shown in figure 2.

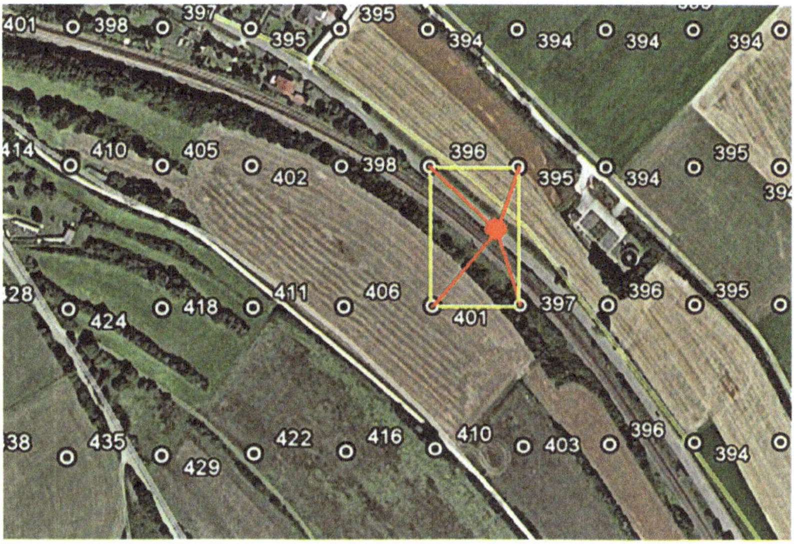


Fig. 2. Interpolation of ASTER elevation data

The mapping of the simulation database with real world coordinates is realized by using the universal transverse Mercator [6] coordinate system in both scenarios Hager [6]. There are two methods available for generating the virtual simulation database. The first one only reproduces the actual route, which is calculated by the navigation algorithm, itself. The second method reconstructs all roads within a bounding box which contains the route. To increase performance, the algorithm can optionally exclude a set of road classes, e.g. highways or regional roads.

### 3.2 Virtual Test Drive with Driving Dynamics Model

The generated road trajectories are being used in a virtual test drive. By using a simulation process consisting of a vehicle model, a driver model and an environment simulation, the test drive can be modeled very realistic. Figure 3 shows the general modules of the complete closed-loop simulation. The components communicate via TCP/IP network connection and compose a closed-loop simulation where the driver and the environment can interact using the vehicle model from which measured data is recorded.

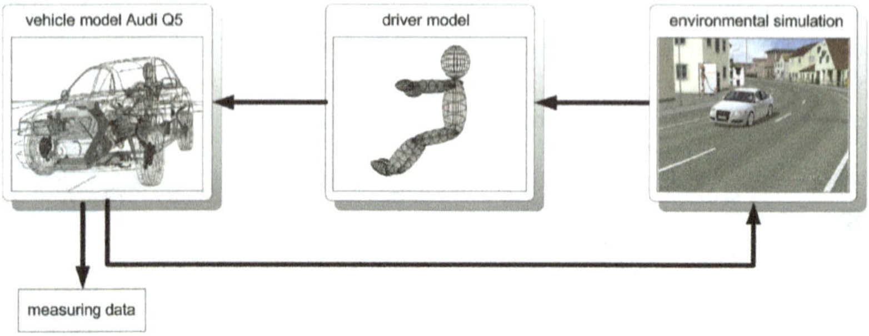


Fig. 3. Modules of the simulation

To provide similar surrounding conditions in the simulation and the test drive, disturbance variables like traffic are being excluded by simply avoiding it in the real test drive. Including traffic would be an important next step in enhancing the quality of the simulation. Parameters like the cycles of traffic lights and speed limits are defined prior to the simulation in the so called scenario. The driver parameters were chosen according to a normal driver as defined by Ebersbach [7].

## 4 Evaluating the Quality of the Environment Simulation

In order to validate the quality of this automatically generated virtual environment, several options would be possible. The algorithm which reconstructs the simulation database out of navigation map data does not deliver a one-hundred percent success rate. Due to the modeling quality of the real world into the navigation database and then to a simulation environment, several problems arise due to logical link errors in the simulation database. Sometimes the algorithm is not able to find the correct predecessors and successors of each road segment. Especially in inner-city situations with roundabouts or multiple lanes present, the algorithm sometimes encounters problems which have to be corrected manually. The success rate however is a first indication for the quality of the algorithm.

Furthermore, a validation method shall be pursued which analysis the simulation database in terms of the data needed for the applications intended to use the simulation. The essential quality criteria are the road gradient in terms of altitude as well as the road curvature in the two dimensional planar surface. The reconstructed simulation environment is benchmarked against data measured from a real test drive. A test vehicle was equipped with an inertial measurement unit, a GPS-receiving unit and data loggers which were attached to the vehicle's CAN bus. All measured data was synchronized and stored into one consistent log file. The GPS receiver was operating at an update frequency of 5 Hz which delivers an absolute accuracy of about 30 m. The focus of this investigation is relative correctness, thus the absolute error of a GPS-signal has no significant influence on the road gradient and road curvature.

## 4.1 Comparison Between Test Drive and Simulation

A test drive was performed in the greater area of the city of Ingolstadt. Due to the main interest in altitude information, a route leading into Altmühltal was chosen due to the hilly landscape leading to altitude differences of about 100 m in total. The results can be seen in figure 4. It can be asserted that the relative accuracy of the two GPS paths in the two dimensional plane are very high. It was found that the maximum deviation of the simulation and the test drive was less then 25 m which lies well within the area of the GPS-receiving-unit's accuracy. Simulated vehicles which now drive in the virtual environment are not expected to experience significant deviations from a real test drive. All attached vehicle systems that use the location of the vehicle are expected to operate exactly like they would do on a real test drive. This result is more or less surprising since the two-dimensional map data provided by NAVTEQ already has very high quality which can be transformed to the simulation with almost no loss off quality.

Due to the lack of altitude information in the map database on the chosen test drive, the ASTER elevation data had to be used in the simulation. As described in the previous paragraph, the missing information was conflated to one simulation database in the end. In figure 4, the developing of altitude on the test drive and simulation can be seen. At first glance, a high correlation between test drive and virtual test drive can be seen. The absolute deviation between the two signals had a maximum of 29.68 m with a mean value of 8.81 m. The deviation in altitude was below 5 m about 40% of the length of the test drive. An accuracy of better than 10 m was achieved over 60% of the length. A closer look at the data makes the reasons for the large deviations apparent. The first three areas of large deviation at kilometers 1, 3, 6 and 14 along the route are lead back to the presence of surrounding houses and woodland. The conflation algorithm treated the elevation data as if it was referring to the earth's surface when it was actually referring to houses and woodland. Starting at road kilometer 8, the highest ascent starts with a maximum inclination of about 20%. This is also when the other phenomenon can be observed. The used interpolation method for finding the corresponding ASTER elevation to a base point assumed planar surfaces. Especially on mountainsides, the different elevation of the four surrounding data points leads to incorrect results.

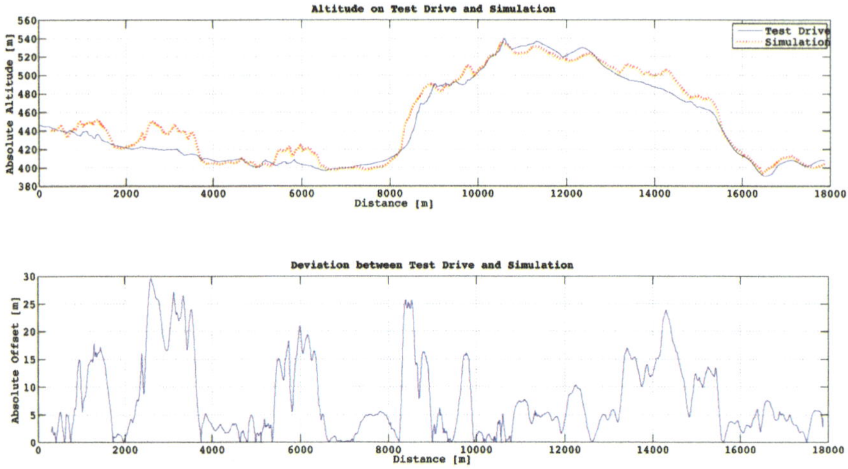


Fig. 4. Comparison of test drive and virtual test drive

## 4.2 Comparison of Alternative Routes

Three alternative driving routes between Manching and Gaimersheim were investigated in terms of the fuel necessary to drive along those routes. The fast route takes quite a detour following the highway A9 whereas the short route follows the straight-line distance as much as possible. A high precision vehicle model provided by AUDI AG was integrated into the simulation to evaluate the fuel consumption. Table 1 shows the results of the comparison between the simulation and the test drive. The first indication was that the characteristics of each alternative route could be preserved in the simulation as well as the real test drive. Not surprisingly, the short route was indeed the shortest compared to the other routes in all cases whereas the fast route had the shortest duration. The eco route also had the least amount of fuel in all cases, but the accuracy for this value is much less compared to the other two. The short route and the eco route had a relative accuracy of 6.7% and 5.9% which is acceptable for the scope of this project. The fast route however had a deviation of almost 20%. Looking at the relevant data leads to the conclusion that besides of the altitude profile, also the speed profile created by the driver model was responsible for the error. The fast route had a very significant portion of highway driving. Whereas the test-driver was limited by the traffic and could only drive a maximum speed of 130 km/h, the simulation went up to 160 km/h. At such high speeds, the wind resistance has the biggest influence to the vehicle's fuel consumption, so the large deviation can indeed be lead back to an inappropriate maximum highway speed in the simulation. Another source for errors is the definition of the simulation scenarios. In a real test drive, the driver has to stop the vehicle on various occasions. Not all of them can be modelled correctly, for instance multiple stops at traffic lights. At times with high traffic congestion, vehicles sometimes need to wait two or three cycles until they can pass the intersection. This was not modelled in the simulation and is possibly another reason for the deviation in fuel.

| | *Short route* | *Fast route* | *Eco route* |
|---|---|---|---|
| ***Deviation in length*** | *+ 1.9 %* | *+ 6.3 %* | *+ 6.7 %* |
| ***Deviation in duration*** | *+ 0.2 %* | *- 0.3 %* | *+ 0.9 %* |
| ***Deviation in fuel*** | *+ 6.7 %* | *+ 19.4 %* | *+ 5.9 %* |

Tab. 1: Relative accuracy of real test drive and simulation with respect to different route options

## 5 Discussion and Future Work

The paper shows that a simulation consisting of a virtual vehicle, a driver model and an environment simulation can be used to simulate the vehicle's movement in the real world. The process of generating the environment simulation can be automated by using map data from navigation systems to generate the simulation database. The quality of this environment simulation has very good accuracy in the two-dimensional plane which could be proven with a validating test drive. The elevation of the environment situation is more challenging to be automatically recreated because today's maps do not have sufficient elevation information, especially for smaller roads and cities. Using freely available elevation data from the ASTER space shuttle mission, the missing elevation data could be conflated into the simulation database with an adequate quality. The simulation process was then used to validate the results of an innovative routing algorithm which aims at minimizing the amount of fuel needed to drive between two locations. It was found that the main properties of the alternative routes (fast, short, eco) were preserved in the simulation as well as in the test drive.

One of the key-factors necessary for a good simulation database is sufficient elevation information. During the past years, map providers started to integrate this information into their databases, but this process is only finished for a small portion of the roads so far. Other companies selling elevation data emerge on the market today. When the technical- and also perhaps commercial difficulties are overcome, much better elevation data could be available. This could also contribute too many applications that directly rely on predictive navigation data that contains altitude information. Especially for all automotive applications relevant for electric vehicles, the altitude is a very significant factor for predicting the energetic costs of routes.

## References

[1] Neumann-Cosel, K., Dupuis, M., Weiss, C., Virtual Test Drive provision of a consistent Tool-Set for [D,H,S,V]-in-the-loop, In Proceedings of the Driving Simulation Conference Monaco, 2009.
[2] Barth, M., Boriboonsomsin, K., Vu, A., Environmentally-Friendly Navigation, 10th International IEEE Conference on Intelligent Transportation Systems, 2007.
[3] Bartsch, P., Fahrtoptimierung im Energiemanagement-Kontext, Dissertation an der Universität Hannover, 2009.
[4] Al-Harbi, S., Tansey, K., The accuracy of a DEM derived from ASTER data using differential GPS measurements, University of Leicester, 2007.
[5] OpenDRIVE Format Specification V1.2, http://www.opendrive.org, 2007.
[6] Hager, J., The Universal Grids: Universal Transverse Marcator (UTM) and Universal Polar Stereographic (UPS), Defense Mapping Agency, 1986.
[7] Ebersbach, D., Entwurfstechnische Grundlagen für ein Fahrerassistenzsystem zur Unterstützung des Fahrers bei der Wahl seiner Geschwindigkeit, Dissertation an der Technischen Universität Dresden, 2006.

**Andreas Lamprecht**
Technische Universität München
Boltzmannstr. 18
Garching bei München
Germany
lamprecht@ini.tum.de

**Thomas Ganslmeier**
Audi Electronics Venture GmbH
Sachsstr. 18
85040 Gaimersheim
Germany
thomas.ganslmeier@audi.de

**Keywords:** environmental simulation, hardware in the loop, driver in the loop, software in the loop, eco navigation, virtual test drive

# Priority System and Information Workflow for BHLS Services: An Interregional Project Between Italy and Germany

S. Gini, G. Ambrosino, MemEx srl
P. Frosini, Resolvo srl
F. Schoen, University of Engeneering of Florence
H. Kirschfink, Momatec GmbH

### Abstract

In order to make public transport more reliable and attractive for potential users in terms of performances and competitiveness, public authorities (municipality, mobility planning agency and transport operators, etc.) can act at infrastructural / network level (facilities design, dedicated lanes, priority infrastructures, etc.) and / or at traffic lights level with the management of the priority for public transport vehicles in a mixed context. The TIPS&INFO4BRT project, approved in late 2009 under the ERA-NET scheme, addresses the definition, implementation, tuning and evaluation of two cross-related sw modules to provide public vehicle with priority and to manage information workflow for road work. These two modules will be tested in a common simulated environment built on the transport network of Livorno (Tuscany) and Aachen (North Rhine-Westphalia).

## 1 Introduction

Different EU actions and initiatives (from the Lisbon agenda to the last White Paper on Mobility 2010 and the Green Paper "Towards a new culture for urban Mobility" issued in 2007) pushed the concept of "co-modality" as key tool for reducing the traffic congestion, improving the overall urban mobility governance and making the overall public transport system more efficient in terms of performances and more attractive for potential users. From this perspective EU recognised that the coordination of the different actors is mandatory in order to guarantee the quality and the integration of the transport chain as a whole.

European urban and metropolitan areas are facing heavy levels of traffic congestion and related environmental degradation (noise, pollution, vibration, etc.)

due to the high rate of private cars use and the lack of optimisation in the management of "conflicting needs" of the traffic flows sharing the same network (public transport services, goods distribution, waste collection services, tourists buses, etc.). Other face of this generalised degradation is the inability of public transport to increment its quota in the mobility demand definitely.

The implementation of this approach moved public administrations and authorities (municipalities, mobility agencies, public transport companies, shires and regional administrations) to operate at two different levels. From one side they enlarge the transport offer aiming at differentiating the mobility transport modalities between main axes and secondary feeders (implementation of tramways and / or BRT services, demand responsive transport schemes) and at customizing them to the different users requirements (flexible services, car sharing, car polling, etc.). From the other side, on the basis of the constraints ranging from urban infrastructures and environmental issues to regulations, costs and community budget limits, local authorities look for ITS systems/ tools as supporting tool for the efficient coordination of the overall transport chain: mobility and transport modes have been evolving towards cooperation schemes managed under an integrated ICT umbrella (mobility management and control system).

Focusing the analysis to priority measures for public transport the implemented interventions must be grouped into two major categories:

- Measures that are based on facility design that usually consists of exclusive lanes for buses on arterials as well as infrastructure design that facilitates the movements of the public transport vehicles;
- Measures that rely on traffic control and range from changes to fixed-time signal settings so that they favour the movements of public transport vehicles, to signal priority locally or network-wide to assist their movements in real-time.

Fig. 1. A cut-through roundabout used by the BusWay® in Nantes [1]

Fig. 2. Example of traffic light "abuse" in the urban area.

## 2 An Overlook to BRT Implementations Over Europe

Even though efforts seem to be turning towards bus-based systems in Europe, BRT has emerged as an attractive solution to Europeans who adapt it to their diverse contexts through the progressive implementation of BHLS (Bus High Level Service, as the BRT are going to be indicated in the European Context followed the French terminology). BHLS relies on the advantages of the tramway (speed, regularity, comfort, image) but its overall cost, capacity and flexibility place it somewhere between the "regular bus" and the tramway.

The term "level of service" covers a host of factors, expressed in terms of objectives regarding frequency, service span, reliability, journey time, comfort, accessibility, image and ease of use. The word "high" here refers to high performance; this, of course, is somewhat subjective and will depend on local contexts and objectives, but in all cases this means performance levels that are higher than those of conventional buses.

Achieving this "high" level of service requires the implementation of a certain number of measures. A list of quality indicators which define BHLS concepts have been proposed by CERTU [1]. The objectives behind the implementation of a BHLS system primarily concern the organisation and management of transport aiming at limiting the use of private vehicles, encouraging a modal shift to alternative forms of transport and at increasing satisfaction among existing public transport users. The BRT / BHLS potentials for further development is great, both in large urban areas seeking to lend a hierarchical structure to their network and in smaller urban areas for which BHLS represents a rapid transit system well suited to their size and capable of accompanying their urban development.

Just like BRT, the BHLS concept remains generic and can be integrated into any type of infrastructure configuration. In 2005, the French workgroup headed by CERTU, wanted to define its own concept of BRT based on initial local experiences (the "new town" of Évry in the 1970s, the Trans-Val-de-Marne system for the Greater Paris region starting in 1993, TEOR in Rouen in 2001). French concept is based on the use of on-street exclusive lanes as fundamental component allowing for the increase of speed and regularity. In the case of BHLS, its use can be limited to congested zones, such as city centres (for instance, Rouen's TEOR). BHLS is also present in the Netherlands (Amsterdam in 2001, Eindhoven in 2005), England (Leeds in 1998, Cambridge in 2009), Sweden (Gothenburg in 2003) and in Spain (Castellón in 2008). While Full-BRT is not present in Europe, as it has not been chosen to fulfil the mass transit function, numerous systems could be classified as BRT-Lite. Among the most well-known systems there are the blue buses of Stockholm (Sweden) since 1999, the Lianes of Dijon (France) since 2004 and the Linea Alta Mobilità in Italy (Prato, Brescia, Pisa).

In Europe's largest cities (over 1 million inhabitants), the subway and the tramway are already widely developed for the main arterial movements. BHLS allows for a complementary hierarchical organization of the bus network (Madrid's MBus projects, Lyon's Cristalis lines, Stockholm's blue buses, Hamburg's MetroBusse, Helsinki's Jokeri-line, Amsterdam's Zuidtangent). In the large urban areas (300 thousand to 1 million inhabitants), BHLS can be integrated into the main rapid transit system obtaining a status equivalent to that of a tramway (the BusWay® in Nantes, TEOR in Rouen, Gothenburg) or it can constitute a feeder service for a heavier rapid transit system to serve the inner suburbs (Toulouse's exclusive bus lanes). In smaller urban areas (Castellón, Nîmes, Metz, Jönköping, Lund, Utrecht, etc.), BHLS constitutes the kernel axes of the public transportation system, allowing the renovation of urban area and creating a "rapid transit effect" that benefits the entire system/city.

The analysis of different BHLS solutions and their classification in terms of role in the transport scheme, ITS applications (fleet monitoring and priority systems, users information systems, etc.) and service performance indicators has been achieved by European workgroup COST TU 0603 [2].

### 2.1 BRT and Priority Measures / Systems

Analysing the different solutions implemented to guarantee priority measures in BRT services in Europe, it is possible to identify:

- The use of two-way central dedicated lane, raised or level bridged, delimited by a light coloured border, with or without a central island;

Just like BRT, the BHLS concept remains generic and can be integrated into any type of infrastructure configuration. In 2005, the French workgroup headed by CERTU, wanted to define its own concept of BRT based on initial local experiences (the "new town" of Évry in the 1970s, the Trans-Val-de-Marne system for the Greater Paris region starting in 1993, TEOR in Rouen in 2001). French concept is based on the use of on-street exclusive lanes as fundamental component allowing for the increase of speed and regularity. In the case of BHLS, its use can be limited to congested zones, such as city centres (for instance, Rouen's TEOR). BHLS is also present in the Netherlands (Amsterdam in 2001, Eindhoven in 2005), England (Leeds in 1998, Cambridge in 2009), Sweden (Gothenburg in 2003) and in Spain (Castellón in 2008). While Full-BRT is not present in Europe, as it has not been chosen to fulfil the mass transit function, numerous systems could be classified as BRT-Lite. Among the most well-known systems there are the blue buses of Stockholm (Sweden) since 1999, the Lianes of Dijon (France) since 2004 and the Linea Alta Mobilità in Italy (Prato, Brescia, Pisa).

In Europe's largest cities (over 1 million inhabitants), the subway and the tramway are already widely developed for the main arterial movements. BHLS allows for a complementary hierarchical organization of the bus network (Madrid's MBus projects, Lyon's Cristalis lines, Stockholm's blue buses, Hamburg's MetroBusse, Helsinki's Jokeri-line, Amsterdam's Zuidtangent). In the large urban areas (300 thousand to 1 million inhabitants), BHLS can be integrated into the main rapid transit system obtaining a status equivalent to that of a tramway (the BusWay® in Nantes, TEOR in Rouen, Gothenburg) or it can constitute a feeder service for a heavier rapid transit system to serve the inner suburbs (Toulouse's exclusive bus lanes). In smaller urban areas (Castellón, Nîmes, Metz, Jönköping, Lund, Utrecht, etc.), BHLS constitutes the kernel axes of the public transportation system, allowing the renovation of urban area and creating a "rapid transit effect" that benefits the entire system/city.

The analysis of different BHLS solutions and their classification in terms of role in the transport scheme, ITS applications (fleet monitoring and priority systems, users information systems, etc.) and service performance indicators has been achieved by European workgroup COST TU 0603 [2].

### 2.1 BRT and Priority Measures / Systems

Analysing the different solutions implemented to guarantee priority measures in BRT services in Europe, it is possible to identify:

- The use of two-way central dedicated lane, raised or level bridged, delimited by a light coloured border, with or without a central island;

- The use of alternating one-way central dedicated lane: when there is not enough space, this system allows the bus to have priority when approaching intersections, as the other direction is with the flow of traffic;
- In case of use of lanes shared with normal traffic, priority is guaranteed by the traffic lights signal at the approach to intersections (TVM Paris – Ilê de France, etc.);

These different measures can also coexist as in Nantes or Rouen.

From the technological / architectural point of view, vehicles can be equipped with transponders (as in Nantes, Lorient, etc.) that signal loops detect first at the approach to intersections (triggering red lights to stop traffic) and then at intersection exits (turning off the red lights so that traffic can resume its normal flow). In Lorient bus lanes that are not in conflict with other traffic flows benefit from an amber flashing light (i.e. go if safe) instead of a red stop light. In other cases (Rouen) traffic light priority system is based on via radio signal transmitted from the vehicles to the traffic light controllers. In other cases (Strasbourg, Stockholm, etc.) the detection of public transport vehicle is carried out on the basis of GPS signals and priority system is managed as AVL/ AVM functionalities.

Different algorithms can be used to give priority at traffic signals (prolonging the current green phase, placing the ordinary green phase earlier, using an "extra green phase", going back to green shortly after normal green phase).

## 3 Urban Traffic Control: Priority Systems and State of the Art

UTC systems are usually based on the identification of a "minimun cost" control policy based on a certain combination of the following three elements:

- The minimization of lost time in queue and/of travel time for private cars;
- The increase of performances of public transport;
- The environmental constraints (minimization of queues and stops in order to reduce pollution, avoidance of congestion by traffic gating).

Different objectives may be used to manage the priority: criteria may be used for granting priority to public transport vehicles include schedule adherence, minimization of travelling time, minimization of waiting time, headway adherence, number of passengers on the vehicle, etc. However, such criteria can only be used, if the required corresponding information is available in real time.

Priority rules aim to provide priority to public transport vehicles by modifying locally in a suitable way the network-wide signal settings produced through UTC system.

### 3.1 Review of Priority Strategies

The main methods adopted include:

- Green extension and stage recall (extension of green time when a bus approaching towards the end of the green time, or the recall of a stage if the signal is on red) commonly used where the detection is relatively close to the priority junction and subject to constraints like maximum extension time, minimum green-time for non-priority stages, etc.;
- Queue jump providing priority to the bus not only against traffic moving in conflicting to the bus routes but also against traffic moving in parallel with the bus;
- Stage re-ordering which is alternative and stronger form of priority modifying the normal sequence;
- Allocation of a special stage for public transport vehicles and introduction of this stage into the normal sequence at the first available opportunity in order to serve a received priority request;
- Stage skipping allowing the omissions of one or more stages from the normal sequence;
- Real-time optimisation based on a traffic model use and the definition of a performance index.

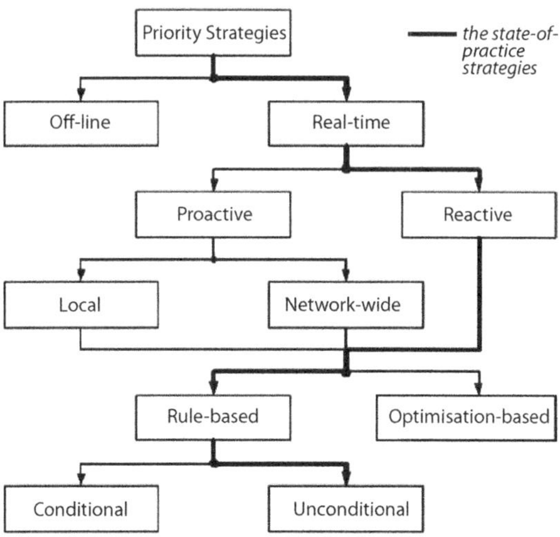


Fig. 3. Classification scheme for UTC systems [3]

### 3.2 Classification of Priority Systems

Classification of UTC systems can be articulated using the following criteria: off-line/real-time, proactive (local or network wide) / reactive (rule based or optimisation based).

A classification scheme proposed in [3] is depicted in figure 3.

### 3.3 Architectural and Technological Options for Bus Priority Implementation

To implement UTC and priority systems functionalities, the following hardware devices are generally required: detectors collecting occupancy measurements (alternatively the numbers of vehicles within the junction approaches are measurable via a video detection system), detectors (or video detecting system, or any other relevant device) for detecting the approach of public transport vehicles (GPS, beacon etc.) and / or providing information on the arrival time, local controllers at each junction, a central computer able to receive the real-time measurements from the detectors (or the video detection system) and to send the signal settings to the local controllers. The devices can be integrated in a physical architecture which can be either fully or partially centralised.

In the first case, a fully centralised control system, all the measurements collected from the network are transmitted to the central computer which implements all the control laws of UTC and sents the final signal settings to the local controllers for implementation. In this case, a continuous communication of the central computer with the local controllers is necessary to allow UTC to receive, process and serve priority requests of public transport vehicles. As a consequence, high-speed communication networks will avoid delays in the transmission of data and decisions.

In the second case, a partially centralised control system, the local controllers receive the settings from the central computer, as well as data that relate to the local movements of public transport vehicles. Given this information, each local controller either implements the received settings or modifies them appropriately in order to serve locally received priority requests. In other words, in this implementation case, the priority logic is implemented locally, in each relevant controller. Such an architecture reduces substantially the performance requirements (and cost) for communications.

Signal controlled junctions that are located and operated independently are known as isolated junctions. This form of control is selected when traffic arrivals at the junction are largely unaffected by any neighbouring traffic signals. These signals, which may still be linked to a traffic control centre (i.e. for fault monitoring), are more common in suburban areas where road junction density is lower. When signal controlled junctions are more closely spaced, and traffic interactions occur, co-ordinated control is implemented. Operations at a junction are then influenced by operations at one or more neighbouring junctions.

## 4 TIPS&INFO4BRT: European Project on Priority Modules

TIPS&INFO4BRT is an European project including for partners from Italy and Germany: small and medium sized enterprises supporting public authorities in the tendering, implementation and management of ITS system and, in general, in the area of technological transfer and research institutes. The project started on October 2009 and lasts 18 months. The project answered the fifth EraSME call funded within the ERA-NET scheme of FP7 for Research and Technological Development launched by the European Commission.

The project addresses the definition and implementation of a priority module to be used in corridor environment. Beside the development of the junction bus priority control module, it is important to offer solutions for whole stretches. In the context, road works and other predictable accidents are very often a big problem as they disturb the predefined schedules. Therefore a second module "information workflow" will be developed in order to the construction site area. The service needs to interpret in order to enhance traffic light control systems at road works controlling variable one-direction traffic to ensure bus priority in coordination with normal traffic lights at intersections.

The TIPS&INFO4BRT project aims to specify, develop and validate two specific interrelated modules for improving the bus priority at the junction and for manage the road information workflow and to estimate the impact of predictable accidents on the bus transit lines.

In details, the project activities are the following:

- To analyse the reference context and the state of the art of the implementation and management of traffic lights control (in particular the analysis will address bus priority management implemented in corridors). A part of the survey will deal with the analysis of optimisation algorithms available in the literature;
- To specify, develop and validate two specific interrelated modules for improving the bus priority at road junctions and to manage the road information workflow in order to estimate the impact of predictable

accidents on bus transit lines. The validation of the modules will be carried out under a common simulation environment (AIMSUN) tuned on road network models built up on two different typical European cities (Livorno in Tuscany and Aachen in North Rhine-Westphalia) which are typical of middle-small urban areas. Technical synergies and exchanges of experiences will be guaranteed at different steps of the project in order to foster interregional fertilisation;

- To define market recommendations for the product, involving stakeholders, representatives of potential users and international experts working in the reference market;
- To disseminate the results of the project at national and international level.

First results produced by the analysis of the state of art are available. The description of the main outputs will follow in the next section.

## 5 Outputs of State of the Art Produced by the TIPS&INFO4BRT Project

The review has been carried out in terms of: analysis of the objectives of the priority functions (maximize the speed of the vehicle, minimize the waiting time, maximize the regularity of the service), adopted architecture (centralized / decentralized control, isolated / coordinated control, demand responsive), technological solutions for vehicle detection and for the communication between the traffic light controller and the coordination centre (if any), ITS (AVM/AVL) supporting systems, strategies for bus priority implementation (management of conflicting directions, optimisation, etc.); procedures for the testing and the evaluation of the impacts of the introduction of priority systems, operational issues (adaptation capability to different traffic condition, configurability, resources and procedure for the management and the maintenance of the system). Current systems are characterised as real-time, reactive, rule-based that can be conditional or unconditional.

The real-time strategies attempt to overcome the disadvantages of fixed settings via real-time operation. They require the ability to detect or identify in real-time public transport vehicles approaching signalised junctions. Reactive strategies are applied at each junction separately of the others and provide isolated output to the received priority requests independently of whether the priority request has been received from an isolated or co-ordinated controlled junction. These strategies require only local communication between the public transport vehicle and the traffic signal. The optimisation-based strategies attempt to provide public transport priority based on the optimisation of some performance criterion, primarily the delay. Unconditional strategies provide priority regardless of the status of the public transport vehicle.

From the technological point of view GPS based solutions are mostly used even if beacon system is still present. Decentralized systems (London, Geneva, Malmoe, Prague, Vienna, etc.) and centralized ones (Aalborg, Glasgow, Stockholm, Toulouse, etc.) are about equally present.

## 6 State of the Art and Approach Promoted by the TIPS&INFO4BRT Project

Currently traffic light control systems / junctions controllers offered by the market present specific priority function and algorithms (ready to be applied) and / or measures that can be customised to specific situations. Anyway each type of systems has their own capabilities and suit to a different extent of traffic conditions and characteristics. However there are various factors which affect the performance of these control systems and bring into evidence their limits as traffic management tools; main relevant ones are: the overloading of junctions, the capability to respond to sudden and unforeseen changes in traffic flows, the poor interaction of the Public Transport (PT) control systems, the lack well consolidated traffic light priority and work information management functions. In conclusion the implementation of such systems can be fruitful if the system offers the appropriate level of configurability in order to facilitate the adaptation of the system to a wide range of traffic conditions. Only if all these aspects are taken into account and faced in the bus priority control module the use of these techniques can be extended and long-lasting.

On the other hand some difficulties in the management of traffic priority systems seems to arise from the lack of preliminary result assessment procedures and of related costs / benefits since, generally, the mobility actors activated on the same network have different objectives: local councils wish to ensure fluid traffic flows, while the Public Transport Operators wish to reduce journey times and improve schedules and the single driver aims to optimise its travel time independent of the other situations. In this sense it must be underlined that a clear definition of the objectives of the systems and the appropriate definition of methodologies for costs / benefits analysis are two other main issues to be considered in the implementation, tuning and evaluation of priority systems.

From the market potential and in particular for the traffic light system and controller producers the TiPS&Info4BRT project could open a market segment not fully covered by the current products which are unable for better cope the critical traffic situations (the core of TiPS&Info4BRT and one of the main directions of the actual research trends).

## References

[1] Centre for the Study of Urban Planning, Transport and Public Facilities (CERTU), Buses with a high level of service (French Bus Rapid Transit): choosing and implementing the right system, 2009.

[2] European workgroup COST TU 0603 – www.bhls.eu.

[3] SMART NETS IST-2000-28090 Signal Management in Real Time for urban traffic NETworkS - Deliverable 7: Public Transport Priority in TUC.

[4] Gardner, K., D'Souza, C., Hounsell, N., Shrestha, B., Bretherton, D., Technical Cluster Deliverable 1 Review of Bus Priority at Traffic Signals around the World, Working Program Bus Committee 2007-2009 – Extra-vehicular technology, UITP WORKING GROUP: Interaction of buses and signals at road crossings.

**Saverio Gini, Giorgio Ambrosino**
MemEx srl
Via Cairoli 30
57123 Livorno
Italy
saverio.gini@memexitaly.it

**Paolo Frosini**
Resolvo srl
Via Giardini della Bizzaria 12
50127 Firenze
Italy
paolo.frosini@resolvo.info

**Fabio Schoen**
Università di Firenze
Via Santa Marta 3
50139 Firenze
Italy
fabio.schoen@unifi.it

**Heribert Kirschfink**
Momatec GmbH
Weiern 171
52078 Aachen
Germany
heribert.kirschfink@momatec.de

**Keywords:** BHLS, corridors, traffic lights control, public transport priority system

# iTETRIS - A System for the Evaluation of Cooperative Traffic Management Solutions

D. Krajzewicz, German Aerospace Center
R. Blokpoel, Peek Traffic bv
F. Cartolano, Comune di Bologna
P. Cataldi, EURECOM
A. Gonzalez, O. Lazaro, Innovalia Association
J. Leguay, Thales Communication France
L. Lin, Hitachi Europe Ltd.
J. Maneros, CBT Comunicación & Multimedia
M. Rondinone, Universidad Miguel Hernández de Elche

## Abstract

V2X communication - communication between vehicles (V2V) and between vehicles and infrastructure (V2I) - promises new methods for traffic management by supplying new data and by opening new ways to inform drivers about the current situation on the roads. Currently, V2X cooperative systems are under development, forced by both the industry and by the European Commission which supports the development as a part of its Intelligent Car Initiative. Within this publication, "iTETRIS", a new system for simulating V2X-based traffic management applications is described which aims on high-quality simulations of large areas. This is achieved by coupling two well-known open source simulators. The sustainability of the project is guaranteed by making the whole also available as an open source tool.

## 1 Introduction

Mobility has had an increasing impact on worldwide economical and social development over the last decades, which is highlighted by the fact that up to 40% of World Bank loans have been used on transport projects [1]. However, current road traffic yields in high congestion and pollution and both will increase with the predicted increase of road traffic amount. One goal towards avoiding these problems and achieving sustainability in road traffic is to improve traffic management by intelligent information and communication technologies (ICT). Here, the wireless vehicular communications ("V2X communication" or "V2X" for short) technology gained a major interest in the recent years and became part of the EC's Intelligent Car Initiative [2]. Cooperative systems which use this technology will be able to assist the driver with additional knowledge gained through the dynamic exchange of messages between vehicles and between vehicles and infrastructure. A wide range of possible applications which use V2X are under development. They range from safety-related applications, such as collision warning or pre-crash warnings, up to applications which unfold their potentials in large areas, such as broadcasts of travel times and route guidance based on this information.

The European research activities with regard to cooperative systems started under the 6th Framework program with projects such as CVIS [3], Safespot [4] or Coopers [5]. To validate and better estimate the impact of cooperative systems, Field Operational Tests (FOTs) have been launched under the 7th Framework Programme. These FOTs will provide a needed feedback on the impact, performance, and predicted adoption of cooperative systems. Nevertheless, they are limited in size, time and the number of involved vehicles. For traffic management a view on large areas and different numbers of vehicles is needed to answer questions the new systems pose: What infrastructure is needed to support a certain traffic management strategy? Does a new strategy bring a benefit which justifies the costs? How does the system behave under different rates of vehicles equipped with V2X systems? And, above all: Will autonomous systems disturb centralized traffic management policies?

These questions can only be answered using simulations. Such simulations must be able to resemble a large area's road traffic, possible incidents, and the communication between a variable amount of vehicles equipped with V2X devices, optionally including a variable amount of installed communication infrastructure. Additionally, the traffic management strategies that shall be evaluated must be modelled and incorporated into the simulation. Several approaches which allow such simulations exist and are well described [6-10]. Unfortunately, many of these simulations are not available to the public. Also, being developed in academic context, most of them lack a real support for potential users and get unsupported after the thesis in which context they were developed ends.

The European FP7 iTETRIS (Integrated Wireless and Traffic Platform for Real-Time Road Traffic Management Solutions, http://ict-itetris.eu/) project aims at implementing an open-source, integrated wireless and traffic simulation platform. It will allow testing and optimising V2V and V2I communications and cooperative traffic management strategies for large-scale scenarios. The sustainability will be guaranteed by continuing its usage and dissemination after the project's end, as well as by using modern simulators and the possibility to exchange them by newer developments.

Within this publication, the results of this project's first year are presented, starting with descriptions of real-world needs. Then, a set of strategies developed within the project are described. Afterwards, the simulation system architecture is presented, together with extensions of the used simulators the system consists of. We will close with a short outlook on the project's next steps.

## 2 Investigation of a City's Problems

In order to develop a tool which addresses real world problems, the first step was to analyze the needs of traffic management within a real city. The municipality of Bologna, one of iTETRIS project partners, evaluated typical traffic scenarios in the city. This analysis has taken into account different factors, such as urban structure, commuting, public transport offer, traffic control systems, etc. This preliminary work defined traffic flow parameters describing various large scale traffic test sites from the mobility point of view. In addition, specific congestion events were studied and classified, in order to collect a wide range of cases and to identify the possible causes for unexpected or recurring traffic jams. In the following, these results will be presented. At first, a description of the current mobility situation of the city of Bologna is given. Then, the generated traffic scenarios which describe the spatiotemporal shape of problematic zones as input to traffic simulations are presented.

### 2.1 The Mobility in the Urban Area of Bologna

The city of Bologna, located in the middle-north of Italy, constitutes an important crossroad for national mobility flows over north-south and east-west axis. Moreover, it has a very evocative historical centre that attracts many tourists during the year. The city has an important fair centre and one of the biggest universities in Italy (and the oldest continually operating one in Europe). The resident population in the city today is about 374 thousand (the municipality area is 140.85 $km^2$), while the metropolitan area, which extends to the neighboring towns, and more pertinent from the mobility's point of view, counts nearly 600 thousand residents. There are about 180 thousand local workers in the municipal area, while the daily city users are around 300 thousand. The result of this demographic structure is that traffic flows within the network are very heavy and include very short trips within the city, short distance trips among the main city and the towns belonging to the same province, medium distance trips taking place within the regional network of cities, and long distance through traffic with areas outside the region.

Like other medium-sized European cities, Bologna suffers because of the regularly high levels of urban traffic. In the last few years, traffic to and from the areas surrounding the city has increased. On the contrary, following new mobility management policies, there has been less traffic in the city itself. During a medium work-day approximately 2 million movements are produced in Bologna; in particular, almost half of these movements is completely within the administrative city area, while the other half is equally distributed between city crossing trips and exchange ones with other cities. Private car use predominates in the exchange (approximately 70%) and city crossing (approximately 90%) movements. Considering only the inner movements, the private car use percentage is less than 40% because in urban travels people generally use public transportation, two wheeled vehicles, walk or bike.

## 2.2 Recognized Bottlenecks and Possible Solutions

Three areas were chosen as being exemplary for a city's problematic zones. These three areas were used to set up scenarios, each defining the area they take place within, a description of the reasons of the problems occurring within this area, and possible strategies to solve these problems. Besides textual descriptions, the scenarios were modelled in Vissim or VISUM. These models, which include both the road network, and the traffic demand, could be transferred into a description usable by the traffic simulator SUMO which is used within the iTETRIS project for traffic modelling. In the following, the scenarios will be presented.

The first scenario describes an area affected by recurring events taking place at the city stadium, such as football matches or concerts. The chosen area also includes a hospital as well as a cemetery. The goal to achieve in this scenario is to manage the traffic in an area that offers few alternative routes, and where the stadium often generates a significant and extraordinary increase in traffic flow that affects adjacent areas of the traffic network. Possible solutions may include adapting traffic light timings, postponing driver departure times by offering information about current state in this part of the network, recommending routes around the affected area, or changing traffic regulation, e.g. by allowing private vehicles to use the bus lanes located in this area.

Being an old city, Bologna is surrounded by an inner city ring-way, which draws a line between the Bologna-centre and the remaining urban city. The ring-way is equipped with plenty of traffic lights and detection loops. It can be covered clockwise or anti-clockwise. It is the only way to cross the city, because the area within it is a Limited Traffic Zone (LTZ) and only few vehicles are allowed to enter it. The only exception is "Via Irnerio", a street that represents a shortcut connection to the ring-way and has no restrictions in being passed. Via Irnerio has a strong public transportation offer and a big open market on weekends which attracts a lot of vehicles and pedestrians. Here, the goals of traffic management are to reduce congestions on the roads which result from the high amount of traffic that crosses the city, from broken induction loops that make the installed control not to work, or from strikes or other events involving a large number of pedestrians who block the street. Solutions proposed for solving these problems are to adapt traffic light timings to the current situation, to detect and replace malfunctioning induction loop detectors by other traffic surveillance methods, and to disseminate routing information allowing drivers to pass the area more quickly.

The third investigated area is the highway located to the north of the city. An "Orbital" road runs beside the highway and, since 4 out of the 5 highway exits are connected directly to the Orbital, the Orbital is often chosen by drivers as an alternative route to the highway. As a result, the Orbital is often congested. The traffic management goals are to reduce the travel times and solve congestions of the Orbital. The proposed solution is to establish a routing support for drivers proposing them a certain exit depending on their destinations and the current traffic situation on the Orbital and on the highway, in terms of travel times.

Fig. 1. iTETRIS areas; from left to right: city stadium area, inner city ring, highway area

## 3 V2X Management Strategies

Traffic management, in general, means optimizing the supply of road infrastructure for a given traffic demand. This process is based on the knowledge about the situation on the considered road network that is mainly described by the number of vehicles and their velocities. Additionally, traffic management involves informing road users about accidents, construction work, road closures or rerouting advices etc.

Conventional traffic surveillance systems, such as induction loops, are relatively expensive (some thousands Euros each). This usually disallows an area-wide coverage of the investigated road network. Furthermore, induction loops have a limited operational reliability over the time. Besides induction loops, other surveillance systems exist, such as cameras which detect and track vehicles or GSM-based level-of-service determination. However, it should be stated that such systems are not yet used on a large scale. All these techniques measure traffic at discrete points or along certain routes, only. Network-wide considerations of the actual traffic situation are not possible using nowadays methods. In conventional traffic management, road users are either informed by radio broadcasts of traffic reports or by variable message signs (VMS). Traffic reports are updated seldom and therefore tend to be outdated in a considerable number of cases. VMS are only situated on larger roads and their number is limited. For these reasons, only a part of road users can be provided with correct traffic information. Newer possibilities for supplying information are navigation devices of the third generation with an updated traffic situation.

Cooperative traffic management tries to close the gap of lacking data about the state of the considered road network by exchanging information between vehicles and between vehicles and infrastructure. In dependence to the equipment rate of vehicles and the number of installed road side units ("RSUs" or "ITS Roadside Stations") a high coverage, ranging up to a complete knowledge about all vehicles (cars, busses) on the road network is theoretically achievable. This knowledge offers new ways of managing traffic efficiently. I2V and V2V communication also promises to offer new possibilities to disseminate information about a road network's current and predicted state, reaching all equipped vehicles. Gathered or generated information can be both presented to a driver directly, or being processed by in-vehicle devices, generating only relevant information.

Within iTETRIS, traffic management methods based on vehicular communication were developed and formally described. They were mainly derived from the traffic management proposals from the work on evaluation Bologna's traffic, presented in the last section. The results were descriptions of the 15 methods which can be combined into complex traffic management strategies. These methods can be grouped into four topics, as presented in the following:

- **Surveillance:** "Distributed traffic jam ahead detection", "Travel time estimation", "Identification of malfunctioning loop detectors", "Induction loop replacement", "(Decentralized) Floating car data"

- **Navigation and Guidance:** "Bus lane management", "Limited access warning", "Request-based personalized navigation", "Postpone departure time for road network balancing", "Decentralized route recommendation based on travel time estimation", "Routing in traffic light controlled network"
- **Traffic Lights:** "Emergency vehicle prioritization", "Traffic light adaptation by traffic management centre"
- **Assistance Systems:** "Regulatory and contextual speed limit information", "Event based traffic condition notification"

Each method was described in a high detail, including a) a textual descriptions, b) information about actors, used messages, communication mode and required network, expiry time, and transmission frequency, c) a diagram showing the data processing within the involved actors if the method would be implemented in reality, d) a sequence diagram showing the data flow for simulating the method, e) a diagram and descriptions of the structures needed for simulating the method, and f) the derived requirements to the simulation system and to the involved simulators.

## 4 Simulation System

The developed strategies will be evaluated using a simulation system which is the major result of the iTETRIS project. The simulation system couples two well-known simulators, ns-3 [11] for networking simulation, and SUMO [12] for traffic simulation. Both simulation applications are available as open source. Furthermore, instances of the strategies to evaluate are run as external applications. In the following, the design of the overall simulation system is presented, first, including the reasons for choosing it. Then, extensions to the used simulators implemented within the iTETRIS project are outlined.

### 4.1 System Architecture

Different approaches for simultaneous, coupled simulation of communication and traffic were used in the past: a) "Integrated Simulation" where the simulation is performed using a single program which includes modules for the communication, the application, and the traffic simulation [9, 10], b) "Direct Simulator Coupling" where two simulators (SUMO and ns-2 for example) are directly coupled, for example using a socket connection and the application simulation is built into one of the simulators, usually into the communication simulation [7], c) "Three-Blocks Architecture" where two simulators are

tion instances and for synchronizing them. In addition, it contains structures enriching and storing information obtained from both simulators making the whole simulation system compliant with the COMeSafety-architecture [13].

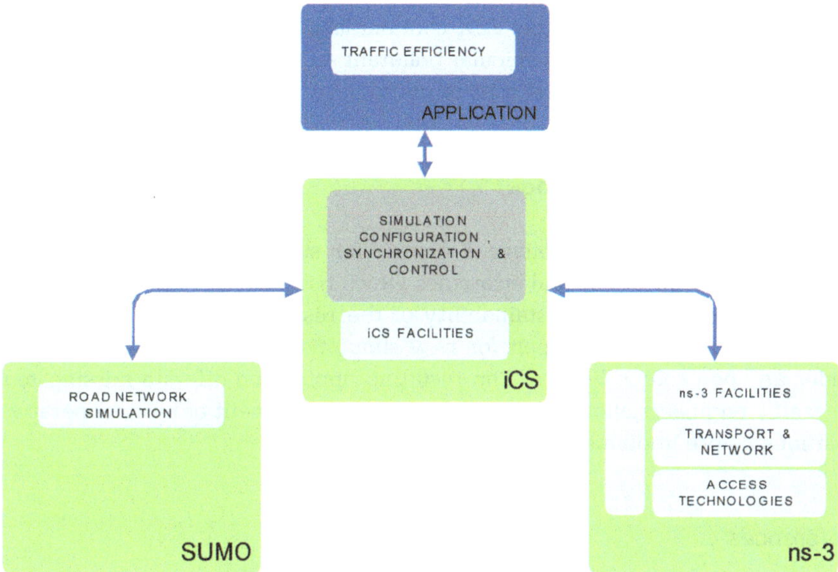


Fig. 2. Chosen simulation system architecture

### 4.2 Extensions to the Involved Simulators

Both used simulators, ns-3, and SUMO had to be extended for being suitable to model the questions addressed by the iTETRIS project. ns-3 had no support for V2X wireless communication simulation and had therefore to be extended. A new module was created to emulate radio transmissions over ITS G5A, the European version of the IEEE 802.11p communication technology (also known as WAVE), being in turn an amendment of the 802.11 standard allowing communications in vehicular environments. A similar work was made for the creation or adaptation of specific modules for the other radio technologies considered in the project: UMTS, WiMAX and DVB-H. Moreover, in order to be compliant with the COMeSafety ITS Communication Architecture, new modules and functionalities are being added to represent all the layers of the communication stack needed for wireless communication in an ITS scenario (see figure 2). Additionally, a communication module named inCI (iCS to ns-3 connection interface) for coupling ns-3 to the iTETRIS Control System was implemented.

The communication module of SUMO was extended in order to support the simulation of traffic management algorithms, by allowing to access and manipulate simulated traffic lights, induction loops, and vehicles. Because a very strong attention is put on developing strategies which show the ecological benefits of strategies based on V2X communication, SUMO was extended in order to be able to compute vehicular pollutant – CO, $CO_2$, $PM_x$, HC, $NO_x$ – and noise emissions as well as fuel consumption.

## 5 Conclusions and Outlook

The iTETRIS project will deliver a system for simulating large-scale traffic management applications and strategies which use V2V and V2X communication in high quality. The sustainability of the resulting simulation system is guaranteed by keeping it open for new simulation tools and by releasing it under an open source licence. The resulting application offers a missing tool for traffic engineers allowing them to measure the benefit of new cooperative solutions before implementing them in reality.

## References


[1] Department for Transport, UK Government, Transport and the economy: full report (SACTRA), 1999.

[2] European Commission, Intelligent Car Initiative Web Portal, http://ec.europa.eu/information_society/activities/intelligentcar/index_de.htm, last visited on 10.02.2010.

[3] CVIS project home page, http://www.cvisproject.org/, last visited on 10.02.2010.

[4] Safespot project home page, http://www.safespot-eu.org/, last visited on 10.02.2010.

[5] Coopers project home page, http://www.coopers-ip.eu/, last visited on 10.02.2010.

[6] Piórkowski, M., Raya, M., et al, TraNS: Realistic Joint Traffic and Network Simulator for VANETs, SIGMOBILE Mob. Comput. Commun. Rev., Vol. 12, No 1, ISSN 1559-1662, pp. 31-33, 2008.

[7] Wegener, A., Hellbrück, M., et al, VANET Simulation Environment with Feedback Loop and its Application to Traffic Light Assistance, Proceedings of the 3rd IEEE Workshop on Automotive Networking and Applications, New Orleans, LA, USA, 2008.

[8] Queck, T., Schuenemann, B., Radusch, I., Runtime Infrastructure for Simulating Vehicle-2-X Communication Scenarios, Proceedings of the Fifth ACM international Workshop on Vehicular inter-Networking, San Francisco, California, USA, September 15, VANET '08, ACM, pp. 78, New York, NY, USA, 2008.

[9] Kerner, B., Klenov, S., Brakemeier, A., Testbed for wireless vehicle communication: A simulation approach based on three-phase traffic theory, Proceedings of the IEEE Intelligent Vehicles Symposium (IV'08), pp. 180–185, 2008.
[10] Krajzewicz, D., Teta Boyom, D., Wagner, P., Untersuchungen der Performanz einer auf C2C-Kommunikation basierenden, autonomen Routenwahl bei Stauszenarien, Heureka '08, 03.05.2010-03.06.2010, 2007.
[11] ns-3 home page, http://www.nsnam.org/, last visited on 10.02.2010.
[12] SUMO home page, http://sumo.sf.net/, last visited on 10.02.2010.
[13] COMeSafety Project, European ITS Communication Arctitecture, Overall Framework, Proof of Concept Implementation, version 3.0, 2010.

**Daniel Krajzewicz**
German Aerospace Center
Rutherfordstr. 2
12489 Berlin
Germany
Daniel.Krajzewicz@dlr.de

**Robbin J. Blokpoel**
Peek Traffic bv
Basicweg 16
Postbus 2542
3800 GB Amersfoort
The Netherlands
Robbin.Blokpoel@peektraffic.nl

**Fabio Cartolano**
Comune di Bologna
Piazza Liber Paradisus,10
40129 Bologna
Italy
Fabio.Cartolano@comune.bologna.it

**Pasquale Cataldi**
EURECOM
2229 route des crêtes
06560 Sophia-Antipolis cedex
France
pasquale.cataldi@eurecom.fr

**Ainara Gonzalez, Oscar Lazaro**
Innovalia Association
Rodriguez Arias, 6
48008 Bilbao
Spain
aigonzalez@innovalia.org
olazaro@innovalia.org

**Jérémie Leguay**
Thales Communication France
160, boulevard de Valmy
92706 Colombes Cedex
France
jeremie.leguay@fr.thalesgroup.com

**Lan Lin**
Hitachi Europe Ltd.
1503 Route des Dolines
06560 Valbonne - Sophia Antipolis
France
lan.lin@hitachi-eu.com

**Julen Maneros**
CBT Comunicación & Multimedia
Carretera Asua, 6
48930 Getxo
Spain
jmaneros@cbt.es

**Michele Rondinone**
Universidad Miguel Hernández de Elche
Avda de la Universidad, s/n
03202, Elche (Alicante)
Spain
mrondinone@umh.es

**Keywords:** traffic management, V2V communication, simulation, traffic management strategies

# A Comprehensive Simulation Tool Set for Cooperative Systems

T. Benz, PTV AG
R. Kernchen, University Surrey
M. Killat, KIT
A. Richter, DLR
B. Schünemann, FOKUS

## Abstract

PRE-DRIVE C2X is a project funded within the 7th Framework Programme of the EU to support the development and introduction of cooperative systems. One major part of this work is the creation of a comprehensive tool set of simulation models integrating all fields of expertise involved. The objective is to create, test and apply an integrated simulation tool set that allows to simulate and evaluate the interaction between vehicle traffic, vehicular communication and co-operative applications. Each of these three areas is treated by dedicated models. Additionally the environmental effects are modelled by a separate modelling approach with detailed algorithms for vehicle dynamics and engine behaviour. The process to arrive at such a model combination from user requirements over pre-existing know-how and other significant steps are described as well as the current status of the work.

## 1 Introduction

Cooperative Systems for future traffic will constitute a major development stage in the evolution of road transport. Only by such an exchange of information between vehicles and with the infrastructure, new levels of safety and efficiency can be achieved. The EU funded project PRE-DRIVE C2X was set-up to further support the development, test and introduction of cooperative systems. Due to the importance of simulation tools in the development process, a significant part of the work in PRE-DRIVE C2X is dedicated to the development of a comprehensive simulation tool set which combines different proficiencies like communication modelling and traffic flow modelling.

The tool set is based on the available know-how about modelling in the different areas. However, the complete tool must satisfy all user requirements resulting from the technical and systematic development work in other parts of the project. It is therefore necessary to analyse both, the state of the art in modelling and in cooperative system development. The whole process contains the following distinct but strongly interlinked individual steps:

- User needs analysis
- Requirements for a comprehensive tool set
- Evaluation of existing tools
- Overall simulation tool set architecture
- Simulation of communication
- Simulation of traffic and safety effects
- Simulation of environmental effect from traffic
- Integration and validation

## 2 User Needs and Requirements

Integrated simulations for cooperative ITS systems can answer many questions related to the large scale behavior of communication (e.g., message distribution, latency, cumulated noise, congestion control) and more importantly estimate the impact and benefit of cooperative applications in terms of traffic and fuel efficiency, increased safety through fewer accidents and security. They provide support for validation and standardizations. Cooperative systems are complex ecosystems involving a number of different stakeholders such as communication experts, user case owners (including application engineers), traffic engineers, system/component developers, simulation tool providers, car manufacturers, infrastructure providers and evaluators. Each of them may have different user needs based on their expectations what simulations should provide, but together with some common goals.

Within PRE-DRIVE C2X common user needs for integrated simulations have been identified. First of all user friendly and understandable tools (w.r.t. scope, limitations, method, model, etc.) with a good HMI are needed. Experience in use, insight in simulation methods, and expertise to evaluate results are required in order to develop and perform simulations. Well defined interfaces for data exchange, integration, and a consistent high level architecture approach must be available with regard to integrated simulations.

Therefore simulation tools have to solve various problems and must satisfy several requirements, which are strongly depending on the class of use cases and on the use cases itself. PRE-DRIVE C2X chose a classification of applications with simulation needs in the area of safety and traffic efficiency based on an extensive use case description and selection process as described in [1]. All the selected use cases have been analyzed for their specific needs towards simulation [2] in the first step. Secondly an extensive requirements collection process took place. The objective was set on enabling combined traffic-communication-application scenarios. Simulation should be performed on microscopic and macroscopic levels. Microscopic simulations are modeled in a way that results support on macroscopic levels. The requirements are selected in several categories, traffic, driving, communication and applications simulators have been identified as the main categories. Nevertheless in order to support integrated simulations, requirements for a simulation integration platform, its functionalities and interface requirements between simulators have to be considered. Requirements are serving as the basis for the evaluation of existing tools and as preparation for the application simulation scenarios. A requirements template has been developed to capture all requirements in a consistent manner.

| *Code* | *Type* | *Description* | *Applicable Element* | *Simulation model* | *Priority* | *Dependency* | *Justification* |
|---|---|---|---|---|---|---|---|
| *REQ. SAR.007* | *F* | *The simulation architecture shall allow an overall scenario minimum length that considers the use case relevant region.* | *Traffic simulator* | | *M* | | *The length of the scenario has to be enough to cover the use case relevant region.* |

Tab. 1. Requirements template and requirements example

Table 1 shows an example of the requirements template and an example of a requirement (REQ) on the simulation architecture (SAR) with number 7. Each of the template categories has been defined. In the case of the example the requirement (explained from left to right) has a unique code "REQ.SAR.007", a type "functional (F)" a description, the applicable element of the integrated simulation architecture "Traffic simulator", no applicable simulation model, the priority level identified "mandatory (M)", there are no dependencies to other requirements and a justification for the requirement. A detailed explanation of the categories is given in [2]. With this method around 30 general requirements on the simulator interfaces, 50 against the simulation architecture functionalities and in average 16 requirements towards each selected application scenario were identified.

## 3 Evaluation of Available Tools

This evaluation identified the simulation tools publicly available and used in other studies. For this purpose state-of-the-art commercial and non-commercial simulation tools are assessed for their applicability to the envisaged simulation approach. In order to simulate traffic, communication, and the actions resulting from these subparts' interactions, a complete C2C/C2X simulation system must couple the traffic and the communication simulators. Also, the actions resulting from exchanged messages must be computed. This is mainly done using so-called "application simulations". Furthermore, driving simulators are taking into regard, in order to allow the evaluation of the HMIs which present exchanged information – or any other actions derived from such information – to the driver. Additionally further tools exist which enhance the system by simulating environmental impacts such as pollutant emissions, or allow to model environment behavior, such as weather conditions.

After collecting a lot of different simulation tools, six traffic simulators (such as VISSIM, SUMO, V2XMS etc.) seven communication simulators (such as ns-2, OPNET, JiST/SWANS etc.) and three application simulators (PRESCAN, V2XMS and VSimRTI) were analyzed in detail by the PRE-DRIVE C2X partners. Also five architecture tools (such as VSimRTI, iTETRIS etc) were evaluated, too. Characteristics of these tools were checked against the requirements mentioned in the previous chapter.

All investigated traffic simulators are covering at least 60% of all requirements. Assuming that partial met requirements can fulfilled with slight improvements, nearly 80% of all requirements are met. Major gaps were found in the online road modification (change of used infrastructure) and in the driver model (especially pre-crash scenarios and driver behavior). The majority of the communication simulators can fulfill more or less only about 50% of all requirements. Here the coupling capability and access to the message related information is an issue where improvement is needed. Another critical point is that almost every communication did not take environmental effects into account. The three evaluated application simulators do not make much difference, they cover more than 80% (with assumed improvements up to 95%) of all requirements. Last but not least, the architecture tools meet about 60% of all requirements. Most of them are not capable of dealing with different driver models.

As result of the evaluation one major need seems to be the ability to use different models for simulating the communication and for simulating the traffic. There are several reasons for forcing it. A system which allows exchanging the used models will support:

- The simulation of the system at different granularities: starting with coarse simulations for rough parameter optimization and ending at fine granular simulations.
- More reliable results: using different simulators of the same granularity will provide results less dependent on the individual simulator's particularities.
- Replacing the traffic simulation by a driving simulation.

Such an open system requires a "loose" coupling of the traffic and the communications simulations. On the side of the traffic simulation, a common representation of the environment to simulate, including the road infrastructure and the demand would be necessary. Furthermore, the simulations would have to support the message exchange used by the C2C/C2X simulation system, though translation modules could be used as long as the simulation to couple supports any kind of interaction with external modules. The application logic should move from the simulators to keep the simulations "clean" – free of structures that do not belong to the simulation's original duty. A recommendation how to couple simulators is given in the following chapter – together with all the pros and cons of different possibilities.

## 4 Simulation Tool Set Architecture

In general, existing C2X simulation architectures can be divided into three different areas: An integrated simulator covering all C2X domains, a fixed coupling of different simulators, and a flexible simulator coupling realized by a simulation runtime infrastructure.

### 4.1 Integrated Simulator Covering Traffic, Communication, and Applications Domains

An integrated simulator covers the three domains important for C2X simulations, i.e. vehicular traffic, communication, and application simulations. Only one clock for the time management exists and no further synchronization is necessary. Here, the integration of different simulation domains is a major challenge for the simulator developers. In the history of computer simulations, vehicular traffic and wireless communication have been divergent domains without intersections. In general, an expert in developing traffic simulators does not necessarily have detailed knowledge about the simulation of wireless communication and vice versa. Thus, the development of such a tool with a high accuracy in traffic, communication, and application simulation has proven to be difficult. As a result, existing integrated simulators are rather suited for high-level simulations. In the PRE-DRIVE C2X project, only the simulator V2XMS belongs to the category of integrated simulators.

### 4.2 Fixed Coupling of Different Simulators for the Different Domains

Fixed simulator couplings are couplings of independent simulators where each simulator is specified for one of the domains, i.e. vehicular traffic, communication, or application simulation. Each simulator has an own clock for its time management. Thus, additional synchronization mechanisms have to be implemented in the coupling mechanism to ensure that all events of the coupled simulators are processed in the correct order. The coupling component is adapted to the used simulators and integrated simulation tools cannot be exchanged. A disadvantage of the fixed coupling is that a re-implementation is not only necessary if a simulator is to be exchanged, but also the integration of new versions of one of the coupled simulators mostly requires to make adjustments. So, a fixed simulator coupling only works well as long as all simulation scenarios have similar requirements that are fulfilled by the existing coupling. The following fixed couplings of simulators have been evaluated in PRE-DRIVE C2X: VISSIM + ns-2, iTETRIS (SUMO + ns-3), TraNS (SUMO + ns-2), and VISSIM + VCOM.

### 4.3 Simulation Runtime Infrastructure (RTI) with Common Interfaces for Flexible Simulator Coupling

If requirements vary depending on the simulated scenarios, it is not satisfying to use simulator couplings that are adapted to specific simulators and cannot be exchanged. To master this challenge, a simulation runtime infrastructure (RTI) with common interfaces allows the integration of arbitrary discrete event-based simulators. The coupling via common interfaces provides the flexibility to exchange simulators according to the specific requirements of a simulation scenario. A central management is provided by the simulation runtime infrastructure and offers services to handle synchronization, communication among the coupled simulators as well as lifecycle management of each component. A solution can be inspired by the IEEE Standard for Modeling and Simulation (M&S) High Level Architecture (HLA) [3]. Thus, attaching a simulator only requires to implement the interfaces of the simulation runtime infrastructure and to realize the commands specified within. The internals of the underlying implementation are hidden. In the PRE-DRIVE C2X project, the VSimRTI [4] belongs to the category of simulation runtime infrastructures for a flexible simulator coupling. The VSimRTI allows varying the composition of coupled simulators depending on the specific requirements of a scenario. Hence, an arbitrary composition of the traffic simulators VISSIM and SUMO, the communication simulators JiST/SWANS, OPNET, and OMNeT++, the application simulator VSimRTI_App, and further simulation and visualization tools is possible.

## 5 Simulation Models for Communication, Traffic and Environment

The challenge of a combined simulation study for ITS systems results from the difficulty of determining a suitable combination of the respective simulation models out of the potential large set founded in the Cartesian product. The consequences of inappropriate model selection are at least twofold: on the one hand, too abstract chosen models may provide incompatible input for dependent models causing the simulation results to become useless in the end. On the other hand, the selection of too detailed models unnecessarily demands computational effort which reflects in time consuming simulation studies. As a consequence, an integrated simulation tool set inevitably depends on recommendations which simulation purpose is met by which combination of simulation models. The following presents the modelling background of the research disciplines communication, traffic, and environment and further elaborates on a proper selection of combined simulation models.

### 5.1 Modelling of Communication

Research on communication comprises activities in various distinctive areas. Focusing on wireless communication as used in ITS systems, research challenges arise from physical to application layer. A rough classification indentifies three blocks dividing communication efforts into a) bit level, b) packet level, and c) information level. The bit level typically deals with physical issues like the design and position of antennas, the influence of obstacles on the radio wave propagation, or the characteristics of the chosen frequency and modulation scheme. The large number of influencing factors and expensive computations make an investigation on bit level time consuming which also gives reason for the often restricted scope to few transmitter-receiver-pairs. Simulation studies on bit level are typically conducted in ray tracers which provide as an output metrics like bit error rate, for instance. The results of the bit level serve as an input to the adjacent upper packet level. The packet level abstracts from transmitted rays and received bits and considers the failed or successful reception of an entire packet as atomic processing unit. The applied abstraction significantly saves computation time, however, simultaneously the scope of the studies is enlarged to many communicating nodes in networks by which computational effort rises again. Prominent representatives of simulation tools on packet level are NS-2/3, OPNET, or OMNeT++. The evaluation of the network performance is specific to the simulation purpose; often chosen objectives are the probability of packet reception or the delay of message delivery, for instance. Both latter mentioned measures are suitable inputs to the information level. Self-describing, the information level takes again an abstraction step and limits communication to the information which is available at a specific point in time to a communicating node. The question for the impact of received information, i.e., how information is used and which consequences are inferred, is specific to the application of interest. In ITS systems, for example, received information may change the driver's selected route or trigger a braking manoeuvre due to a hazardous situation ahead. Thereby, the output of the information level typically reflects in a different research discipline, like vehicular traffic.

Beyond the presented distinction of bit, packet, and information level, further abstractions can be applied in each level individually which allows to extend the discussion on the trade-off between accuracy and performance of the simulation study. Independent of the processing unit, the basis for the argumentation is found in the methodology of computing a level's output. A prime example is, for instance, given on packet level where the network simulator NS-2 and the VCOM module balance the level of accuracy and the required runtime. VCOM is a statistical model which instantaneously provides the probability of packet reception derived from numerous simulation traces generated in NS-2. For applications which operate on packets but are not sensitive to specific receptions, the time-intensive event-discrete simulation in NS-2 might be an overkill and VCOM a proper match. Exemplarily, one may think of radio-disseminated speed advices to increase traffic flow for which a statistical number of correctly behaving vehicles fits the simulation needs. Other simulation purposes require a more detailed representation of the networking behaviour making VCOM an unfavourable choice. All in all the simulation purpose determines which communication level is concerned and on which level of detail communication is simulated.

## 5.2 Modelling of Traffic

Just like in the communication domain, traffic simulations vary in the level of detail with which they treat the vehicular flows in a street network. While macroscopic models only use parameters like traffic flow (vehicles per hour) and average speed, microscopic approach involve individual vehicles; the macro-models may serve for up-scaling effects found on a microscopic level. Clearly, when the effects of ITS determine the driving behaviour and thus the movement of individual vehicles, such aspects must be treated on a microscopic level. Such models vary in the degree to which they describe the movement patterns of the vehicles. While agent-based approaches show advantages in run-time performance, their relatively simply movement patterns do not allow to model ITS influences with the degree of detail required. On the other hand, more detailed models with a more elaborate driver-vehicle model still show a run-time performance which is more than adequate when compared to the communication models. Several models on this level of detail were identified as being suitable for the envisaged tool set. These included models like the ITS-Modeller, SUMO or VISSIM. Each of the identified models shows distinct advantages (and draw-backs) which, in the end, result from their primary purpose they were developed for.

### 5.3 Modelling of Environmental Issues

One important aspect when assessing cooperative systems are environmental issues. There are several different models available to evaluate exhaust emissions from vehicular traffic. Generally, these models use the output of a traffic model as input and add the estimation of emission based of the traffic description. In this step, too, there are different levels of detail possible, depending on the time-scale and area for which the evaluation is performed. Cooperative systems usually change the way a vehicle is operated at a level, where a microscopic traffic flow simulator is the appropriate choice. Within PRE-DRIVE C2X, the combination of VISSIM and PHEM [5] was tested and has provided good results.

## 6 Conclusion

The number of available simulation models from each research discipline enables far more model combinations for a joint simulation study. Many of those potential combinations, however, do not seem to be reasonable, at least, if the perspective of a specific application of interest is taken. The user of the envisaged integrated tool set should thus be aided by a recommendation which model combinations fit which simulation purposes. A specific application of interest is thereby not necessarily linked to a single recommendation since the perspective of the simulation study may require to change the (modelling) focus. As an example, one may think of an emergency message disseminated in a hazardous traffic situation. For an application designer, the primary interest is on a suitable reaction behaviour depending on the remaining time to collision. For an antenna expert, however, the perspective of the use case changes, likely putting less emphasis on an advanced modelling of the reaction procedure but on a detailed configuration of the antenna model. At the first instance, the recommendation list depends on the foreseen use cases and requires coordination among the involved research disciplines.

The approach followed within PRE-DRIVE C2X to set up a tool set of simulations from different areas has proven to be a worthwhile task. It brings together the know-how of various disciplines and thus greatly enhances the understanding of such complex systems as Cooperative Systems of ITS. Although the process of creating a mutual understanding between experts with different backgrounds sometimes seems tedious, the "end product" incurred is valuable and, in the end, indispensable.

## Acknowledgement

The work described here was performed within the joint research project PRE-DRIVE C2X, which is funded by DG Infso of the European Commission within the 7th Framework Programme. For further information please refer to http://www.pre-drive-c2x.eu/.

## References

[1] Pre-Drive C2X Consortium, Detailed description of selected use-cases and corresponding technical requirements, Deliverable D4.1, 2008.

[2] Pre-Drive C2X Consortium, Description of user needs and requirements, and evaluation of existing tools, Deliverable D2.1, 2009.

[3] Institute of Electrical and Electronics Engineers, IEEE standard for modeling and simulation (M&S) High Level Architecture (HLA)–federate interface specification, IEEE Standard 1516.1, IEEE, New York, USA, 2000.

[4] Naumann, N., Schünemann, B., Radusch, I., VSimRTI - Simulation Runtime Infrastructure for V2X Communication Scenarios, Proceedings of the 16th World Congress and Exhibition on Intelligent Transport Systems and Services (ITS Stockholm 2009), 2009.

[5] Rexeis, M., Hausberger, S., Zallinger, M., Kurz, C., PHEM and NEMO: Tools for micro- and meso-scale emission modelling, 6th International Conference on Urban Air Quality, Cyprus, 27-29 March 2007.

**Thomas Benz**
PTV AG
Stumpfstr. 1
76131 Karlsruhe
Germany
thomas.benz@ptv.de

**Ralf Kernchen**
University Surrey
GU2 7XH Guildford
Surrey
UK
r.kernchen@surrey.ac.uk

**Moritz Killat**
Karlsruhe Institute of Technology (KIT)
Zirkel 2
76131 Karlsruhe
Germany
moritz.killat@kit.edu

**Andreas Richter**
German Aerospace Center (DLR)
Lilienthalplatz 7
38108 Braunschweig
Germany
andreas.richter@dlr.de

**Björn Schünemann**
Fraunhofer Institute - FOKUS
Kaiserin-Augusta-Allee 31
10589 Berlin
Germany
bjoern.schuenemann@fokus.fraunhofer.de

**Keywords:** cooperative systems, simulation, communication, traffic, environment

# GuideWeb: A New Paradigm for Navigation Support based on V2V Communication

B. X. Weis, A. Sandweg, BlackForestLightning


## Abstract

GuideWeb is a support network for vehicle navigation and guidance. It is constituted by the cooperation of a multitude of autonomous MapSynthesisers located in the vehicles. A MapSynthesiser, being the autonomous constitutional core of GuideWeb, receives over radio communication traffic flow information via information-enhanced maps (called map syntheses) from other GuideWeb participants' MapSynthesisers. It creates from received map syntheses and the information of its own travel route a new map synthesis, which is then broadcasted. MapSynthesiser provides timely and accurate information on traffic flow, density and trend on how traffic will develop as well as traversability everywhere within a radius of approx. 100 km to a navigation system for driver assistance. MapSynthesiser cooperation is based on short range radio communication e.g. Wireless Local Area Network (WLAN) according to IEEE 802.11 standards.


## 1 Introduction and Background

There is an increasing demand for individual traffic and route information and navigation. The existing systems are typically based either on elaborate infrastructures for traffic measurement or on traffic patterns derived e.g. from individual mobile phone movements. Both concepts have their drawbacks.

The first one requires substantial investment into setting up the infrastructure, its maintenance incurs cost and in the case of a privately run enterprise should result in some profits. Therefore, this service for the end user can't be free of charge. Furthermore, the measurement infrastructure is typically only available on motorways.

The second one collects data by tracking individual user-movements e.g. mobile phones. By processing this data and matching it to maps allows the derivation of traffic flow estimations. This process can't ensure privacy as patterns of movement of individuals are recorded. The potential misuse of this data mandates a rigid supervision of the process by independent auditors. Again, the effort to put these concepts into place incurs capital expenditures in computing infrastructure and operational cost.

V2V communication offers a remedy for these drawbacks. Firstly, V2V communication allows in addition to navigation a variety of different other applications e.g. communicating to other vehicles in dangerous situations [1, 2, 3] and/ or mobile Internet [4, 5] with their corresponding issues of security [6] etc. V2V communication will be based on IEEE 802.11p standard. A German study [7] indicates that V2V communication networking will not be in place before 2020 (see Fig. 1), though the European Union supports a variety of projects with this target. Very often, especially in Europe, the propensity of customers to spend money for services on a recurrent basis is rather low; they prefer a one-time payment if necessary. Finally, the quality of data sets the price customers are willing to pay for this service.

This sets the scene for the concept presented here. The requirements derived from above are:

- to create most up-to-date traffic flow information and navigation support everywhere – city, over-land or motorway,
- to offer most up-to-date information on temporarily traversable/non-traversable routes not provided by the maps of a navigation system,
- to ensure privacy and data security,
- to avoid recurrent service charges.

The implications of these requirements are manifold. The most important one is: No investments into infrastructure and cost for system maintenance: Making the reasonable assumption of there being no public subsidy, the direct structural implication is that no centralized processing, evaluation or distribution unit must be required for the functioning of the process.

The concept GuideWeb presented in the following fulfils these requirements. GuideWeb is a systemic process, in which a multitude of participants are involved, for transmitting and receiving information of vehicles, using a radio transmitter and receiver and, more particularly, based on a device, which also processes the information received and enables the presentation of synthesized information to the user. A similar concept has been presented in [8]. Furthermore, GuideWeb communication technology is simple and easy to implement, and can therefore serve as an intermediate introductory step in the full-fledged V2V communication networking (see Fig. 1) that is technologically and commercially viable.

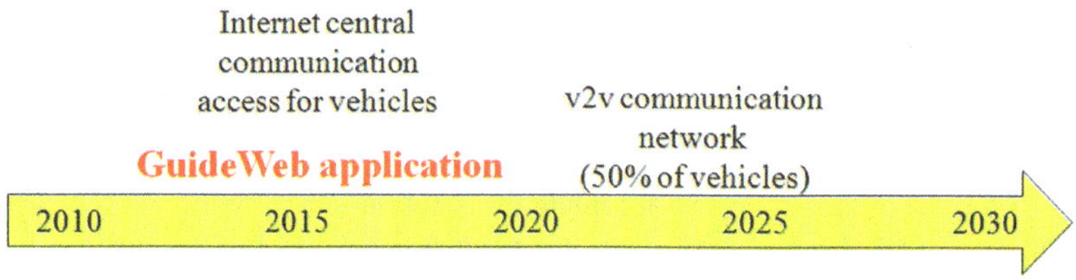


Fig. 1. Time line of V2V communication introduction: GuideWeb positioning

This paper describes the functioning and implementation of GuideWeb. First the concept of GuideWeb and the autonomous MapSynthesiser as its constituting element is introduced followed by a description of how GuideWeb/ MapSynthesiser fits into a driver assistance environment. The next two sections present the MapSynthesiser and its map processing capabilities. Simulation results that quantify penetration requirements and a conclusion complete the paper.

## 2 GuideWeb

The principle of GuideWeb is intriguingly simple – one gets information, processes it, uses it and distributes the processed information. Information is freely offered and everyone can take it (give and take). GuideWeb is constituted by a multitude of participants where the participants continuously broadcast their information and knowledge about current traffic and environment in the form of map syntheses. A map synthesis is a data compressed form of the synthesized map (NowMap) derived from the aggregation of all routes GuideWeb participants have traversed including averaged speed and traffic density information of route segments. Thus, the individual map synthesis is built up from every participant's best knowledge of traffic flow and environment. In order to facilitate this process each participant is equipped with a device called MapSynthesiser of which the functional diagram is shown in Fig. 2.

The MapSynthesiser is typically a software application (executing together with a navigation application), which autonomously enables the functionalities of the GuideWeb only requiring connectivity from the communication platform and some basic vehicle information. MapSynthesiser provides timely and accurate information on traffic flow and density as well as traversability of roads everywhere within a radius of approx. 100 km. This information can be further processed and displayed to the driver e.g. by a navigation system. The MapSynthesiser generates the NowMap, the representation of the current information and knowledge about the environment, from which the broadcasted map synthesis is derived.

Conceptually GuideWeb is self-organizing and self-contained and independent of any infrastructure. Functionally GuideWeb rests entirely on the cooperation of the multitude of autonomous MapSynthesiser nodes

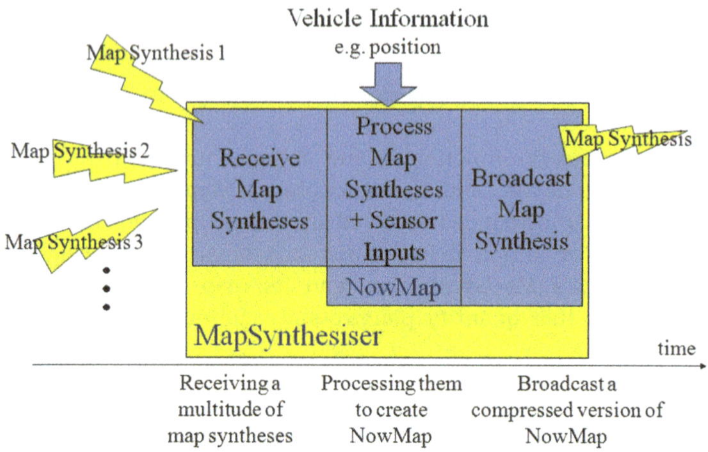


Fig. 2. Functional diagram of the MapSynthesiser

## 3 Environment for MapSynthesiser Operation

MapSynthesiser typically operates in a driver assistance environment and Fig. 3 shows how MapSynthesiser fits into this. MapSynthesiser exploits the V2V communication platform to send and receive messages. Two message types have been defined:

Synthesis messages contain a data condensed version of NowMap, and additional information, e.g. road conditions, radar traps, etc. when available and appropriate. Synthesis messages comprise on the incoming side the data compressed NowMaps of other GuideWeb participants' MapSynthesiser and on the outgoing side of its own NowMap. From the incoming synthesis messages the corresponding map data are extracted and made available for further processing.

Presence messages contain a minimum of information indicating the presence of a MapSynthesiser. Presence messages are very short and from their reception together with the reception of synthesis messages an estimate of traffic density can be derived. Furthermore, in phases of very low MapSynthesiser density, i.e. traffic density, presence messages serve as a trigger to broadcast

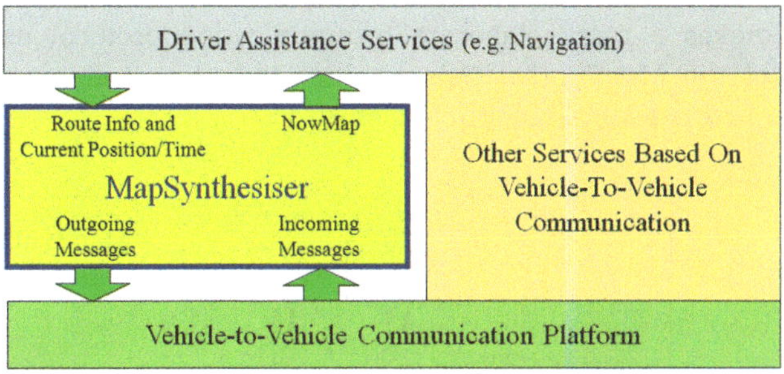


Fig. 3. MapSynthesiser in the driver assistance environment

synthesis messages; if no MapSynthesiser is in radio reach to receive synthesis messages they will not be transmitted. Further, MapSynthesiser uses driver assistance services for positioning as well as time information and optionally for a map of the environment of the navigation system. This map may then serve as the basis for the map processing in the MapSynthesiser. The map representation in MapSynthesiser is based on Earth coordinates to ensure independence from the map representation of the navigation system (9). As a result MapSynthesiser provides the navigation system with a map representation that comprises attributes for route segments – most prominently traffic flow and traffic flow trend estimation as well as traffic density. These attributes are time stamped.

## 4 Outline of MapSynthesiser

The MapSynthesiser comprises the functionalities as described in the following based on the block diagram shown in Fig. 4. The MapSynthesiser broadcasts its messages in two modes:

- In Broadcast Mode 1 it broadcasts a synthesis message,
- In Broadcast Mode 2 it broadcasts a presence message.

In addition to the flow control mechanisms of the V2V communication platform stochastic processes control the broadcast modes to ensure a maximum throughput of messages.

The V2V communication platform sends the messages on some carrier frequency typically in the 2,4 GHz or 5 GHz band reserved for short-range applications (WLAN – Wireless Local Area Network). In order to minimize the expenses of the radio devices it is most cost effective if all radio subsystems transmit and receive information on the same carrier frequency. Furthermore, this choice makes a more elaborate communication protocol unnecessary. When each of the MapSynthesisers employs the same frequency selection algorithm for broadcasting, reception is ensured.

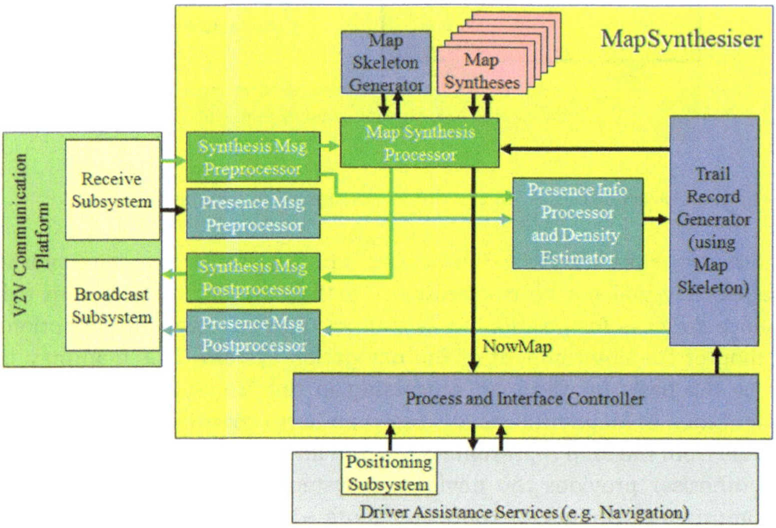


Fig. 4. MapSynthesiser block diagram

# 5 Processing Map Syntheses

## 5.1 Map Synthesis Attributes

The map synthesis processor generates the map synthesis. From the received synthesis messages the relevant map synthesis information and additional information is extracted in a preprocessing stage. An example of an extracted map synthesis is shown in Fig. 5. On the right the map skeleton is presented, which summarizes the retrieved knowledge about routes in the chosen environment. On the left the map synthesis is shown with the attributes for the route segments. For each segment of the map skeleton – typically of length 100 m – the six values indicate average speed (AS), average density (AD) and traffic flow trend (T) for each direction of the road as determined. Together with a weight indicating the relevance of the information, i.e. the older the information the less relevant it becomes, they form the route segment attribute.

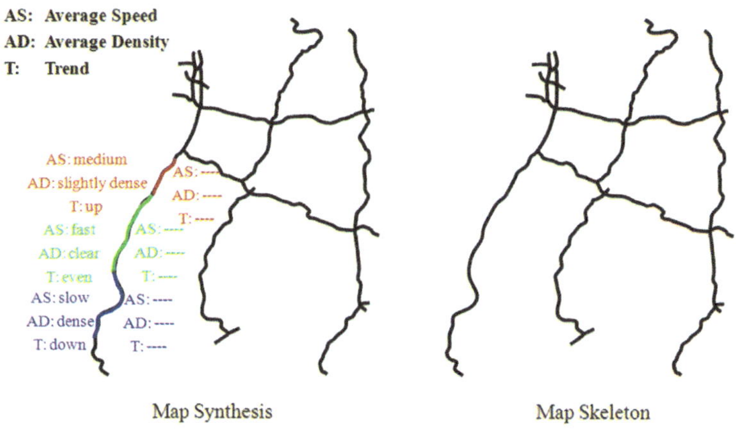


Fig. 5. Example of a map synthesis

## 5.2 Map Synthesis History

To generate the NowMap the map synthesis processor collects and processes all received map syntheses for a defined time period, and keeps the history of the information of each route segment in memory as shown in Fig. 6.

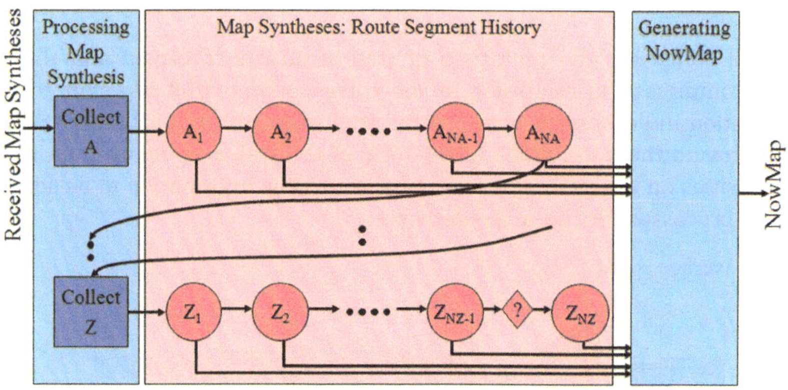


Fig. 6. Route segment history

The weights of the route segment attributes provide the time references to create this route segment history. The route segment history is used for computing the actual route segment attributes as well as for trend calculation. The history is stored in different levels e.g. in level $A$ the time difference of the attributes $A_i$ and $A_{i+1}$ is 1 min, in level $B$ $B_i$ and $B_{i+1}$ is 5 min, and so on. The number of levels and the corresponding timing is a matter of map synthesis processor configuration.

Keeping the route segment history enables the map synthesis processor to perform plausibility checks to detect and eliminate unrealistic attribute values (e.g. introduced to the system by a malicious user) as well as to compute the attributes for NowMap taking into account historical attributes. Furthermore, the historical attribute data allows a trend computation that indicates how traffic has evolved over time and what is to be expected.

### 5.3 Generating Trail Record

The most important ingredient to create a map synthesis is the trail record. A trail record is generated according to the following mechanism. When the process is started in a vehicle, the position is recorded with respect to the information received from the positioning subsystem e.g. a GPS receiver, and the time. For a certain distance e.g., 100 m, called a trail segment, the instantaneous speeds of the vehicle are determined and their average is computed resulting in an average speed for this trail segment and forms the trail segment record together with traffic density information that is computed by counting the received presence and synthesis messages and inferred from this number and the speed indications. Then the process is started again.

The trail comprises the multitude of trail segments traversed and the trail record comprises the multitude of the corresponding trail segment records. The position indications of points of the trail segments, in particular the start point, provide the information of the form of the trail and the average speeds and densities on the trail segments provide suitable information to be advantageously processed for the map synthesis.

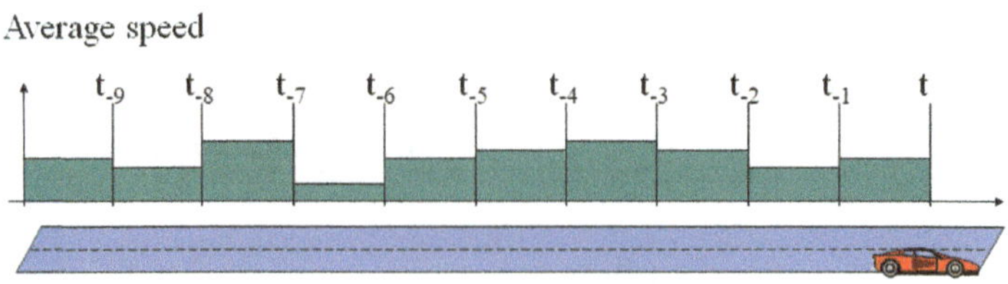


Fig. 7. Trail record contribution to map synthesis

The vehicle under consideration generates a trail record (see Fig. 7), which it contributes to its present map synthesis. Initially i.e. without having received a synthesis message from another participating vehicle, a map synthesis consists of just the trail record. If speed is low, a time-out terminates and restarts this process after e.g., 1 min. In the case where the vehicle is not actively participating e.g., a vehicle with switched-off engine, it may cease to contribute.

## 6 Conclusion

In this paper the concept GuideWeb for vehicle navigation support based on V2V communication has been presented. It has been shown that GuideWeb is a very suitable candidate for a commercially viable introduction of V2V communication imposing a minimum requirement on networking. By using a broadcast communication concept, important difficulties in V2V communication are overcome or circumvented, and due to its coordinate-based exchange format it is independent of map suppliers and easily integrated in any driver assistance system. GuideWeb finds its application window beginning now until all the challenges of the V2V networking capabilities are resolved.

Further, the insights derived from GuideWeb deployment and its behavior allows learning about V2V communication system performance. Keeping historical data of the route segment attributes adds a new level of complexity – the systemic memory. However, the advantages of historical data (detection of implausible attributes or malicious users, computing trends ...) trade off favorably with the implications of the systemic memory as shown by simulation. Furthermore, by concept the penetration level required for GuideWeb to function is much lower than in other V2V communication systems. The processing methodology allows that information about specific vehicles can neither be extracted nor traced, i.e. that privacy is ensured.

## References

[1] Schnaufer, S., et al., Vehicular Ad-Hoc Networks: Single-Hop Broadcast is not enough, Proceedings of 3rd International Workshop on Intelligent Transportation (WIT), Hamburg, Germany, pp. 49-54, 2006.

[2] Torrent-Moreno, M., Inter-Vehicle Communications: Assessing Information Dissemination under Safety Constraints, 4th Annual IEEE/IFIP Conference on Wireless On Demand Network Systems and Services (WONS), Obergurgl, Austria, 2007.

[3] Baldessari R. et al., NEMO meets VANET: A Deployability Analysis of Network Mobility in Vehicular Communication, , Proceedings of 7th International Conference on ITS Telecommunications (ITST 2007), Sophia Antipolis, France, pp. 375–380, 2007.

[4] Bechler, M., et al., Mobile Internet Access in FleetNet , in Proceedings of KiVS 2003, Leipzig, Germany, 2003.

[5] Baldessari, R., et al., Flexible Connectivity Management in Vehicular Communication Networks, Proceedings of 3rd International Workshop on Intelligent Transportation (WIT), Hamburg, Germany, pp. 211-216, 2006.

[6] Harsch, C., et al., Secure Position-Based Routing for VANETs, Proceedings of IEEE 66th Vehicular Technology Conference (VTC Fall), Baltimore, MD, USA, 2008.

[7] Zukunft und Zukunftsfähigkeit der Informations- und Kommunikationstechnologien und Medien, Internationale Delphi-Studie 2030, Münchner Kreis e.V., EICT GmbH, Deutsche Telekom AG, TNS Infratest GmbH, 2009.

[8] Wischhof, L., et al., SOTIS – A Self-Organizing Traffic Information System, Proceedings of 57th IEEE Semiannual Vehicular Technology Conference VTC 2003-Spring, Jeju, South Korea, 2003.

[9] ISO, Intelligent Transport Systems (ITS) – Location Referencing for Geographic Databases – Part 3: Dynamic Location References (Dynamic Profile), Draft Standard 17572-3, 2009.



**Bernd X. Weis, Astrid Sandweg**
Haeberlinstr. 29 b
70563 Stuttgart
Germany
bernd.weis@blackforestlightning.de
astrid.sandweg@blackforestlightning.de


**Keywords:** traffic management, driver assistance, car-to-car communication, dynamic routing

# Road Course Estimation Using Local Maps and Digital Maps

M. Konrad, K. Dietmayer, Universität Ulm
M. Szczot, Daimler AG

## Abstract

This paper presents a novel approach for road course estimation using a multilayer laser scanner and a digital road map, which is similar to that used for navigation purposes. If the correct position of the host vehicle on the digital map is known, the road course can be predicted. But because of inaccurate GPS measurements this position is not determined satisfyingly. Therefore, the road course is extracted from the digital map and is matched to an occupancy grid which is a representation of a local map. Based on the correct position as the result of this matching, the road course is extracted from the digital map and can be predicted even in great distances where no sensor measurements exist.

## 1 Introduction

Many common and future driver assistance systems are based on information about the environment of the vehicle. The road course as basic environmental information is useful for many assistance systems such as Adaptive Cruise Control, pedestrian-, or obstacle detection. An important task in recent research is not only to detect an object but to rate its relevance and hence to establish a sophisticated treatment. A pedestrian detected on the road yields much more risk potential than one detected aside the road and should result in a higher warn level. A road course prediction in great distances (above 80 meters) is desired because the earlier the relevance of an object can be determined, the earlier an appropriate reaction can be deduced. In general, road course estimation is a wide field in research. There are many approaches which differ in sensors and appropriate features of their measurements, the road model, and the kind of prediction. Only a few methods are based on laser scanner measurements. In [1] raw laser points are used and in [2] the echo pulse width is additionally taken into account. The conventional occupancy grid [3] was extended in [4] and was used to estimate the road course. Digital road maps, which consist of waypoints that describe the road network, are only used in a few publications [5, 6]. This paper presents a road course estimation approach based on a digital road map to provide a prediction greater than 80 meters. Because of inaccurate GPS positions an occupancy grid is built by integrating the laser measurements over time. Afterwards the digital map is matched to this grid, which leads to an estimation of the correct position and the road width. To solve this matching problem, a particle filter is applied. Based on the correct position, the road course is extracted from the digital map.

## 2 Local Map

In general, a local map is built by integrating sensor measurements (e.g. from radar, sonar, or lidar). Thus, the influence of noise and uncertainties are decreased and the probability of mapping relevant data is increased. Here, the occupancy grid mapping is chosen for building a local map, which is described below.

### 2.1 Occupancy Grid

The Occupancy Grid Mapping is presented in [3] and was extended by [4] for automotive applications. The main idea is to divide the environment of the vehicle into grid cells using a defined resolution, e.g. 0.5m per pixel. Each cell $m_i$ holds an occupancy likelihood $p(m_i|z_{1:t}, u_{1:t})$, that contains the history of all corresponding laser measurements $z_{1:t}$ regarding the movements $u_{1:t}$ of the vehicle up to time $t$. A likelihood of $p(m_i|z_{1:t}, u_{1:t}) = 0$ stands for a free and $p(m_i|z_{1:t}, u_{1:t}) = 1$ for an occupied cell. Three steps are executed for every new laser measurement: a new pose of the vehicle is determined, a measurement grid is built and the occupancy grid is updated.

**Determine the Pose of the Vehicle:** The pose of the vehicle is estimated using an Extended Kalman Filter with a Constant Turn Rate and Velocity model (CTRV). The measurement vector consists of the four wheel speeds and the yaw rate. The state vector holds the position and orientation related to the grid.

**Building the Measurement Grid:** The following Inverse Sensor Model maps a laser scan to the so called measurement grid. Initially, all cells of the measurement grid are set to 0.5 because it is not known whether they are occupied or not. Afterwards, each laser point, related to the laser scanner coordinate system, is transformed to the measurement grid. It is uncertain in its position caused by an inaccurate measurement of distance and angle on the one hand and errors based on the estimation of the movement of the vehicle on the other hand. These uncertainties are modeled for each point. Finally, a free space function is applied. If there is an occupied cell, the region behind it is hidden and that cells keep their initial likelihood. It can be assumed that the area between the laser scanner and this occupied cell is free. Additionally, if no object is detected in a certain scan direction, this region seems to be free, as well. There are different approaches for handling these regions. For example the likelihoods can be set to a low value [3, 7]. Nevertheless, because of discrete scan directions, the probability to miss an object increases with its radial distance. Therefore, a function which depends on the radial distance was proposed in [8]. Here another free space function is presented for two reasons. The first reason originates from the mounting position and scanner characteristics. For example: in a great distance, a laser beam hits a curb which is inserted into the grid. If the vehicle moves closely to the curb, no laser beam is able to detect it, because all beams measure above that object. Thus, the corresponding cells are wrongly assumed to be free. Therefore, the likelihoods increase starting at 0.5 in a small distance interval close to the vehicle. The second reason is that the laser scanner has a variable angle resolution. It gets coarser to the sides, resulting in a higher probability of missing an object. This is modeled by a Gaussian function, which standard deviation increases linearly over the longitudinal direction and is combined with the linear decreasing over the radial distance from [8]. Figure 1 shows the proposed free space function as three-dimensional plot as well as the resulting measurement grid.

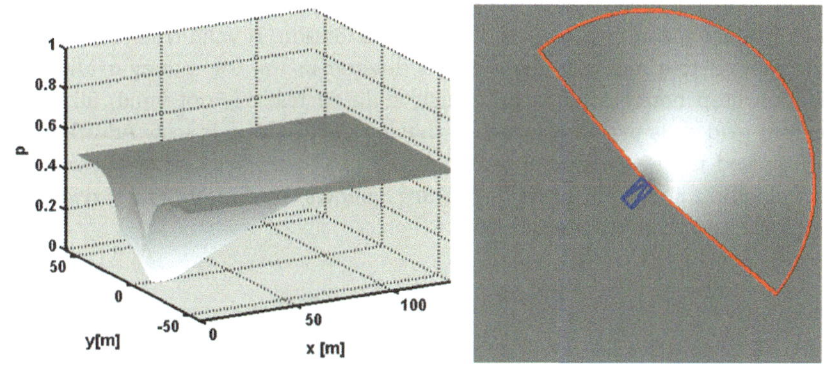


Fig. 1. The proposed free space function as schematic 3D plot and the corresponding measurement grid.

**Updating the Occupancy Grid:** The problem of calculating the occupancy grid is broken down into a number of estimation problems. Each cell $m_i$ is considered as a binary problem with static state, which is solved using a Binary Bayes Filter. Therefore, the formulation deduced in [4] is used. These equations describe the update of the likelihoods $p(m_i z_{1:t}, u_{1:t})$ in the current occupancy grid regarding the likelihoods $p(m_i|z_t,u_t)$ of its corresponding cells in the measurement grid and the last occupancy grid $p(m_i|z_{1:t-1},u_{1:t-1})$. Figure 2 depicts an overview over the whole mapping algorithm for the shown scene.

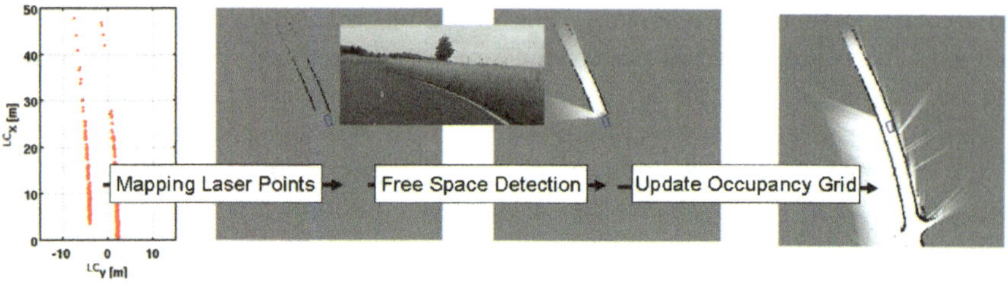


Fig. 2. Overview over the mapping algorithm. The laser points for the shown scene are mapped to the grid, the free space detection is applied and the occupancy grid is updated.

## 2.2 Detecting Moving Objects

A moving object remains as kind of "tail" in the occupancy grid (see figure 3, left), which is disadvantageous for road detection. To eliminate the corresponding cells, a so called difference grid is determined between the current and the previous occupancy grid. Afterwards, a segmentation algorithm [9] yields potential moving objects. The centroids of these segments are associated to existing tracks or lead to new tracks. Characteristics like speed, direction, size and length are taken into account for track validation. A valid track is assumed to correspond to a moving object and is deleted in the occupancy grid. If the validation step notices that a previously deleted track is not valid, all corresponding grid cells are reinserted so that no information is lost. Additionally, the proposed method yields approximations of the trajectories of the objects, their velocities, and their direction. Figure 3 shows the elimination of two moving objects and their trajectories.

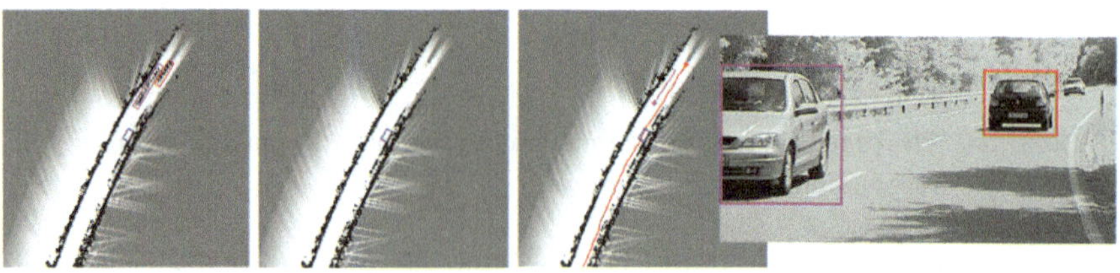

Fig. 3. Moving object detections in the shown scene. Left, the original occupancy grid with two moving objects. Middle, the occupancy grid without moving objects. Right, the trajectories of the deleted objects.

## 3 Road Course Estimation

The following road course estimation is based on the match of digital map feature to so called road border features which are extracted from the occupancy grid.

### 3.1 Extracting Road Border Feature Points

Occupied cells, which describe the road borders, are searched based on a center line $C = (C_b, C_f)$. Thereby, $C_b$ contains the previous positions of the vehicle. Additionally, a linear interpolation adds points to get a trajectory without gaps. $C_f$ is a line that points in driving direction. Starting at the position of the vehicle, $C_f$ ends if an occupied cell is hit or a certain distance is exceeded. Additionally, every $c_i = (x_i, y_i)$ in $C$ gets a direction information $o_i$ which is the number of the octant in which the straight line between $c_{i-1}$ and $c_i$ points. The first octant denotes the interval from 0 to 45°, the second from 45° to 90° and so on. Afterwards, $d_i$ is used to determine the left and right road border feature points. For example: if $o_i = 1$ the left road border is above $c_i$ and the right one below. A counter $k_l$ is increased until $(x_i-k_l, y_i)$ addresses an occupied cell which is now a feature point for the left road border $l_i = (x_i-k_l, y_i)$. No left feature point is found for $c_i$ if $|x_i - (x_i-k_l)| > d_l$, where $d_l$ denotes a threshold depending on the road width. Analog, $k_r$ is increased until $(x_i+k_r, y_i)$ addresses a feature point $r_i = (x_i+k_r, y_i)$ for the right border or a distance $d_r$ is exceeded. For each $c_i$ in $C$ left and right feature points are searched. Figure 4 schematically illustrates this algorithm which leads to left and right feature points $L = \{l_i\}$ and $R = \{r_i\}$.

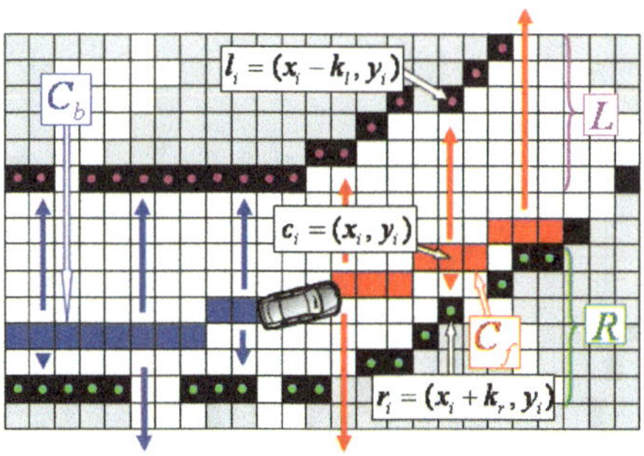


Fig. 4. Schematic illustration of the extraction of road border feature points

### 3.2 Extract Digital Map Features

The digital map consists of GPS waypoints that describe the road network. At every on-road GPS position of the host vehicle, a history and an expected future of such waypoints are known, but only that ones are considered which lie in a distance interval around the car (e.g. -250m to 50m). A *Hermite spline* interpolates these extracted points, which leads to a road course estimation. Afterwards, the spline is sampled using a certain sampling rate considering computational costs and desired accuracy. These sampled points are used as digital map features.

### 3.3 Matching the Digital Map to the Occupancy Grid

Matching the digital map to the occupancy grid is an optimization problem. Therefore, a criterion $e$ which quantifies the match for a hypotheses $(N,E,\varphi,b)$ is defined, where $N$, $E$, $\varphi$ denote a GPS position in UTM coordinates and $b$ is the road width. The calculation of the criterion for a hypothesis is realized as follows, which is additionally illustrated in figure 5.

- Transforming the digital map features to the occupancy grid using $N$, $E$, $\varphi$ and the corresponding position in the grid
- Translating the digital map features to a left and a right road border considering the road width $b$ of the hypotheses which leads to left and right digital map features.
- For every left road border feature point the corresponding straight line between two left digital map features is determined and their distance is calculated. Analog, a distance for every right road border feature point is established. All of these distances are combined to a sum of squared errors.

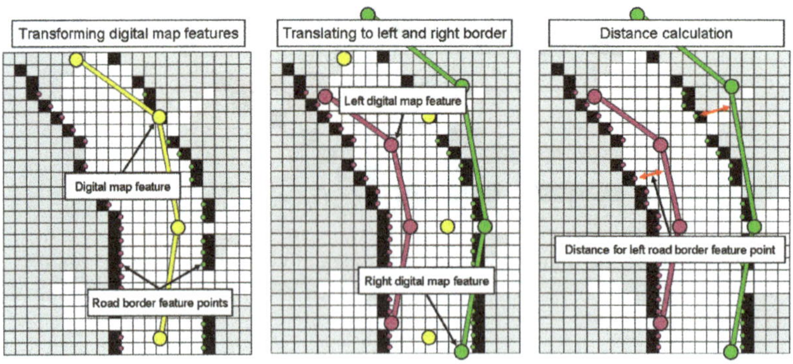


Fig. 5. Schematic illustration of the calculation of the criterion

Additionally to the above mentioned algorithm, an error weighting is introduced. It is assumed that there are no feature points between the road borders, which leads to a higher weighting of points between the borders. Furthermore, there are scenarios, which provide more left road border feature points than right ones or vice versa. To balance their influence to the criterion, the sum of errors for each border is weighted in relationship to the ratio of left and right feature points.

The optimization problem is solved using a particle filter [10], which additionally provides a filtering over time. Each particle denotes a hypothesis $(N,E,\varphi,b)$ which is predicted using the vehicle movement since the last step while the road width stays constant. Furthermore, uncorrelated Gaussian noise is added. The innovation step assigns a weight to each particle according to the calculated criterion value. The weighted mean over all particles is calculated as best estimation result. For initialization of the process, the particles are uniformly distributed around the initial measurement of the GPS sensor and their road width is set to a certain value related to the type of the current road (e.g. 6m for a rural road).

## 4 Results

Evaluating an estimated road course is difficult because it is hard to get a reliable ground truth. To get an impression of the performance of the proposed method, three different scenarios are shown in figure 6. Furthermore, to quantify the presented approach, the road course of the urban sequence was labeled. For that, the occupancy grids were built and, for one frame per second, the road borders were marked manually. Afterwards, the labeled points were compared to the estimated road course which leads to a mean error of about one meter. It can be determined, that there is no significant variation between errors in different distances. This is caused by the fixed road course extracted from the digital road map which is matched to the occupancy grid. Nevertheless, the error increases a little in higher distances as expected. The road width can be determined with a mean error of 0.5 meters. High errors are caused by the assumption of a constant width for the whole distance, which is used for the matching process. Figure 7 shows the results.

Fig. 6. Three basically different scenes mapped to occupancy grids (300m x 300m) and the estimated road course. Top: highway scenario, middle: winding street in a small village, bottom: rural road with sparse road border information.

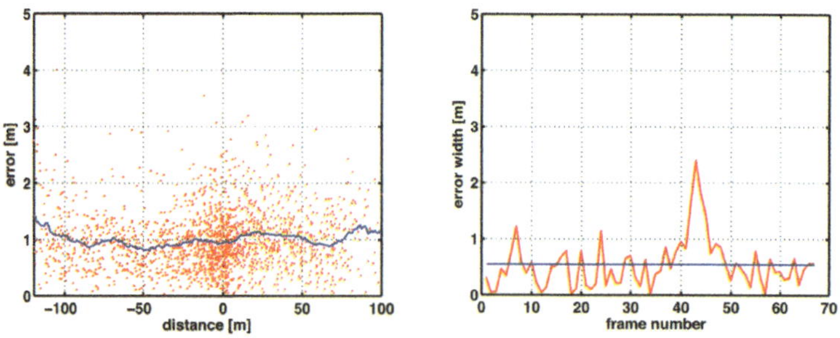


Fig. 7. Left: Every red point depicts an error between a labeled point and the estimated road course. The blue line illustrates the mean error according to a distance. Right: The red curve shows the errors between labeled and estimated road widths. The blue line depicts the mean error. In reality the road was smaller in the frames 43-45 which leads to a high error.

## 5 Conclusions and Future Works

This paper proposed a road course estimation method based on digital road maps. Matching the map to an occupancy grid results in a correct position on the map, which is necessary to estimate a reliable road course. The matching step is based on road border feature points extracted from the occupancy grid and digital map features. A particle filter was applied to filter the result over time. In general, the presented approach yields good road course estimations even in great distances. The main source of errors is the estimation of the vehicle movement, which can cause an inaccurate occupancy grid. Using SLAM (Simultaneous Localization and Tracking) techniques or attaching more accurate odometry sensors, especially for measuring the yaw rate, could minimize this effect. Furthermore, reflecting posts describe the road course very well. Detecting and incorporating them as features could increase the performance of the proposed method.

## Acknowledgement

This work was supported by Daimler AG, Department Environment Perception, Ulm, Germany.

## References

[1] Kirchner, A., Heinrich, T., Model based detection of road boundaries with a laser scanner, Proceedings of IEEE International Conference on Intelligent Vehicles, pp. 93-98, 1998.

[2] Kibbel, J., Justus, W., Fürstenberg, K., Lane estimation and departure warning using multilayer laserscanner, Proceedings of IEEE International Conference on Intelligent Transportation Systems, pp. 777–781, 2005.

[3] Thrun, S., Burgard, W., Fox, D., Probabilistic Robotics, MIT Press, Cambridge, 2005.

[4] Weiss, T., Schiele, B., Dietmayer, K., Robust driving path detection in urban and highway scenarios using a laser scanner and online occupancy grids, Proceedings of IEEE Intelligent Vehicles Symposium, pp. 184–189, 2007.

[5] Serfling, M., Schweiger, R., Ritter, W., Road course estimation in a night vision application using a digital map, a camera sensor and a prototypical imaging radar system, Proceedings of IEEE Intelligent Vehicles Symposium, pp. 810–815, 2008.

[6] Tsogas, M., Polychronopoulos, A., Amditis, A., Using digital maps to enhance lane keeping support systems, Proceedings of IEEE Intelligent Vehicles Symposium, pp. 148–153, 2007.

[7] Vu, T., Aycard, O., Appenrodt, N., Online localization and mapping with moving object tracking in dynamic outdoor environments, Proceedings of IEEE Intelligent Vehicles Symposium, pp. 190–195, 2007.

[8] Weiss, T., Dietmayer, K., Automatic detection of traffic infrastructure objects for the rapid generation of detailed digital maps using laser scanners, Proceedings of IEEE Intelligent Vehicles Symposium, pp. 1271–1277, 2007.

[9] Haralick, M., Shapiro, L., Computer and Robot Vision, Addison-Wesley Longman Publishing Co., Boston, MA, USA, 1992.

[10] Isard, M., Blake, A., CONDENSATION - conditional density propagation for visual tracking, International Journal of Computer Vision, pp. 5-28, 1998.

**Marcus Konrad, Klaus Dietmayer**
Universität Ulm
Albert-Einstein-Allee 41
89081 Ulm
Germany
marcus.konrad@uni-ulm.de
klaus.dietmayer@uni-ulm.de

**Magdalena Szczot**
Daimler AG
Wilhelm-Runge-Str. 11
89081 Ulm
Germany
magdalena.szczot@daimler.com

**Keywords:** driver assistance, laser scanner, occupancy grid, digital map, matching

# Galileo GNSS Based Mobility Services

R. Uhle, SAP AG


## Abstract

Fleet managers more and more want to have a process oriented approach integrated into their fleet management applications. In most cases it is not enough anymore to know that a vehicle passed a trip from location A to B in a certain timeframe with a certain number of transactional activities (such as parcel deliveries etc.). Furthermore process oriented information such as detailed routing information, number of stops at traffic lights, number of left turns (normally time consuming in right lane traffic areas), number of gear changes, time wasted in traffic jams, average speed, max speed, altitude profile of the trip, number of car door openings, $CO_2$ emission, etc. is more and more timely available through modern powerful Broadband Wireless Access (BWA) technologies and is waiting for exploitation through sophisticated analytical and business applications in order to reduce the total cost of ownership for businesses processes with moving assets involved. The article will show how a business case could look like that covers all needs to turn technical information into a business that is able to realize a win-win situation for all parties involved into the mobility process.


## 1 A New Industry Based on New Technologies

### 1.1 The Internet of Things and Services

The Internet of Things more and more brings Information and Communication Technologies into the private life of the human being. Location Based Services on SMART Phones are a good example how Geographic Information Systems (GIS), Global Navigation Satellite Systems (GNSS) and Business Applications can be combined to give tourists guidance in typical situations answering questions such as 'Where is a Point of Interest (POI) or a Sushi Bar next to my current location?' From the business point of view this kind of 'Real Life Awareness' opens up a new dimension for Internet Services with incredible opportunities. In our example the tourist – in his car or by feet – is a moving asset, his sightseeing trip is a process that gives the opportunity to be exploited in offering services for the process management such as guidance, information, quality assurance, or security etc. The quality of these internet services can be optimized, the more process data can be tracked and stored in backend systems as a basis for analytical applications that allow strategic, operational and predictive analyses. The introduction of Business Intelligence principles into this open platform will be discussed in detail below.

Fig. 1. Vision of a web-based service economy the internet of things and services, Source: SAP TechEd 2009 Session BI301.

## 1.2 Process Oriented Data Acquisition for Business Intelligence Needs

In their Seventh Framework Program (FP7) for research and technology development the European Commission (EC) defined a special Program for Information and Communication Technologies (ICT). One challenging Work Program (WP) where huge research resources are currently bundled especially in Europe is the area of Intelligent Transportation Systems (ITS) where concepts for vehicular communications are discussed and developed in funded projects (e.g. PreDrive [1]). Based on the availability of powerful Broadband Wireless Access (BWA) technologies such as UMTS, WiMAX, or LTE, communication concepts for the integration of vehicles with business applications (Vehicle-to-Business,) are more and more thinkable.

The integration of vehicles and business systems needs well-defined and standardized integration platforms that allow for a reliable and scalable interconnection between vehicles and business systems. A major challenge that needs to be tackled by such an integration platform concerns the intermittent connectivity of vehicles caused by their high mobility and the incomplete network coverage of existing BWA technologies. While the platform needs to deal with the technical connectivity, it also needs to offer dedicated functionality for priority-based queuing and scheduling of message transmission (Quality of Service) [1]. The following figure shows how a vehicle integration platform has to be designed that enables the full quality of service in regards to a business application connectivity:

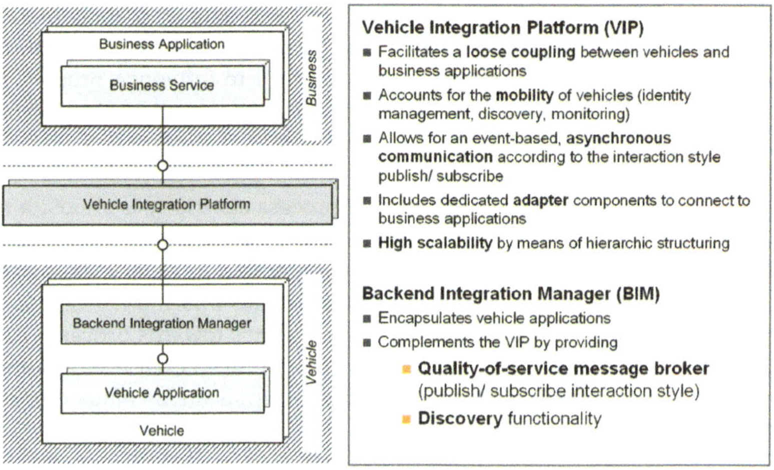


Fig. 2. Vehicle-to-business integration architecture: 2010 [1]

Another research area that is complementary to the technical integration between motor vehicles and business applications concerns the processing and enrichment of the raw technical data transmitted by vehicles. A provider that mediates between vehicles and business applications needs to receive the raw technical data, process it and enrich it with business logic in order to provide the business data to interested business applications. The mediation process for floating car data typically has to be run in an operational, process-oriented mode, which reduces the accepted latency time frames to a maximum of several minutes, whereby latency has to be seen in two ways: On the one hand, GNSS (Global Navigation Satellite System) position information such as GPS, EGNOS, or in the near future Galileo tracked from the in-vehicle GNSS receiver can be used easily and very fast to visualize the position of the particular vehicle on a geographical map in the in vehicles' car navigation system. If the position of the vehicle has also to be surveyed in a computer center management system of a backend application system, the accepted latency time would be 60-100 seconds. If – on the other hand, but still in the same example – the application now detects an error within a transportation process for dangerous goods, where e.g. a road type has been chosen by the driver that does not satisfy the security service level agreements for the related transport contract, the application wants to inform the truck driver by firing an event within a few seconds again in order to correct the transport process in changing the route planning of the On-Board-Unit application.

Exactly this way of near-real-time, bidirectional communication between moving assets and backend business applications leads to the term Process-Centric Business Intelligence (PCBI). PCBI has to turn technical data into transactional data by adding a business flavor to it. It has to aggregate transactional data to business process-related information. It has to analyze this information along dedicated algorithms and finally it has to be able to influence processes for which the analysis of the generated information on business level has led to the conclusion that the underlying process is running in a wrong way.

Especially the second aspect of this way of process centricity is almost a terra incognita and has to be investigated from scratch. Of course modern Business Intelligence and Data Warehousing solutions are able to collect transactional data from business applications and to analyze the information in business related dashboards, but in most cases the capabilities towards process-orientation, real-time capabilities, business transformation needs, service-orientation, model driveness, load and analysis performance, application connectivity, and interoperability are insufficient.

### 1.3 Service Oriented Architecture (SOA) Based Business Applications

As explained beforehand the PCBI platform has to take over the role of a mediator between Telematics and the Business Applications. The mediation process so far has been explained in the short form 'Turn data into business'. The cornerstone of the PCBI platform is a data warehouse with a dedicated data persistency in an underlying database accompanied with the necessary technologies for mass data volumes in direction of in –memory processing, Massive Parallel Processing (MPP) databases or Near-line Storage (NLS) support. The data warehouse architecture is service oriented and model driven which allows to implement typical data management and analysis processes as a kind of applications for data warehousing. In addition a data warehouse implementation methodology describes different layers for data acquisition, harmonization and consolidation, business transformation, propagation, reporting and operational needs and the dedicated modeling objects and implementation steps that have to be done per layer type.

In our case the telematic data typically has to enter the data warehouse through the data acquisition layer, then it has to pass different business transformations that route the resulting information into business related areas of the propagation and reporting layers, where they are accessible for business application needs. The most important part has to be managed during the business transformations that have to massage the incoming, granular and technical data in a way that it is understandable and useful for business application purposes. Of course business transformations can be time consuming but sophisticated caching mechanisms and look up APIs allow to optimize the system performance during this important step of the enrichment process of the data.

The most important part is the data acquisition itself. In our case a standard system-to-system communication has to be established in order to implement an open system of shared applications. Therefore the data acquisition layer offers the implementation of WebServices, in order to allow Internet communication using standardized protocols, Uniform Resource Identifiers (URI), XML based communication and data exchange, WebService Description Language for parameters, message encapsulation using Simple Object Access Protocol (SOAP) and central libraries for Webservices based on the technology of Universal Description, Discovery and Integration (UDDI). This technology will allow external applications to directly push data into the acquisition layer of the data warehouse. On the other hand the data warehouse can use WebServices also to pull data from external applications if necessary or to push data into external applications. The handover from the data acquisition layer to the business transformation layer and top level propagation and reporting layers can be totally automated. Therefore Daemon processes in the data warehouse can permanently start data processing activities for incoming data based on a reliable delta handling which is based on an exactly once in order service level of the WebService framework.

Regarding the above listed features and capabilities SAP NetWeaver Business Warehouse BW™ is more or less predestinated to take over the leading role of a data warehouse component that will cover all aspects of turning technical data into business relevant transactional data. Thus BW is closing the gap between the backend part of the vehicle integration platform and the possible business applications in SAP's broad Business Suite™ that offers solutions for most of the thinkable Mobility Services such as Enterprise Asset Management (EAM), Electronic Toll Collect (ETC), Fleet Management, Transportation, Financials with pricing and billing engines etc. As the nature of a data warehouse is to provide a single place of truth for data from many different source systems, BW offers various data acquisition capabilities. BWs sound and mature WebService interface is the ideal basis to include car floating data from moving assets into the information management processes of the data warehouse as the data acquisition is real-time enabled and highly automated on a minute wise processing periodicity. Due to such process oriented capabilities of BW incoming data can be handed over directly to the layer of business transformations and consolidation in order to aggregate and consolidate technical raw data into business relevant information for BI consumers or into transactional data for business applications sitting on top of BW. Electronic Toll Collect applications, for instance deal with objects such as customers, vehicles, contracts, pricing segments and invoices. The most important transactional data for the entire business comes from the vehicles with key figures such as 'driving kilometers per road type', 'amount of up- and downhill kilometers', 'Average Speed', '$CO_2$ Emission', 'exact routing descriptions' etc. which are generated in the BW environment. This transactional data can easily be handed over to the ETC application using standardized Business Application Programming Interfaces (BAPIs). Therefore the BW is mastering Online Transactional Processing (OLTP) needs of the scenario in the same way as it satisfies Online Analytical Processes (OLAP) demands. In short form: Real world process data from sensor networks can immediately be integrated into business applications thanks to state of the art real-time and in memory processing technologies – and this is what the Internet of Things and Services is finally all about. But the challenge of the given use case is not only a question of data warehousing capabilities for semantics and business transformations. It is also a question of managing massive data volumes regarding data load times as well as ad-hoc reporting needs. In this category the BW again shows advantages due to the fact that it can be deployed in the same way as an application over different application servers on top of different operating systems and database systems. Classical Relational Database Management Systems (RDBMS) are also an option as Massive Parallel Processing (MPP) solutions (planned for 2010) and column based in-memory indices for reporting in the speed of thought. Beside that the BW is the only data warehouse solution on the market that provides an archiving interface to 3rd party Near-line Storage solutions. All these capabilities can be bundled and open up the space for petabyte dimensions of car floating data.

## 2 The Mobility Services Industry – Use Case: Pay-As-You-Drive for Rental Cars

Having all these sophisticated solutions at hand we see the way paved for a new industry which we would like to give the name 'Mobility Services'. This new industry will have millions of users consuming services such as kilometer and driving behavior based road tolling (in a similar way to our today's mobile phone invoicing), usage based asset maintenance management, fleet management as a service for hundreds of thousands of private households, pay back options for consumers mobility behavior (drive your car in the right way, at the right time, in the right area and you will save money). The role of the Galileo GNSS standard may not be underestimated under these circumstances. Only the integrity and the legislation of the signal can finally make the before mentioned scenario possible. All partners of the Mobility Services Industry will ask for such standardizations and specifications in order to make their business service offerings reliable and compliant. Assuming the existence of all these standards certificates for particular mobility behavior patterns are thinkable that lead into various additional business scenarios.

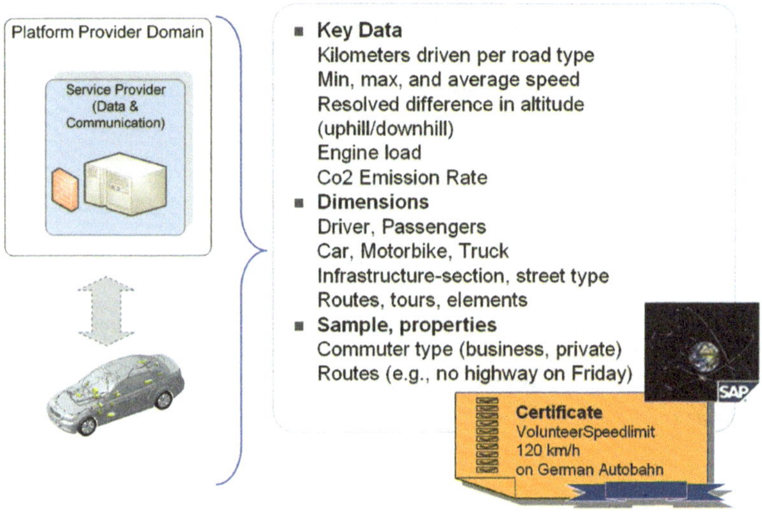


Fig. 3. Telematic Business Intelligence – Use Case Galileo GNSS based Mobility Services Source SAP TechEd 2009, Session BI301

Based on these functions and capabilities explained above the following use case can be implemented accordingly. In this use case, a rental car company offers different contract types with different tariffs and rates reflecting different usage types or typical customer behavior. Therefore dedicated tariffs exist for customers who typically drive long distances on motorways (motorway ratio over 80%) or in the opposite way for customers who mainly use their rental cars in urban regions. The tariffs can be combined and enriched with special agreements such as volunteer speed limits, using dedicated filling and service stations or pay back offers at certain POIs, such as holiday parks, fast food chains, museums, and parking service providers. Depending on the customers behavior during the rental period the rental car business provider can optimize his fleet management costs due to cooperation with other business partners from different industries. A small part of these savings is made available for customers who are willing to adjust their habits to the contract tariffs.

Therefore, the use case involves vehicles that are tracked and one fleet management application that uses the collected data to calculate invoices for car rental contracts. To collect the raw technical data, a V2B integration platform is used and the data needs to be processed into a BW PCBI platform. Typical processing steps include the matching between GNSS position data and GIS map data to determine the type of used roads (as this influences rental costs premium rates). Based on the GNSS data PCBI platform algorithms can also provide key figures such as average speed per road type, road type ratios per rental period, resolved difference in altitude (uphill, downhill), $CO_2$ Emission, fuel consumption estimations, park ticket verification etc. During the rental period it is of course possible for the customer at any time to ask the back end system for guidance and statistical information over the rental process such as the before mentioned key figures. Also this kind of conversation would be handled over the PCBI platform framework.

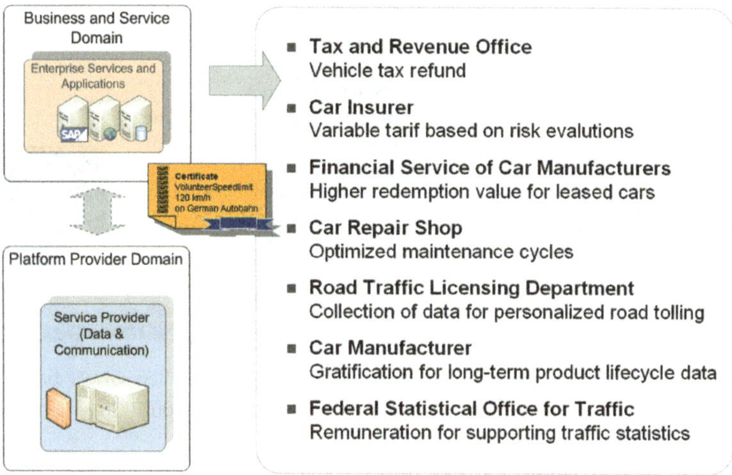


Fig. 4. Pay Back Systems in the area of mobility management
Source: SAP TechEd 2009, Session BI301

## 3 'Vision impossible?'

Against the explanations above the following story doesn't sound that visionary any more: It is the year 2015. A new industry called Mobility Services Industry offers internet based services for fleet management based on state of the art Car-to-X Communication Standards, Telematic and Business Application Platforms for floating car data, the newly available European Galileo GNSS standards and many Business Applications on top of this architecture.

I am heading one of 30 Mio private households in Germany with statistically 1.5 cars. My household has only one car and I am the manager for this tiny car fleet. To optimize the costs for my car fleet I use the 'FleetMan as a Service' offering of one of the largest German automotive club. The service bundles a chipset that is able to receive the Galileo GNSS signal, the chipset is part of a receiver-recorder that can store the Galileo information, the recorder is a black box in my car similar to flight recorders in planes. The Galileo black box of my car is certified so that the data cannot be manipulated by me or anyone else. The black box is sending the Galileo information to my individual Galileo Box hosted by my Internet Provider. I am the owner of that private Galileo data and I allow my Internet Provider to derive key figures out of that Galileo data pool using appropriate, again certified Galileo Applications (Analytical Transformation Services). Such a key figure set could look like this: Rainer Uhle drove 6.000 km on European roads during the last 3 months with a maximum speed of only 130 km/h. Such a service represents a huge semantic: It reflects my way of mobility behavior, it proves my environmental engagement (most people use to drive much faster in Germany, blowing out much more $CO_2$ than I do), and finally it shows my style and personality. In former days my friends laughed at me for my environmental engagement that very often brought me disadvantages. But nowadays the advantage is on my side, because I can sell this information to parties like my car insurance or the German government (Bundesverkehrsministerium). My car insurance agent puts this personal key figure into his Risk Management Modules and will find my crash risk on motorways reduced by 40%. He will offer me a 10% reduction of my car insurance fee and he will take the rest of the money to increase his margin. If the insurance contract is related to the car everything works already. If an authentication of the Galileo signal to a person is needed my car would need a fingerprint scanner in the steering wheel and an iris scanner in the rear-view mirror etc. All this is available on the market for me if I need it. And my car insurance agent will ask me another important question: 'On which GNSS standard is your car black box working, we only accept the Galileo standard.' Galileo is outstanding, the key differentiator is the legislation and the compliance that comes with the signal and all service layers built on top of it. Therefore Galileo has become European standard for GNSS tracking due to the acknowledged legislation behind that is widely accepted over all Industries and Businesses. My Internet Provider is offering a steadily growing key figure catalogue for me, which helps me to dive into a wide range of Business Scenarios offering numerous Pay Back options to me. I do not pay taxes for my car and the fuel any more. I only pay a road toll based on the kilometers I drive and which road categories I use at which time. I have chosen for this option because this pricing model brings additional benefits for me compared to the classical one based on taxes per engine volume and fuel consumption. It is a kind of a Galileo based Pay Back System. I am getting pay back points whenever I drive my car at the right time, at the right point, in the right way, or I use the subway instead. I can make my own money based on my own data, but only if I am willing to. My private data is hosted from an Internet Provider in a trustworthy way. I am acting autonomously - I am not an object or a victim in an Orwell scenario. I am actively investing in Internet Services that reduce my Total Cost of Ownership (TCO), increase my life quality and help to protect the environment.

## References

[1] Miche, M., Bohnert, T. M., The Internet of Vehicles or the Second Generation of Telematic Services, ERCIM News, 2009.

**Rainer Uhle**
SAP AG
Dietmar-Hopp-Allee 16
69190 Walldorf/Baden
Germany
rainer.uhle@sap.com

**Keywords:** internet of things, internet of services, telematic platforms, mobility services, fleet management, intelligent traffic systems, traffic management, galileo GNSS, broadband wireless access (BWA), business intelligence, process centric business intelligence (PCBI)

# Appendix A

## List of Contributors

## List of Contributors

# Appendix B

## List of Keywords

## List of Keywords

The manufacturer's authorised representative in the EU is Springer Nature Customer Service Centre GmbH, Europaplatz 3, 69115 Heidelberg, Germany. If you have any concerns regarding our products, please contact ProductSafety@springernature.com

Printed and bound by CPI Group (UK) Ltd, Croydon, CR0 4YY
07/01/2026
02030927-0019